WELL
PRODUCTION
PRACTICAL
HANDBOOK

NEW EDITION EXPANDED

Henri CHOLET, Editor

Engineer, Project Manager at IFP

WELL PRODUCTION PRACTICAL HANDBOOK

NEW EDITION EXPANDED

2008

t
Editions TECHNIP   5 av. de la République, 75011 PARIS, FRANCE

ISBN 978-2-7108-0917-3

Foreword

In order to meet the increasing world energy demand, the oil and gas industry is developing new technologies and improving the existing ones to better produce conventional fields while developing more and more challenging fields. It has led to significant evolutions in the field of well production covered by this handbook.

This book is dedicated to any engineer dealing with well productivity and thus intends to provide people involved in production and reservoir engineering or even in drilling and cementing with a common tool thanks to a comprehensive coverage of a broad scope of topics.

The first chapter provides general informations: mathematical formulae, chemical symbols, physical properties of various materials and fluids, stratigraphic scale. The handbook is then organized along the following main topics:

- Well technology (chapters B to D): casings and tubings, coiled tubing, packers.
- Well productivity estimate and control (chapters E to I): flluid flow and pressure losses, well tests, formation damage control and testing, sand control.
- Stimulation (chapter J).
- Horizontal and multilateral wells (chapter K).
- Production management (chapters L to V): water management, heavy oil production, enhanced oil recovery, artificial lift, pumps, gas lift, multiphase pumping, deposits and their inhibitors, well servicing.
- Cased-hole logging and imaging (chapter W).
- Financial formulae (chapter X).

For each topic, relevant information is provided on the main parameters and physical laws, on current equipment and industrial practices including new, recently designed technologies and practical formulae. A list of references for further reading is also provided for those who need more details or want to investigate the theoretical aspects.

Maurice Boutéca
Director, Exploration-Production
Technology Business Unit
IFP

Preface

Well production techniques have considerably progressed in recent years. Therefore, we thought that a new handbook on the subject might well prove valuable to the practicing engineer.

This work is designed to give a complete, comprehensive overview of field development and well production, providing a wealth of practical information. It is intended as a reference guide for petroleum engineers and oilfield operators, yet also provides readily-available solutions to practical problems. The user will find the guidelines, recommendations, formulas and charts currently in use, as it covers most of the cases encountered in the field. Even when a problem has been contracted out to a service company, reference to this handbook will help the oilfield manager to better monitor outsourced work and current operations.

The information used in this book to choose a solution to a given problem in oilfield development is based on a wide variety of petroleum industry documents and reviews, and on the many papers presented at SPE and other international conferences. Numerous references, both at the end of each chapter and in the captions to figures and tables, will enable the interested reader to study a specific topic in greater detail.

Many examples are given throughout to facilitate the use of the formulas. Also, measurements are frequently expressed in both metric and U.S. units. The symbols used for these units conform to the recommendations of the SPE Board of Directors.

This publication will therefore serve both as a guide and as a handbook, in which the operator will find answers to his questions, along with quick and easy solutions to most of the problems that occur in field development.

Henri Cholet

Acknowledgments

Well production is a very broad area that covers the analysis of reservoirs and of well productivity, activation and stimulation techniques, one-phase and multiphase production in horizontal, deviated and multilateral wells, problems due to water control and gas influx, deposit treatment, well reconditioning, measurements in cased holes, standards and economics. One cannot be a leading specialist in all these areas. I thus greatly appreciated the help of my former colleagues at IFP, oil and service company engineers and petroleum consultants.

Works published by professors of the IFP School provided me with a wealth of practical information. The *Drilling Data Handbook* was particularly useful in preparing the first two chapters of this work.

At SPE and AFTP conferences, I had the opportunity to discuss with and benefit from the advice of speakers, session chairpersons and organizing committee members.

I offer my hearty thanks to them and Editions Technip for helping bring this book into being.

The preparation of this work, which involved many long hours, would have been impossible without the support and understanding of my wife.

Henri Cholet

Contents

Contents

Contents

Contents

Contents

A

General Data

General Data

A1 SI UNITS SYSTEM

The International System of Units is customarily termed the SI Units system.

Conversion by the United States petroleum industry to the SI system is proceeding slowly.

- The SI base quantities and units are set forth in Table A1.1.
- Table A1.2 lists some of the more common derived units.
- Table A1.3 lists the prefixes used with the SI system to indicate multiplication by factors ranging from 10^{18} to 10^{-18}.
- Table A1.4 lists some of the SI units and conversion factors that are frequently used by engineers and operators in many companies.

Table A1.1 SI base quantities [1, 2].
(*Source*: SPE, The SI Metric System of Units and SPE Metric Standard)

Base quantity	SI unit	SI unit symbol	SPE letter symbol for mathematical equations
Length	meter	m	L
Mass	kilogram	kg	m
Time	second	s	t
Electric current	ampere	A	I
Thermodynamic temperature	kelvin	K	T
Amount of substance	mole	mol	n
Luminous intensity	candela	cd	

Table A1.2 Some common derived SI units [1, 2].
(*Source*: SPE, The SI Metric System of Units and SPE Metric Standard)

Quantity	Unit	SI unit symbol	Formula
Acceleration	meter per second squared	…	m/s^2
Angular acceleration	radian per second squared	…	rad/s^2
Angular velocity	radian per second	…	rad/s
Area	square meter	…	m^2
Celsius temperature	degree Celsius	°C	K
Density	kilogram per cubic meter	…	kg/m^3
Electric capacitance	farad	F	$A·s/V (= C/V)$
Electric charge	coulomb	C	$A·s$
Electrical conductance	siemens	S	A/V
Electric inductance	henry	H	$V·s/A$
Electric potential	volt	V	W/A
Electric resistance	ohm	Ω	V/A
Energy	joule	J	$N·m$
Entropy	joule par kelvin	…	J/K
Force	newton	N	$kg·m/s^2$
Frequency	hertz	Hz	$1/s$
Power	watt	W	J/s
Pressure	pascal	Pa	N/m^2
Quantity of electricity	coulomb	C	$A·s$
Quantity of heat	joule	J	$N·m$
Specific heat	joule per kilogram kelvin	…	$J/(kg·K)$
Stress	pascal	Pa	N/m^2
Thermal conductivity	watt per meter kelvin	…	$W/(m·K)$
Velocity	meter per second	…	m/s
Viscosity, dynamic	pascal second	…	$Pa·s$
Viscosity, kinematic	square meter per second	…	m^2/s
Volume	cubic meter	…	m^3
Work	joule	J	$N·m$

Table A1.3 SI unit prefixes [1, 2].
(*Source*: SPE, The SI Metric System of Units and SPE Metric Standard)

Multiplication factor	SI prefix	SI prefix symbol
1 000 000 000 000 000 000 $= 10^{18}$	exa	E
1 000 000 000 000 000 $= 10^{15}$	peta	P
1 000 000 000 000 $= 10^{12}$	tera	T
1 000 000 000 $= 10^{8}$	giga	G
1 000 000 $= 10^{4}$	mega	M
1 000 $= 10^{3}$	kilo	k
100 $= 10^{2}$	hecto	h
10 $= 10$	deca	da
0.1 $= 10^{-1}$	deci	d
0.01 $= 10^{-3}$	centi	c
0.001 $= 10^{-3}$	milli	m
0.000 001 $= 10^{-6}$	micro	μ
0.000 000 001 $= 10^{-9}$	nano	n
0.000 000 000 001 $= 10^{-12}$	pico	p
0.000 000 000 000 001 $= 10^{-15}$	femto	f
0.000 000 000 000 000 001 $= 10^{-18}$	atto	a

Table A1.4 Recommended SI units [1, 2].
(*Source*: SPE, The SI Metric System of Units and SPE Metric Standard)

Quantity and SI unit	Customary unit	Metric unit		Conversion factor* Multiply customary unit by factor to get metric unit	
		SPE preferred	Other allowable		
Space					
Length m	mile	km		1.609 344*	E + 00
	ft	m		3.048*	E – 01
			cm	3.048*	E + 01
	in.	mm		2.54*	E + 01
			cm	2.54*	E + 00
Area m^2	sq mile	km^2		2.589 988	E + 00
	acre	m^2		4.046 856	E + 03
			ha	4.046 856	E – 01
	ha	m^2		1.0*	E + 04
	sq ft	m^2		9.290 304*	E – 02
			cm^2	9.290 304*	E + 02
	sq in.	mm^2		6.451 6*	E + 02
			cm^2	6.451 6*	E + 00
Volume, capacity m^3	acre-ft	m^3		1.233 489	E + 03
			ha·m	1.233 489	E – 01
	bbl (42 U.S. gal)	m^3		1.589 873	E – 01
	cu ft	m^3		2.831 685	E – 02
		dm^3	L or ℓ	2.831 685	E + 01
	U.S. gal	m^3		3.785 412	E – 03
		dm^3	L or ℓ	3.785 412	E + 00
	liter	dm^3	L or ℓ	1.0*	E + 00

* An asterisk indicates that the conversion factor is exact using the numbers shown;
all subsequent numbers are zeros.

(to be continued)

Table A1.4 *(cont'd)* Recommended SI units [1, 2].
(*Source*: SPE, The SI Metric System of Units and SPE Metric Standard)

Quantity and SI unit		Customary unit	Metric unit		Conversion factor*
			SPE preferred	Other allowable	Multiply customary unit by factor to get metric unit
Mass, amount of substance					
Mass	kg	U.S. ton (short ton)	Mg	t	9.071 847 E − 01
		U.S. cwt	kg		4.535 924 E + 01
		lbm	kg		4.535 924 E − 01
Calorific value, heat, entropy, heat capacity					
Calorific value (mass basis)	J/kg	Btu/lbm	MJ/kg KJ/kg	J/g (kW·h)/kg	2.326 E − 03 2.326 E + 00 6.461 112 E − 04
		cal/g	kJ/kg	J/g	4.184* E + 00
		cal/lbm	J/kg		9.224 141 E + 00
Calorific value (mole basis)	J/mol	Btu/lbm mol	MJ/kmol kJ/kmol		2.326 E − 03 2.326 E + 00
Calorific value (volume basis-gases)	J/m^3	cal/mL	kJ/m^3	J/dm^3	4.184* E + 03
		kcal/m^3	kJ/m^3	J/dm^3	4.184* E + 00
		Btu/ft^3	kJ/m^3	J/dm^3 (kW·h)/m^3	3.725 895 E + 01 1.034 971 E − 02
Specific entropy	J/kg·K	Btu/(lbm·°R)	kJ/(kg·K)	J/(g·K)	4.186 8* E + 00
		cal/(g·°K)	kJ/(kg·K)	J/(g·K)	4.184* E + 00
		kcal/(kg·°C)	kJ/(kg·K)	J/(g·K)	4.184* E + 00
Specific heat capacity (mass basis)	J/kg·K	kW·hr/(kg·°C)	kJ/(kg·K)	J/(g·K)	3.6* E + 03
		Btu/(lbm·°F)	kJ/(kg·K)	J/(g·K)	4.186 8* E + 00
		kcal/(kg·°C)	kJ/(kg·K)	J/(g·K)	4.184* E + 00
Molar heat capacity	J/mol·K	Btu/(lbm mol·°F)	kJ/(kmol·K)		4.186 8* E + 00
		cal/(g mol·°C)	kJ/(kmol·K)		4.184* E + 00
Temperature, pressure					
Temperature (absolute)	K	°R	K		5/9
		°K	K		1.0* E + 00
Temperature (traditional)	K	°F	°C		(°F − 32)/1.8
		°C	°C		1.0° E + 00
Pressure	Pa	atm (760 mm Hg at 0°C or 14.696 lbf/in.^{2})	MPa kPa	bar	1.013 25* E − 01 1.013 25* E + 02 1.013 25* E + 00

* An asterisk indicates that the conversion factor is exact using the numbers shown;
all subsequent numbers are zeros.

(to be continued)

General Data   **A**

Table A1.4 *(cont'd)* Recommended SI units [1, 2].
(*Source*: SPE, The SI Metric System of Units and SPE Metric Standard)

Quantity and SI unit	Customary unit	Metric unit		Conversion factor* Multiply customary unit by factor to get metric unit
		SPE preferred	Other allowable	
Pressure Pa	bar	MPa		1.0* E – 01
		kPa		1.0* E + 02
			bar	1.0* E + 00
	lbf.in.^{2}(psi)	MPa		6.894 757 E – 03
		kPa		6.894 757 E + 00
			bar	6.894 757 E – 02
	mm Hg (0°C) = torr	kPa		1.333 224 E – 01
	dyne/cm^2	Pa		1.0* E – 01
Density, specific volume, concentration				
Density (gases) kg/m^3	lbm/ft^3	kg/m^3		1.601 846 E + 01
		g/m^3		1.601 846 E + 04
Density (liquids) kg/m^3	lbm/U.S. gal	kg/m^3		1.198 264 E + 02
			g/cm^3	1.198 264 E – 01
	lbm/ft^3	kg/m^3		1.601 846 E + 01
			g/cm^3	1.601 846 E – 02
	g/cm^3	kg/m^3		1.0* E + 03
			kg/dm^3	1.0* E + 00
	°API	g/cm^3		141.5/(131.5 + °API)
Density (solids) kg/m^3	lbm/ft^3	kg/m^3		1.601 846 E + 01
Specific volume (gases) m^3/kg	ft^3/lbm	m^3/kg		6.242 796 E – 02
		m^3/g		6.242 796 E – 05
Specific volume (liquids) m^3/kg	ft^3/lbm	dm^3/kg		6.242 796 E + 01
	U.S. gal/lbm	dm^3/kg	cm^3/g	8.345 404 E + 00
Specific volume (mole basis) m^3/mol	L/g mol	m^3/kmol		1.0* E + 00
	ft^3/lbm mol	m^3/kmol		6.242 796 E – 02
Concentration (mass/volume) kg/m^3	lbm/bbl	kg/m^3	g/dm^3	2.853 010 E + 00
	g/U.S. gal	kg/m^3		2.641 720 E – 01
Concentration (volume/volume) m^3/m^3	bbl/acre·ft	m^3/m^3		1.288 923 E – 04
			m^3/ha·m	1.288 923 E + 00
Concentration (mole/volume) mol/m^3	lbm mol/ft^3	kmol/m^3		1.601 846 E + 01
	std ft^3 (60°F, 1 atm)/bbl	kmol/m^3		7.518 18 E – 03

* An asterisk indicates that the conversion factor is exact using the numbers shown; all subsequent numbers are zeros.

(to be continued)

Table A1.4 *(cont'd)* Recommended SI units [1, 2].
(*Source*: SPE, The SI Metric System of Units and SPE Metric Standard)

Quantity and SI unit	Customary unit	Metric unit		Conversion factor* Multiply customary unit by factor to get metric unit	
		SPE preferred	Other allowable		
Flow rate					
Flow rate (volume basis) m^3/s	bbl/D	m^3/d		1.589 873	E – 01
			L/s or ℓ/s	1.840 131	E – 03
	ft^3/D	m^3/d		2.831 685	E – 02
			L/s or l/s	3.277 413	E – 04
	U.S. gal/min	dm^3/s	L/s or l/s	6.309 020	E – 02
Flow rate/ pressure drop (productivity index) $m^3/s{\cdot}Pa$	bbl/(D-psi)	$m^3/(d{\cdot}kPa)$		2.305 916	E – 02
Energy, work, power					
Energy, work J	therm	MJ		1.055 056	E + 02
		kJ		1.055 056	E + 05
			kW·h	2.930 711	E + 01
	hp-hr	MJ		2.684 520	E + 00
		kJ		2.684 520	E + 03
			kW·h	7.456 999	E – 01
	kW-hr	MJ		3.6*	E + 00
		kJ		3.6*	E + 03
	Btu	kJ		1.055 056	E + 00
			kW·h	2.930 711	E – 04
	cal	kJ		4.184*	E – 03
	erg	J		1.0*	E – 07
Power W	million Btu/hr	MW		2.930 711	E – 01
	kW	kW		1.0*	E + 00
	hp (550 ft-lbf/s)	kW		7.456 999	E – 01
	Btu/hr	W		2.930 711	E – 01
Mechanics					
Velocity (linear), speed m/s	mile/hr	km/h		1.609 344*	E + 00
	ft/s	m/s		3.048*	E – 01
Force N	lbf	N		4.448 222	E + 00
	pdl	mN		1.382 550	E + 02
	dyne	mN		1.0*	E – 02
Stress Pa	lbf/in.^{2} (psi)	MPa	N/mm^2	6.894 757	E – 03
	$dyne/cm^2$	Pa		1.0*	E – 01

* An asterisk indicates that the conversion factor is exact using the numbers shown; all subsequent numbers are zeros.

(to be continued)

Table A1.4 *(cont'd)* Recommended SI units [1, 2].
(*Source*: SPE, The SI Metric System of Units and SPE Metric Standard)

Quantity and SI unit		Customary unit	Metric unit		Conversion factor* Multiply customary unit by factor to get metric unit	
			SPE preferred	Other allowable		
Yield point, gel strength (drilling fluid)		lbf/100 ft^2	Pa		4.788 026	E – 01
Coefficient of thermal expansion	m/(m·K)	in./(in.-°F)	mm/(mm.K)		5.555 556	E – 01
Transport properties						
Diffusivity	m^2/s	ft^2/s	mm^2/s		9.290 304*	E + 04
		cm^2/s	mm^2/s		1.0*	E + 02
Thermal resistance	(k·m^2)/W	(°C-m^2·hr)/kcal	(K·m^2)/kW		8.604 208	E + 02
		(°F-ft^2·hr)/Btu	(K·m^2)/kW		1.761 102	E + 02
Heat flux	W/m^2	Btu/(hr-ft^2)	kW/m^2		3.154 591	E – 03
Thermal conductivity	W/(m·K)	(cal/s-cm^2-°C)/cm	W/(m·K)		4.184*	E + 02
		Btu/(hr-ft^2-°F/ft)	W/(m·K)	kJ·m/ (h·m^2·K)	1.730 735 6.230 646	E + 00 E + 00
Heat transfer coefficient	W/(m^2·K)	cal/(s-cm^2-°C)	kW/(m^2·K)		4.184*	E + 01
		Btu/(s-ft^2-°F)	kW/(m^2·K)		2.044 175	E + 01
Surface tension	N/m	dyne/cm	mN/m		1.0*	E + 00
Viscosity (dynamic)	Pa·s	(dyne-s)/cm^2	Pa·s	(N·s)/m^2	1.0*	E – 01
		cp	Pa·s	(N·s)/m^2	1.0*	E – 03
		lbm/(ft·hr)	Pa·s	(N·s)/m^2	4.133 789	E – 04
Viscosity (kinematic)	m^2/s	cm^2/s	mm^2/s		1.0*	E + 02
		ft^2/hr	mm^2/s		2.580 64*	E + 01
		cSt	mm^2/s		1.0*	E + 00
Permeability	m^2	darcy (d)	μm^2		9.869 233	E – 01
		millidarcy (md)	μm^2	10^{-3} μm^2	9.869 233 9.869 233	E – 04 E – 01
Miscellaneous						
Capillary pressure	Pa	ft (fluid)	m (fluid)		3.048*	E – 01
Compressibility of reservoir fluid	Pa^{-1}	psi^{-1}	Pa^{-1}	kPa^{-1}	1.450 377 1.450 377	E – 04 E – 01
Gas-oil ratio	m^3/m^3	scf/bbl	"standard" m^3/m^3		1.801 175	E – 01

* An asterisk indicates that the conversion factor is exact using the numbers shown; all subsequent numbers are zeros.

(to be continued)

Table A1.4 *(cont'd)* Recommended SI units [1, 2].
(*Source*: SPE, The SI Metric System of Units and SPE Metric Standard)

Quantity and SI unit		Customary unit	Metric unit		Conversion factor* Multiply customary unit by factor to get metric unit	
			SPE preferred	Other allowable		
Gas rate	m³/s	scf/D	"standard" m³/d		2.863 640	E – 02
Head (fluid mechanics)	m	ft	m	cm	3.048* / 3.048*	E – 01 / E + 01
Heat exchange rate	W	Btu/hr	kW	kJ/h	2.930 711 / 1.055 056	E – 04 / E + 00
Mobility	m²/Pa·s	d/cp	µm²/mPa·s	µm²/Pa·s	9.869 233 / 9.869 233	E – 01 / E + 02
Oil rate	m³/s	bbl/D / short ton/yr	m³/d / Mg/a	t/a	1.589 873 / 9.071 847	E – 01 / E – 01
Particle size	m	micron	µm		1.0*	
Permeability-thickness	m³	md-ft	md·m	µm²·m	3.008 142	E – 04
Pressure buildup per cycle	Pa	psi	kPa		6.894 757	E + 00
Productivity index	m³/Pa·s	bbl/(psi-D)	m³/kPa·d		2.305 916	E – 02
Pumping rate	m³/s	U.S. gal/min	m³/h	L/s or ℓ/s	2.271 247 / 6.309 020	E – 01 / E – 02
Revolutions per minute	rad/s	rpm	rad/s	rad/m	1.047 198 / 6.283 185	E – 01 / E + 00
Recovery/unit volume (oil)	m³/m³	bbl/(acre-ft)	m³/m³	m³/ha·m	1.288 931 / 1.288 931	E – 04 / E + 00
Reservoir area	m²	sq mile / acre	km²	ha	2.589 998 / 4.046 856	E + 00 / E – 01
Reservoir volume	m³	acre-ft	m³	ha·m	1.233 482 / 1.233 482	E + 03 / E – 01
Specific productivity index	m³/Pa·s·m	bbl/(D-psi-ft)	m³/ (kPa·d·m)		7.565 351	E – 02
Surface or interfacial tension in reservoir capillaries	N/m	dyne/cm	mN/m		1.0*	E + 00
Torque	N·m	lbf-ft	N·m		1.355 818	E + 00

* An asterisk indicates that the conversion factor is exact using the numbers shown; all subsequent numbers are zeros.

A2 TEMPERATURE CONVERSION TABLE

Table A2 Temperature conversion table [3].

$$\left(t°F = \frac{9}{5}(t°C + 32) \qquad t°C = \frac{5}{9}(t°F - 32)\right)$$

Example: The central figures refer to the temperatures either in degrees Celsius or degrees Fahrenheit which require conversion. The corresponding temperatures in degrees Fahrenheit or degrees Celsius will be found to the right or left respectively.

°C		°F
6.67	**44**	111.2

44° Fahrenheit → 6.67° Celsius
44° Celsius → 111.2° Fahrenheit

°C	value	°F		°C	value	°F		°C	value	°F		°C	value	°F		°C	value	°F
−56.7	**−70**	−94.0		−11.10	**12**	53.6		11.1	**52**	125.6		33.3	**92**	197.6		216	**420**	788
−53.9	**−65**	−85.0		−10.00	**14**	57.2		12.2	**54**	129.2		34.4	**94**	201.2		227	**440**	824
−51.2	**−60**	−76.0		−8.89	**16**	60.8		13.3	**56**	132.8		35.6	**96**	204.8		238	**460**	860
−48.4	**−55**	−67.0		−7.78	**18**	64.4		14.4	**58**	136.4		36.7	**98**	208.4		249	**480**	896
−45.6	**−50**	−58.0		−6.67	**20**	68.8		15.6	**60**	140.0		37.8	**100**	212.0		260	**500**	932
−42.8	**−45**	−49.0		−5.55	**22**	71.6		16.7	**62**	143.6		48.9	**120**	248.0		271	**520**	968
−40.0	**−40**	−40.0		−4.44	**24**	75.2		17.8	**64**	147.2		60.0	**140**	284.0		282	**540**	1 004
−37.2	**−35**	−31.0		−3.33	**26**	78.8		18.9	**66**	150.8		71.1	**160**	320.0		293	**560**	1 040
−34.4	**−30**	−22.0		−2.22	**28**	82.4		20.0	**68**	154.4		82.2	**180**	356.0		304	**580**	1 076
−31.7	**−25**	−13.0		−1.11	**30**	86.0		21.1	**70**	158.0		93.3	**200**	392.0		316	**600**	1 112
−28.9	**−20**	−4.0		0.00	**32**	89.6		22.2	**72**	161.6		104.4	**220**	428.0		327	**620**	1 148
−26.1	**−15**	5.0		1.11	**34**	93.2		23.3	**74**	165.2		115.6	**240**	464.0		338	**640**	1 184
−23.3	**−10**	14.0		2.22	**36**	96.8		24.4	**76**	168.8		126.7	**260**	500.0		349	**660**	1 220
−20.6	**−5**	23.0		3.33	**38**	100.4		25.6	**78**	172.4		137.8	**280**	536.0		360	**680**	1 256
−17.8	**0**	32.0		4.44	**40**	104.0		26.7	**80**	176.0		148.9	**300**	572.0		371	**700**	1 292
−16.7	**2**	35.6		5.55	**42**	107.6		27.8	**82**	179.6		160.0	**320**	608.0		382	**720**	1 328
−15.6	**4**	39.2		6.67	**44**	111.2		28.9	**84**	183.2		171.0	**340**	644.0		393	**740**	1 364
−14.4	**6**	42.8		7.78	**46**	114.8		30.0	**86**	186.8		182.0	**360**	680.0		404	**760**	1 400
−13.3	**8**	46.4		8.89	**48**	118.4		31.1	**88**	190.4		193.0	**380**	716.0		416	**780**	1 436
−12.2	**10**	50.0		10.00	**50**	122.0		32.2	**90**	194.0		204.0	**400**	752.0		427	**800**	1 472

Interpolation table

°C	0.56	1.11	1.67	2.22	2.78	3.33	3.89	4.44	5	5.56	6.11	6.67	7.22	7.78	8.33	8.89	9.44	10	10.56	11.11
	1	**2**	**3**	**4**	**5**	**6**	**7**	**8**	**9**	**10**	**11**	**12**	**13**	**14**	**15**	**16**	**17**	**18**	**19**	**20**
°F	1.8	3.6	5.4	7.2	9	10.8	12.6	14.4	16.2	18	19.8	21.6	23.4	25.2	27	28.8	30.6	32.4	34.2	36

A3 CORRESPONDENCE BETWEEN SPECIFIC GRAVITY AND DEGREES API

Table A3 Correspondence specific gravity/degrees API [3]
(at 15.56°C in relation to water at 15.56°C and 760 mmHg).

Specific gravity	Degrees API	Specific gravity	Degrees API	Specific gravity	Degrees API	Specific gravity	Degrees API	Specific gravity	Degrees API	Specific gravity	Degrees API	Specific gravity	Degrees API	Specific gravity	Degrees API	Specific gravity	Degrees API
0.600	104.3	0.650	86.2	0.700	70.6	0.750	57.2	0.800	45.4	0.850	35.0	0.900	25.7	0.950	17.4	1.000	10.0
0.602	103.5	0.652	85.5	0.702	70.1	0.752	56.7	0.802	44.9	0.852	34.6	0.902	25.4	0.952	17.1	1.002	9.7
0.604	102.8	0.654	84.9	0.704	69.5	0.754	56.2	0.804	44.5	0.854	34.2	0.904	25.0	0.954	16.8	1.004	9.4
0.606	102.0	0.656	84.2	0.706	68.9	0.756	55.7	0.806	44.1	0.856	33.8	0.906	24.7	0.956	16.5	1.006	9.2
0.608	101.2	0.658	83.5	0.708	68.4	0.758	55.2	0.808	43.6	0.858	33.4	0.908	24.3	0.958	16.2	1.008	8.9
0.610	100.5	0.660	82.9	0.710	67.8	0.760	54.7	0.810	43.2	0.860	33.0	0.910	24.0	0.960	15.9	1.010	8.6
0.612	99.7	0.662	82.2	0.712	67.2	0.762	54.2	0.812	42.8	0.862	32.7	0.912	23.7	0.962	15.6	1.012	8.3
0.614	99.0	0.664	81.6	0.714	66.7	0.764	53.7	0.814	42.3	0.864	32.3	0.914	23.3	0.964	15.3	1.014	8.0
0.616	98.2	0.666	81.0	0.716	66.1	0.766	53.2	0.816	41.9	0.866	31.9	0.916	23.0	0.966	15.0	1.016	7.8
0.618	97.5	0.668	80.3	0.718	65.6	0.768	52.7	0.818	41.5	0.868	31.5	0.918	22.6	0.968	14.7	1.018	7.5
0.620	96.7	0.670	79.7	0.720	65.0	0.770	52.3	0.820	41.1	0.870	31.1	0.920	22.3	0.970	14.4	1.020	7.2
0.622	96.0	0.672	79.1	0.722	64.5	0.772	51.8	0.822	40.6	0.872	30.8	0.922	22.0	0.972	14.1	1.022	7.0
0.624	95.3	0.674	78.4	0.724	63.9	0.774	51.3	0.824	40.2	0.874	30.4	0.924	21.6	0.974	13.8	1.024	6.7
0.626	94.5	0.676	77.8	0.726	63.4	0.776	50.8	0.826	39.8	0.876	30.0	0.926	21.3	0.976	13.5	1.026	6.4
0.628	93.8	0.678	77.2	0.728	62.9	0.778	50.4	0.828	39.4	0.878	29.7	0.928	21.0	0.978	13.2	1.028	6.1
0.630	93.1	0.680	76.6	0.730	62.3	0.780	49.9	0.830	39.0	0.880	29.3	0.930	20.7	0.980	12.9	1.030	5.9
0.632	92.4	0.682	76.0	0.732	61.8	0.782	49.4	0.832	38.6	0.882	28.9	0.932	20.3	0.982	12.6	1.032	5.6
0.634	91.7	0.684	75.4	0.734	61.3	0.784	49.0	0.834	38.2	0.884	28.6	0.934	20.0	0.984	12.3	1.034	5.3
0.636	91.0	0.686	74.8	0.736	60.8	0.786	48.5	0.836	37.8	0.886	28.2	0.936	19.7	0.986	12.0	1.036	5.1
0.638	90.3	0.688	74.2	0.738	60.2	0.788	48.1	0.838	37.4	0.888	27.8	0.938	19.4	0.988	11.7	1.038	4.8
0.640	89.6	0.690	73.6	0.740	59.7	0.790	47.6	0.840	37.0	0.890	27.5	0.940	19.0	0.990	11.4	1.040	4.6
0.642	88.9	0.692	73.0	0.742	59.2	0.792	47.2	0.842	36.6	0.892	27.1	0.942	18.7	0.992	11.1	1.042	4.3
0.644	88.2	0.694	72.4	0.744	58.7	0.794	46.7	0.844	36.2	0.894	26.8	0.944	18.4	0.994	10.9	1.044	4.0
0.646	87.5	0.696	71.8	0.746	58.2	0.796	46.3	0.846	35.8	0.896	26.4	0.946	18.1	0.996	10.6	1.046	3.8
0.648	86.9	0.698	71.2	0.748	57.7	0.798	45.8	0.848	35.4	0.898	26.1	0.948	17.8	0.998	10.3	1.048	3.5

$$\text{Degrees API} = \frac{141.5}{d} - 131.5$$

d (15.56°C) = Specific gravity (60°F)

Specific gravity	Correction for 1°C	**Approximate temperature correction to obtain gravity at 15°C**	
0.600 to 0.700	0.0009	– add if $t > 15°C$	
0.700 to 0.800	0.0008	– subtract if $t < 15°C$	
0.800 to 0.840	0.00075		
0.840 to 0.880	0.0007		
0.880 to 0.920	0.00065		
0.920 to 1.000	0.0006		

A4 KINEMATIC VISCOSITY CONVERSION

Table A4a Kinematic viscosity conversion [4].

cSt	Engler degrees	Saybolt seconds to:			Redwood seconds to:		
		100°F (37.8°C)	130°F (54.4°C)	210°F (98.9°C)	70°F (21.1°C)	140°F (60°C)	200°F (93.3°C)
2	1.140	32.60	32.65	32.83	30.20	30.95	31.20
3	1.224	36.00	36.07	36.25	32.70	33.45	33.70
4	1.308	39.10	39.17	39.37	35.30	35.95	36.30
5	1.400	42.30	42.38	42.60	37.90	38.45	38.90
6	1.481	45.50	45.59	45.82	40.50	41.05	41.50
7	1.563	48.70	48.79	49.04	43.20	43.70	44.15
8	1.653	52.00	52.10	52.36	46.00	46.35	46.90
9	1.746	55.40	55.51	55.79	48.85	49.10	49.63
10	1.837	58.80	58.91	59.21	51.70	52.00	52.55
11	1.928	62.30	62.42	62.74	54.75	55.00	55.60
12	2.020	65.90	66.03	66.36	57.90	58.10	58.75
13	2.120	69.80	69.73	70.09	61.05	61.30	61.95
14	2.219	73.40	73.54	73.91	64.35	64.55	65.25
15	2.323	77.20	77.35	77.74	67.70	67.95	68.75
16	2.434	81.10	81.25	81.67	71.15	71.40	72.20
17	2.540	85.10	85.26	85.70	74.65	74.85	75.75
18	2.644	89.20	89.37	89.82	78.10	78.45	79.35
19	2.755	93.30	93.48	93.95	81.70	82.10	83.10
20	2.870	97.50	97.69	98.18	85.40	85.75	86.90
21	2.984	101.7	101.9	102.4	89.20	89.50	90.70
22	3.100	106.0	106.2	106.7	92.90	93.25	94.50
23	3.215	110.3	110.5	111.1	96.70	97.05	98.30
24	3.335	114.6	114.8	115.4	100.4	100.9	102.2
25	3.455	118.9	119.1	119.7	104.2	104.7	106.1
26	3.575	123.3	123.5	124.2	108.1	108.6	110.0

cSt	Engler degrees	Saybolt seconds to:			Redwood seconds to:		
		100°F (37.8°C)	130°F (54.4°C)	210°F (98.9°C)	70°F (21.1°C)	140°F (60°C)	200°F (93.3°C)
27	3.695	127.7	127.9	128.6	111.9	112.5	114.0
28	3.820	132.1	132.4	133.0	115.8	116.5	118.0
29	3.945	136.5	136.8	137.5	119.7	120.4	122.0
30	4.070	140.9	141.2	141.9	123.7	124.4	126.0
31	4.195	145.3	145.6	146.3	127.5	128.3	130.1
32	4.320	149.7	150.0	150.8	131.5	132.3	134.1
33	4.445	154.2	154.5	155.3	135.1	136.3	138.1
34	4.570	158.7	159.0	159.8	139.3	140.2	142.2
35	4.695	163.2	163.5	164.3	143.3	144.2	146.2
36	4.825	167.7	168.0	168.9	147.2	148.2	150.3
37	4.955	172.2	172.5	173.4	151.2	152.2	154.2
38	5.080	176.7	177.0	177.9	155.2	156.2	158.3
39	5.205	181.2	181.5	182.5	159.2	160.3	162.5
40	5.335	185.7	186.0	187.0	163.2	164.3	166.7
41	5.465	190.2	190.6	191.5	167.2	168.3	170.8
42	5.590	194.7	195.1	196.1	171.2	172.3	175.0
43	5.720	199.2	199.6	200.6	175.2	176.4	179.2
44	5.845	203.8	204.2	205.2	179.2	180.4	183.3
45	5.975	208.4	208.8	209.9	183.2	184.5	187.5
46	6.105	213.0	213.4	214.5	187.2	188.5	191.7
47	6.235	217.6	218.0	219.1	191.2	192.6	195.8
48	6.365	222.2	222.6	223.8	195.3	196.6	200.0
49	6.495	226.8	227.2	228.4	199.2	200.7	204.2
50	6.630	231.4	231.8	233.0	203.3	204.7	208.3
51	6.760	236.0	236.4	237.6	207.3	208.8	212.5

Table A4b Kinematic viscosity conversion [4].

cSt	Engler degrees	Saybolt seconds to:			Redwood seconds to:		
		100°F (37.8°C)	130°F (54.4°C)	210°F (98.9°C)	70°F (21.1°C)	140°F (60°C)	200°F (93.3°C)
52	6.890	240.6	241.1	242.3	211.3	212.8	216.7
54	7.106	249.9	250.3	251.6	219.3	221.0	225.0
56	7.370	259.0	259.5	260.8	227.4	229.1	233.4
58	7.633	268.2	268.7	270.1	235.5	237.2	241.7
60	7.896	277.4	277.9	279.3	243.5	245.3	250.0
62	8.159	286.6	287.2	288.6	251.5	253.5	258.4
64	8.422	295.8	296.4	297.9	259.6	261.6	266.7
66	8.686	305.0	305.6	307.1	267.7	269.8	275.0
68	8.949	314.2	314.8	316.4	275.8	277.9	283.4
70	9.212	323.4	324.0	325.7	283.9	286.0	291.7
72	9.475	332.6	333.3	335.0	291.9	294.1	300.0
74	9.738	341.9	342.5	344.3	300.0	302.2	308.4
76	9.982	351.1	351.8	353.6	307.7	310.2	316.7
78	10.246	360.4	361.1	362.9	315.8	318.4	325.1
80	10.510	369.6	370.3	372.2	323.9	326.6	333.4
82	10.774	378.8	379.6	381.5	332.0	334.8	341.7
84	11.038	388.1	388.7	390.8	340.1	342.9	350.0
86	11.302	397.3	398.1	400.1	348.2	351.1	358.4
88	11.566	406.6	407.4	409.4	356.3	359.2	366.7
90	11.830	415.8	416.6	418.7	364.4	367.4	375.0
92	12.094	425.0	425.9	428.0	372.5	375.6	383.4
94	12.358	434.3	435.1	437.3	380.6	383.7	391.7
96	12.622	443.5	444.4	446.6	388.7	391.9	400.0
98	12.886	452.8	453.6	455.9	396.8	400.1	408.4
100	13.152	462.0	462.9	465.2	404.9	408.2	416.7
110	14.474	508.2	509.2	511.8	445.4	449.0	458.4
120	15.794	554.4	555.5	558.3	485.9	489.8	500.0
130	17.118	600.6	601.8	604.8	526.4	530.7	541.7
140	18.438	646.8	648.1	651.3	566.9	571.5	583.4
150	19.758	693.0	694.4	697.9	607.4	612.3	625.0
160	21.078	739.2	740.6	744.4	647.9	653.1	666.7
170	22.398	785.4	786.9	790.9	688.3	693.9	709.4
180	23.718	831.6	833.2	837.4	728.8	734.8	750.1
190	25.038	877.8	879.5	884.0	769.3	775.6	791.7
200	26.358	924.0	925.8	930.5	809.8	816.4	833.4
220	28.998	1 016.4	1 018.4	1 023.5	890.8	898.0	916.7
240	31.642	1 108.8	1 111.0	1 116.6	971.8	979.7	1 000.1
260	34.282	1 201.2	1 203.5	1 209.6	1 052.7	1 061.3	1 083.5
280	36.922	1 293.6	1 296.1	1 302.7	1 133.7	1 143.0	1 166.7
300	39.562	1 386.0	1 388.7	1 395.7	1 214.7	1 224.6	1 250.1
350	46.162	1 617.0	1 620.1	1 628.3	1 417.1	1 428.7	1 458.6
400	52.762	1 848.0	1 851.6	1 861.0	1 619.6	1 632.8	1 666.8
450	59.362	2 079.0	2 083.1	2 093.6	1 822.0	1 836.8	1 875.0
500	65.962	2 310.0	2 314.5	2 326.2	2 024.5	2 041.0	2 083.5
550	72.562	2 541.0	2 546.0	2 558.8	2 227.0	2 245.2	2 291.9
600	79.162	2 772.0	2 777.4	2 791.4	2 429.4	2 449.2	2 500.2
650	85.753	3 003.0	3 009.9	3 024.1	2 631.9	2 653.2	2 708.7
700	92.303	3 234.0	3 240.3	3 256.7	2 834.3	2 857.2	2 916.9
750	98.853	3 465.0	3 472.0	3 489.0	3 040.0	3 061.6	3 125.3
800	105.403	3 696.0	3 703.0	3 722.0	3 239.	3 266.0	3 334.0
850	111.940	3 927.0	3 935.0	3 955.0	3 442.0	3 470.0	3 542.0
900	118.480	4 158.0	4 166.0	4 187.0	3 644.0	3 674.0	3 750.0
950	125.050	4 389.0	4 398.0	4 420.0	3 847.0	3 878.0	3 959.0
1 000	131.630	4 620.0	4 629.0	4 652.0	4 049.0	4 082.0	4 167.0
1 100	144.790	5 082.0	5 092.0	5 118.0	4 454.0	4 490.0	4 584.0
1 200	157.960	5 544.0	5 555.0	5 583.0	4 859.0	4 898.0	5 000.0
1 300	171.120	6 006	6 018.0	6 048.0	5 264.0	5 307.0	5 417.0
1 400	184.280	6 468.0	6 481.0	6 513.0	5 669.0	5 715.0	5 834.0
1 500	197.440	6 930.0	6 944.0	6 979.0	6 074.0	6 123.0	6 250.0
2 000	263.260	9 240.0	9 258.0	–	8 098.0	8 164.0	–
2 500	329.080	11 550.0	11 573.0	–	10 123.0	10 205.0	–
3 000	394.890	13 860.0	13 887.0	–	12 147.0	12 246.0	–

A5 NUMERICAL CONSTANTS AND MATHEMATICAL FORMULAS

Table A5 Numerical constants and mathematical formulas [3].

π	3.1415927	$\dfrac{1}{\pi}$	0.3183099	$\dfrac{\pi}{2}$	1.5707963	$\dfrac{\pi}{180}$	0.0174533
π^2	9.8696044	$\dfrac{1}{\pi^2}$	0.1013212	$\dfrac{\pi}{3}$	1.0471976	$\dfrac{\pi}{200}$	0.0157080
π^3	31.0062767	$\dfrac{1}{\pi^3}$	0.0322515	$\dfrac{\pi}{4}$	0.7853982	$\dfrac{180}{\pi}$	57.2957795
$\sqrt{\pi}$	1.7724539	$\dfrac{1}{\sqrt{\pi}}$	0.5641896	$\dfrac{4\pi}{3}$	4.1887902	$\dfrac{200}{\pi}$	63.6619763
$\sqrt[3]{\pi}$	1.4645919	$\dfrac{1}{\sqrt[3]{\pi}}$	0.6827840				

$\sqrt{2}$	1.414214	$\sqrt{3}$	1.732051	$\sqrt{5}$	2.236068	$\sqrt{10}$	3.162278
$\dfrac{1}{\sqrt{2}}$	0.70711	$\dfrac{1}{\sqrt{3}}$	0.57735	$\dfrac{1}{\sqrt{5}}$	0.44721	$\dfrac{1}{\sqrt{10}}$	0.31623

e	2.7182818	$\dfrac{1}{e}$	0.3678794	$\log_{10}e = 0.4342945$	$g = 9.80665 \text{ m/s}^2$

$$\frac{1}{\log_{10}e} = \text{colog } e = \log_e 10 = 2.3025851 \qquad \log_e x = 2.3025851 \log_{10} x \qquad \log_{10} x = 0.4342945 \log_e x$$

Arithmetic progression	a = first term
$a \quad a+r \quad a+2r \quad a+3r \quad \dots \quad a+(n-1)r$ $S_n = \left(\dfrac{a+\ell}{2}\right)n = \dfrac{n}{2}[2a+(n-1)r]$	r = common difference n = number of terms ℓ = last term = $a + (n-1)r$

Geometric progression	a = first term
$a \quad aq \quad aq^2 \quad aq^3 \quad \dots \quad aq^{n-1}$ Si $q \neq 1 \quad S_n = \dfrac{\ell q - a}{q-1} = a\dfrac{(q^n - 1)}{q-1}$	r = common ratio n = number of terms ℓ = last term = aq^{n-1}

A6 TRIGONOMETRIC FORMULAS [3, 4]

6.1 Definition

$$\cos\alpha = \frac{OP}{OM}$$

$$\sin\alpha = \frac{PM}{OM} \quad \cotan\alpha = \frac{1}{\tan\alpha} = \frac{OP}{PM}$$

$$\tan\alpha = \frac{PM}{OP}$$

6.2 Geometric interpretation

$$\overline{OA} = \overline{OM} = R = 1$$

$$\overline{OQ} = \sin\alpha$$

$$\overline{OP} = \cos\alpha$$

$$\overline{AT} = \tan\alpha$$

$$\overline{BT'} = \cotan\alpha$$

6.3 Trigonometric relations

$$\cos^2\alpha + \sin^2\alpha = 1$$

$$\tan\alpha = \frac{\sin\alpha}{\cos\alpha}$$

$$\cotan\alpha = \frac{\cos\alpha}{\sin\alpha} = \frac{1}{\tan\alpha}$$

$$\sin 2\alpha = 2\sin\alpha\cos\alpha$$

$$\cos 2\alpha = \cos^2\alpha - \sin^2\alpha$$
$$= 1 - 2\sin^2\alpha$$

$$\tan 2\alpha = \frac{2\tan\alpha}{1 - \tan^2\alpha}$$

$$\sin(\alpha + \beta) = \sin\alpha\,\cos\beta + \cos\alpha\,\sin\beta$$

$$\cos(\alpha + \beta) = \cos\alpha\,\cos\beta - \sin\alpha\,\sin\beta$$

$$\sin(\alpha - \beta) = \sin\alpha\,\cos\beta - \cos\alpha\,\sin\beta$$

$$\tan(\alpha + \beta) = \frac{\tan\alpha + \tan\beta}{1 - \tan\alpha\,\tan\beta}$$

$$\tan(\alpha - \beta) = \frac{\tan\alpha - \tan\beta}{1 + \tan\alpha\,\tan\beta}$$

6.4 Values of trigonometric functions related to half-angle tangents

$$\tan\frac{\alpha}{2} = t \qquad \cos\alpha = \frac{1-t^2}{1+t^2} \qquad \sin\alpha = \frac{2t}{1+t^2} \qquad \tan\alpha = \frac{2t}{1-t^2}$$

6.5 Relations between sides and angle of any triangle

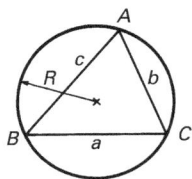

$$\hat{A} + \hat{B} + \hat{C} = \pi$$

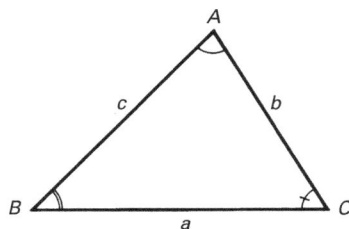

$$\frac{a}{\sin\hat{A}} = \frac{b}{\sin\hat{B}} = \frac{c}{\sin\hat{C}} = 2R$$

$$a^2 = b^2 + c^2 - 2bc\ \cos\hat{A}$$
$$b^2 = c^2 + a^2 - 2ca\ \cos\hat{B}$$
$$c^2 = a^2 + b^2 - 2ab\ \cos\hat{C}$$

A7 GEOMETRY FORMULAS FOR AREAS AND VOLUMES

Table A6   Geometry formulas for areas and volumes [3, 4].

Area	Volume
Triangle $$p = \frac{a+b+c}{2}$$ $$S = \frac{ah}{2} = \frac{abc}{4R} = pr$$	**Regular or oblique prism** $$V = Bh$$
Parallelogram $$S = bh$$	**Right cylinder** $$V = \pi R^2 h = Bh$$ **Hollow cylinder** $$V = \pi(R^2 - r^2)h = \pi(R+r)eh$$
Square: $S = a^2$ Rectangle: $S = ab$	**Right cone** $$V = \frac{\pi R^2 h}{3}$$
Trapezoid $$S = \frac{AB + CD}{2}h = MN \cdot h$$	**Truncated right cone** $$V = \frac{\pi h}{3}(R^2 + r^2 + Rr)$$
Circle $$C = 2\pi R = \pi D$$ $$S = \pi R^2 = \frac{\pi D^2}{4} = \frac{C^2}{4\pi}$$	**Pyramid** $$V = \frac{1}{3}Bh$$
Sector of a circle $$S = \frac{\text{arc } ABC \cdot R}{2} = \frac{\pi R^2 \alpha}{360}$$ (α is the number of degrees of arc ACB) **Segment of a circle** $$S = \frac{\pi R^2 \beta}{360} - \frac{DF}{2}(R-f)$$	**Truncated pyramid with parallel bases** $$V = \frac{1}{3}h(B + b + \sqrt{Bb})$$
Annulus $$S = \frac{\pi}{4}(D^2 - d^2) = \pi(R^2 - r^2)$$ $$= \frac{\pi}{4}(D+d)(D-d)$$ $$= \pi(R+r)(R-r)$$	**Sphere** $S = 4\pi R^2 = \pi D^2$ $$V = \frac{4}{3}\pi R^3 = 4{,}189R^3$$ **Hollow sphere** $$V = \frac{4}{3}\pi(R^3 - r^3)$$
Ellipse a = semi-major axis b = semi-minor axis $$S = \pi ab$$	**Spherical segment with one base** $$1°) \;\; V = \frac{1}{6}\pi h(h^2 + 3\overline{Al}^2)$$ $$2°) \;\; V = \frac{1}{3}\pi h^2(3R - h)$$ **Spherical segment with two bases** $$V = \frac{1}{6}\pi h(3a^2 + 3b^2 + h^2)$$

A8 CONTENT OF HORIZONTAL CYLINDRICAL TANKS [4]

Knowing the characteristics: total volume V and total height H, and measuring the liquid height h, from Fig. A1, the volume of the liquid stocked in the tank can be evaluated.

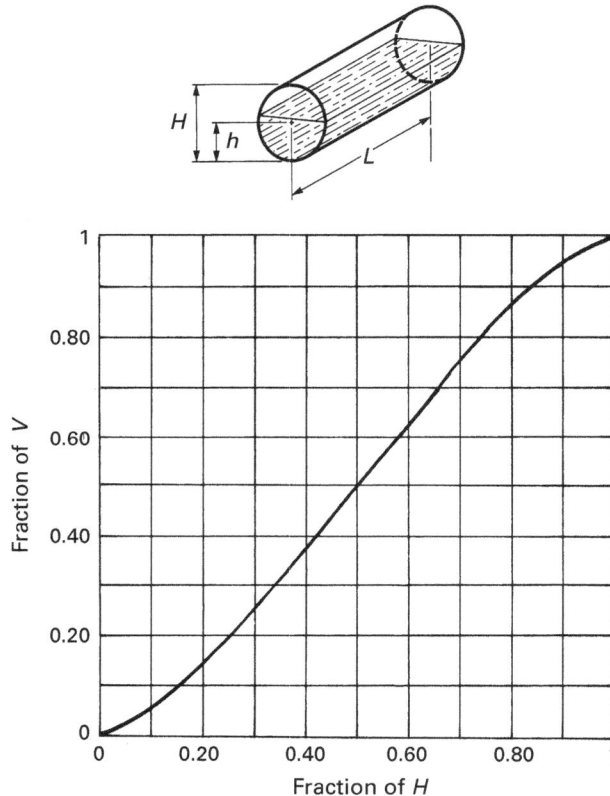

Figure A1 Evaluation of fluid volume in tanks.

▼ Example

Consider a tank of volume $V = 15\ 000$ liters and height $H = 2$ m. Measurements show a height $h = 0.40$ m of liquid in the tank.

How much liquid does the tank contain?

Using Fig. A1, from a fraction of H : $0.4/2 = 0.2$, then a fraction of V of about 0.15 can be evaluated. The quantity of liquid contained in the tank is therefore approximately: $0.15 \times 15\ 000 = 2\ 250$ liters. ▲

A9 MECHANICS AND STRENGTH OF MATERIALS [3, 4]

9.1 Moment of a force about a point. Moment of a torque

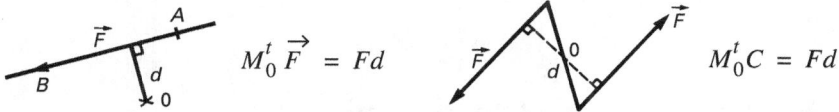

$$M_0^t \vec{F} = Fd \qquad M_0^t C = Fd$$

(M_0^t in newton-meters, F in newtons and d in meters).

9.2 Uniform straight line motion

$$I = I_0 + vt$$

I distance travelled (m)
I_0 initial distance (m)
v velocity (m/s)
t time (s).

9.3 Uniformly-accelerated motion

$$I = I_0 + v_0 t + \frac{\gamma t^2}{2}$$

I distance travelled (m)
I_0 initial distance (m)
v_0 initial velocity (m/s)
t time (s)
γ acceleration (m/s²).

9.4 Uniform circular motion

Angular velocity:

$$\omega = \frac{\alpha}{t} \quad \text{or} \quad \alpha = \omega t$$

(α: angle of rotation during time t).

Angular velocity as a function of revolutions per minute:

$$\omega = \frac{2\pi N}{60}$$

(ω in radians per second and N in revolutions per minute).

Circumferential velocity:

$$v \,(\text{m/min}) = 2\pi RN \quad \text{or} \quad v\,(\text{m/s}) = \omega R = \frac{2\pi RN}{60}$$

(ω in radians per second, R in meters and N in revolutions per minute).

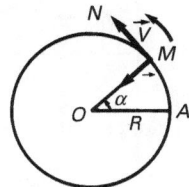

Centripetal acceleration γ_c:

$$\gamma_c = \omega^2 R \quad \text{or} \quad \gamma_c = \frac{V^2}{R}$$

(γ_c in meters per second per second, ω in radians per second, R in meters and V in meters per second).

9.5 Fundamental formula of dynamics

$$F = m\gamma \quad \text{with } m = \text{mass}, \gamma = \text{acceleration}$$

(F in newtons, m in kilograms and γ in meters per second per second).

Specific case of gravity:

$$P = m\vec{g}$$

($\vec{g}$ = gravitational acceleration, approximately 9.81 m/s^2).

9.6 Centrifugal force

$$f_c = m\omega^2 R \quad \text{or} \quad f_c = m\frac{V^2}{R}$$

(f_c in newtons, m in kilograms, ω in radians per second, R in meters and V in meters per second).

9.7 Work of a force

Constant force in quantity and direction displacing its point of application:

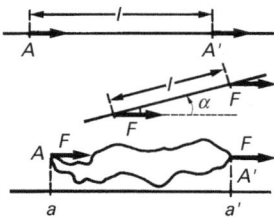

1. On its action line $\quad\quad\quad\quad\quad\quad T = F\ell$
2. On an oblique line to its action line $\quad T = F\ell\cos\alpha$
3. On a curve in its plane $\quad\quad\quad\quad T = Faa'$

(T in joules, F in newtons and I in meters).

Constant force moving tangentially to a circle:

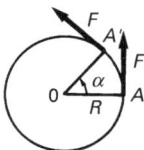

$$T = FR\alpha = M_0^t F\alpha$$
for one rotation $T = 2\pi RF$

(T in joules, F in newtons, R in meters, α in radians and M_0^t in newtons-meters).

9.8 Work of a torque

Torque rotating about an axis perpendicular to its plane:

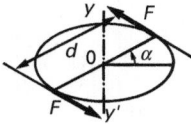

$$T = Fd\,\alpha = M_0^t C\,\alpha$$

for one rotation $T = 2\pi M_0^t C = 2\pi Fd$

(T in joules, F in newtons, d in meters,
α in radians and M_0^t in meter newtons).

9.9 Power

Work produced per unit time:

$$P = \frac{T}{t}$$

(P in watts, T in joules and t in seconds).

Power of a torque rotating at constant speed ω:

$$P = M_0^t C\,\omega \quad \text{or} \quad P = Fd\,\omega = Fd\frac{2\pi N}{60}$$

(P in watts, M_0^t in newtons-meters, ω in radians per second, F in newtons, d in meters and N in revolutions per minute).

9.10 Kinetic energy

$$W = \frac{1}{2}\,mv^2$$

(W in joules, m in kilograms and v in meters per second).

9.11 Strength of materials

Tension and compression:

Stress: $$n = \frac{N}{S}10^{-6}$$

with
- n stress (MPa)
- N tensile or compressive force (N)
- S cross-sectional area (m^2).

Hooke's law: $$n = E\frac{\Delta\ell}{\ell}$$

with
- E   Young's modulus or longitudinal elastic modulus: approximately 200 000 to 220 000 MPa for steel

$\Delta\ell$ elongation $\Big\}$ expressed in the same units.
ℓ length

Torsion:

Torque: $M_t = 2Fr$

(M_t in newton-meters, F in newtons and r in meters).

Torsional unit: $\theta = \dfrac{\alpha}{\ell}$ $\qquad\qquad$ $\dfrac{\alpha}{\ell} = \dfrac{M_t}{GI_0}$

Hooke's law: $t_{max} = Gr\,\theta$ $\qquad$ $t_{max} = \dfrac{M_t}{\dfrac{I_0}{r}}$

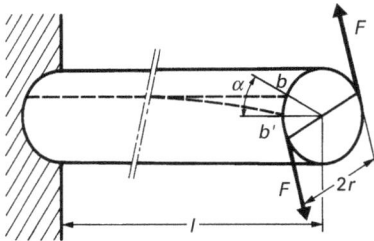

with
- θ torsional unit (rad/m)
- α angle of rotation (rad)
- ℓ length (m)
- t torsional or tangential shear stress (MPa)
- G transverse elastic modulus:
  - $G = 0.4 \times E$ (Young's modulus)
  - $G = 80\ 000$ MPa for steel
- r radius of cylinder (m)
- I_0 polar moment of inertia.

A10 ELECTRICITY. DIRECT CURRENT [3, 4]

10.1 Intensity of current: *I*

Unit: ampere (A).

Constant current which, maintained in two straight parallel conductors of infinite length and negligible circular cross-sectional area, and placed one meter apart in a vacuum, produces a force of 2.10^{-7} newtons per meter of length between these conductors.

10.2 Quantity of electricity: *Q*

Unit: coulomb (C).

Quantity of electricity transmitted in one second by a current of one ampere.

Practical unit: ampere-hour (Ah).

Quantity of electricity transmitted in one hour by a current of one ampere (1 Ah = 3 600 C):

$$Q\,(\text{Ah}) = I\,(\text{A})\,t\,(\text{h})$$

10.3 Potential difference (Voltage): *U*

Unit: volt (V).

Potential difference between two points of conducting wire carrying a constant current of one ampere when the power dissipated between these points is one watt.

10.4 Resistance: *R*

Unit: ohm (Ω).

Resistance between two points of a conducting wire when a potential difference of one volt, applied between these two points, produces a current of one ampere in the conductor, the conductor not being a source of any electromotive force.

Resistivity: ρ (Ω/m/mm^2) at 15°C.

Resistance of a wire one meter long with a cross-sectional area of one square millimeter

	ρ (Ω/m per mm^2)		ρ (Ω/m per mm^2)
Copper	0.017 – 0.0175	Iron	0.11
Silver	0.016 – 0.018	Steel	0.10 – 0.25
Aluminium	0.229 – 0.0175	Nickel/silver	0.36 – 0.39
		(Cu 60%, Zn 20%, Ni 20%)	

$$R = \rho \frac{\ell}{S}$$

(ℓ length of conductor in meters, S: cross-sectional area of conductor in square millimeters).

10.5 Temperature coefficient of a resistance and resistivity

$$R_t = R_0 (1 + \alpha t) \qquad \rho_t = \rho_0 (1 + \alpha t)$$

with

R_t, ρ_t resistance, resistivity at t°C
R_0, ρ_0 resistance, resistivity at 0°C
α temperature coefficient at 15°C.

	α		α
Copper	3.93×10^{-3}	Iron	4.7×10^{-3}
Silver	3.6×10^{-3}	Steel	5×10^{-3}
Aluminium	3.9×10^{-3}	Nickel/silver	3×10^{-4}
		(Cu 60%, Zn 20%, Ni 20%)	

10.6 Resistance connections

1. Connection in series:

$$R = R_1 + R_2 + R_3 \ldots$$
$$U = U_1 + U_2 + U_3 \ldots$$

I constant

2. Connection in parallel:

$$\frac{1}{R} = \frac{1}{R_1} + \frac{1}{R_2} + \frac{1}{R_3} \cdots \quad U \text{ constant}$$

$$I = I_1 + I_2 + I_3 \cdots$$

For two resistances in parallel:

$$R = \frac{R_1 R_2}{R_1 + R_2} \qquad I_1 = I\frac{R_2}{R_1 + R_2} \qquad I_2 = I\frac{R_1}{R_1 + R_2}$$

10.7 Ohm's law

$$U = RI \qquad I = \frac{U}{R} \qquad R = \frac{U}{I} \qquad R\ (\Omega), I\ (A), U\ (V)$$

10.8 Electrical energy: *W,* or quantity of heat: *Q*

Unit: joule (J).

Electrical energy generated each second by a current of one ampere flowing through a resistance of one ohm:

$$\begin{array}{cc} W = R\ I^2\ t & W = U\ I\ t \\ (J)\ (\Omega)\ (A)\ (s) & (J)\ (V)\ (A)\ (s) \end{array}$$

Non SI units:

1. watt-hour (Wh).

Energy expended in one hour by a power of one watt:

$$\begin{array}{cc} W = R\ I^2\ t & 1\ Wh = 3600\ J \\ (Wh)\ (\Omega)\ (A)\ (h) \end{array}$$

2. calorie (cal):

$$\begin{array}{ccc} Q = 0.24\ R\ I^2\ t & 1\ cal = 4.1855\ J & 1\ J = 0.2389\ cal \\ (cal)\ (\Omega)\ (A)\ (h) \end{array}$$

(4.1855 is an experimental value).

10.9 Electrical power: *P*

Unit: watt (W).

Power of one joule per second:

$$\begin{array}{ccc} P = R\ I^2 & P = U\ I & P = \frac{U^2(V)}{R\ (\Omega)} \\ (W)\ (\Omega)\ (A) & (W)\ (V)\ (A) & (W) \end{array}$$

A11   ELECTRICITY. ALTERNATING CURRENT [3, 4]

11.1   General

Voltage-current

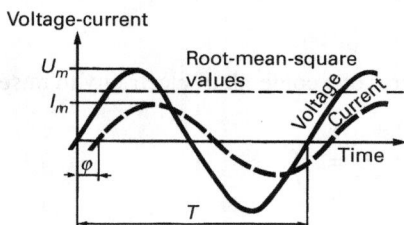

Period $T = \dfrac{1}{F}$

Frequency $F = \dfrac{1}{T}$ (Hz)

Angular frequency $\omega = 2\pi F$ (rad/s)

Instantaneous values:

$u = U_m \cos \omega t$

$i = I_m \cos (\omega t - \varphi)$

φ angle of phase difference between current and voltage

Root-mean-square values (rms values):

$$U = \frac{U_m}{\sqrt{2}} \qquad I = \frac{I_m}{\sqrt{2}}$$

Power:

1. Applied power $S = UI$, in volt-amperes (VA)
2. Active power $P = UI \cos \varphi$, in watts (W)
3. Reactive power $Q = UI \sin \varphi$, in reactive volt-amperes (VAR)

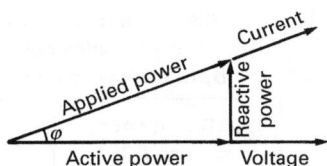

$$S^2 = P^2 + Q^2$$

$$\tan \varphi = \frac{Q}{P}$$

$$\cos \varphi = \frac{P}{S} \text{ (power factor)}$$

11.2   Three-phase system

Phase interconnection (formulas valid with same load for all 3 phases)

Star connection   Mesh or delta connection

$U_v = 1.73\, U_p$ $I = I_p$ $U_v = U_p$ $I = 1.73\, I_p$

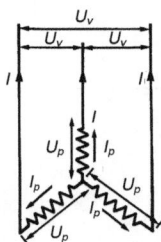

1. Apparent power: $S = UI$ (VA)
2. Active power: $P = 1.73\, U_v I \cos\varphi$ (W)

                  $= 3 U_p I_p \cos\varphi$ (W)

3. Reactive power: $Q = \sqrt{S^2 - P^2}$

                  $= 1.73\, U_v I \sin\varphi$ (VAR)

                  $= 3 U_p I_p \sin\varphi$ (VAR)

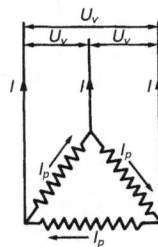

where

U_v voltage in volts between two conductors of the three-phase system
U_p voltage for each phase
I intensity in amperes through each conductor of the three-phase line
I_v intensity per phase
φ phase difference between current and voltage.

Capacitance: *C*

Unit: farad (F). A capacitance of one farad requires one coulomb of electricity to raise its potential one volt:

$$1 \text{ farad} = \frac{1 \text{ coulomb}}{1 \text{ volt}} \qquad C = \frac{Q}{U}$$

Connections of capacitors (or condensers)

Capacitors in parallel:

$$C = C_1 + C_2 + C_3 + \ldots$$

Capacitors in series:

$$\frac{1}{C} = \frac{1}{C_1} + \frac{1}{C_2} + \frac{1}{C_3} + \ldots \quad \text{for 2 capacitors: } C = \frac{C_1 \cdot C_2}{C_1 + C_2}$$

Intensities of current acceptable through conductors

Nominal cross-sectional area (mm²)	Intensity of current Temperature rise = 45°C			For temperature rises different from 45°C multiply the intensities opposite by the following coefficients	
	Number of conductors				
	2	3	4	Temperature rise (°C)	Coefficient
	Intensities of current (A)				
2	20	17	15		
3	27	22.5	21	20	0.67
5	35	31	28	25	0.75
10	53	47	44	30	0.82
16	66	60	55	35	0.88
25	88	81	70	40	0.94
40	110	103	88	45	1
50	130	123	105	50	1.05
75	167	154	132	55	1.10
95	192	184	155	60	1.15

A12 PRINCIPAL CHEMICAL SYMBOLS, ATOMIC NUMBERS AND WEIGHTS

Table A7 Principal chemical symbols, atomic numbers and weights [3, 4].

Name	Symbol	Atomic number	Atomic weight	Name	Symbol	Atomic number	Atomic weight
Aluminium	Al	113	27.0	Mercury	Hg	80	200.6
Antimony	Sb	51	122.0	Molybdenum	Mo	42	96.0
Argon	A	18	40.0	Neon	Ne	10	20.0
Arsenic	As	33	75.0	Nickel	Ni	28	58.7
Barium	Ba	56	137.0	Nitrogen	N	7	14.0
Bismuth	Bi	83	209.0	Oxygen	O	8	16.0
Boron	B	5	11.0	Phosphorus	P	15	31.0
Bromine	Br	35	80.0	Platinium	Pt	78	195.0
Cadmium	Cd	48	112.0	Plutonium	Pu	94	242.0
Calcium	Ca	20	40.0	Potassium	K	19	397.0
Carbon	C	6	12.0	Radium	Ra	88	226.0
Chlorine	Cl	17	35.5	Selenium	Se	34	79.0
Chromium	Cr	24	52.0	Silicon	Si	14	28.0
Cobalt	Co	27	59.0	Silver	Ag	47	108.0
Copper	Cu	29	63.5	Sodium	Na	11	23.0
Fluorine	F	9	19.0	Strontium	Sr	38	87.6
Gold	Au	79	197.0	Sulfur	S	16	32.0
Helium	He	2	4.0	Tin	Sn	50	119.0
Hydrogen	H	1	1.0	Titanium	Ti	22	48.0
Iodine	I	53	127.0	Tungsten	W	74	184.0
Iron	Fe	26	56.0	Uranium	U	92	238.0
Lead	Pb	82	207.0	Vanadium	V	23	51.0
Lithium	Li	3	7.0	Xenon	Xe	54	131.3
Magnesium	Mg	12	24.0	Zinc	Zn	30	65.4
Manganese	Mn	25	55.0	Zirconium	Zr	40	91.0

A13 SPECIFIC GRAVITY OF VARIOUS MATERIALS AND FLUIDS

Table A8 Specific gravity of various materials and fluids [3, 4].

Name	Specific gravity	Name	Specific gravity
Rock:		**Materials:**	
Dry sand	2.6	Baryte (barium sulfate)	4.2 to 4.3
Gypsum	2.3 to 2.37	Compact brick	2.2
Granite	2.4 to 3.0	Compact clay	2.1
Hard limestone	2.4 to 2.7	Concrete	2.25
Marble	2.5 to 2.9	Glass	2.53
Medium-hard limestone	1.9 to 2.3	Portland cement (powder)	3.0 to 3.3
Quartzite	2.2 to 2.8	Portland cement slurry	1.8 to 2.0
Rock salt	2.16	Walnut shells	1.3
Sandstone	1.9 to 2.6	**Gas (at 10°C and 760 mmHg in relation to air):**	
Liquids (at 25°C):		Air	1
Acetone	0.791	Isobutane	2.067
Benzene	0.878	*n*-butane	2.0854
Carbon tetrachloride	1.595	Carbon dioxide	1.529
Chloroform	1.482	Carbon monoxide	0.9671
Ether	0.714	Ethane	1.0493
Ethyl alcohol	0.816	Ethylene	0.9749
Glycerin	1.260	hydrogen	0.06952
Methyl alcohol	0.792	Hydrogen sulfide	1.19
Trichloroethylene	1.455	Methane	0.5544
Water at 4°C	1	Oxygen	1.10527
		Propane	1.554

A14 PHYSICAL PROPERTIES OF METALS

Table A9 Physical properties of metals [3, 4].

Name	Symbol	Specific gravity	Melting point (°C)	Brinell hardness	Mohs scale
Aluminium	Al	2.70	660	16	2.5
Antimony	Sb	6.70	631	–	3.2
Bismuth	Bi	9.75	271	–	2.5
Cadmium	Cd	8.65	321	23	2
Chromium	Cr	7.19	1 890	70-130	9
Cobalt	Co	8.90	1 495	124	–
Copper	Cu	8.94	1 083	–	2.5
Gold	Au	19.32	1 063	–	2.5
Iron	Fe	7.88	1 535	77	4.5
Lead	Pb	11.34	327	4	1.5
Magnesium	Mg	1.74	651	29	2
Manganese	Mn	7.20	1 260	–	5
Mercury	Hg	13.55	– 39	–	–
Molybdenum	Mo	10.20	2 620	150-200	–
Nickel	Ni	8.90	1 455	110-300	–
Platinium	Pt	21.45	1 774	64	4.3
Silver	Ag	10.50	961	–	2.5-7
Tin	Sn	7.30	232	–	1.7
Titanium	Ti	4.50	1 800	–	–
Tungsten	W	19.30	3 370	350	–
Vanadium	V	5.96	1 710	–	–
Zinc	Zn	7.14	419	–	2.5

General Data

A15 STRATIGRAPHIC SCALE

Table A10 Stratigraphic scale [3, 4].

Era	Period	Formations
Quaternary (Anthropozoic)	Holocene (Neolithic) Pleistocene (Paleolithic)	Flandrian Tyrrhenian Sicilian
Tertiary (Cenozoic)	Pliocene	Calabrian (Villafranchian) Astian Plaisancian
	Miocene	Saphelian (Pontian) Vindobonian Burdigalian
	Oligocene	Aquitanian Chattian Stampian Sannoisian
	Eocene	Ludian Bartonian Lutetian Ypresian Sparnacian Thanetian Montian

Era	Period	Formations
Secondary (Mesozoic)	Upper Cretaceous	Danian Senonian Turonian Cenomanian
	Lower Cretaceous (Eocretaceous)	Albian Aptian Barremian (Urgonian) Hauterivian Valanginian
	Upper Jurassic (Malm)	(Purbeckian) Portlandian (Tithonic) Kimmeridgian Sequanian Rauracian Argovian Oxfordian Callovian
	Middle Jurassic (Dogger)	Bathonian Bajocian
	Lower Jurassic (Lias)	Aalenian Toarcian Charmouthian Sinemurian Hettangian Rhetian
	Trias	Keuper Muschelkalk Bunter

Era	Period	Formations
	Permian	Zechstein or Thuringian Saxonian Autunian
	Carboniferous	Coal form. (Stephanian) (Wesphalian) Dinantian (Culm)
Primary (Paleozoic)	Devonian	Fammennian Frasnian Givetian Eifelian Coblencian Gedinnian Downtonian
	Silurian	Gothlandian Ordovician
	Cambrian	Potsdamian Acadian Georgian
	Precambrian (Algonkian)	
	Archean	

REFERENCES

1 *The SI Metric System of Units and SPE Metric Standard* (1984). SPE, Dallas
2 Koederitz LF, Harvey AH, Honarpour M (1989) *Introduction to Petroleum Reservoir Analysis.* Gulf Publishing Company Book Division, Houston, TX
3 Gabolde G, Nguyen JP (1999) *Drilling Data Handbook,* 7th Ed. Editions Technip, Paris
4 Publication de la Chambre Syndicale de la Recherche et de la Production du Pétrole et du Gaz Naturel (1970) *Formulaire du Producteur*. Editions Technip, Paris.

B

Casing and Tubing

Casing and Tubing

B1 INTRODUCTION

The American Petroleum Institute (API) and the International Organization for Standard-ization (ISO) have established standards for oil/gas tubing and casing. Tubing is defined as pipe with nominal diameters from 1.050 to 7.0 in. (ISO 11 960), while casing sizes range from 4 1/2 to 20 in.

  Casing and tubing are classified according to five properties:
- Manner of manufacture
- Steel grade
- Type of joints
- Length range
- Wall thickness (unit weight).

  Standards are presented in Chapter Y.

B2 GEOMETRICAL CHARACTERISTICS OF CASINGS

Tables B1 and B2 introduce dimensions, masses and capacities of casings which are com-monly used after drilling of a well (in U.S. and metric units).

B3 DIMENSIONS AND MASSES OF TUBINGS

Geometric characteristics are introduced in Tables B3 and B4 in relation with grade and nominal weight of tubings from 1.050 in. to 7 in. following the Standard ISO 11 960.

B4 STEEL GRADE AND TENSILE REQUIREMENTS OF TUBINGS

A number of grades of steel and types of tubing connections have been developed to meet demands of greater depths and new completion techniques.

  Numbers in the grade designations indicate minimum yield strength in 1 000 psi. Grade of new pipe can be identified by color bands (see pages 42 to 46).

Table B1 Geometrical characteristics of casings (4 1/2 to 7 5/8 in.).
(*Source*: API Spec 5CT and ISO 11 960)

OD (in. & mm)	Nominal weight		Wall thickness		Inside diameter		Steel Cross section		Capacity		Drift diameter	
	(lbm/ft)	(kg/m)	(in.)	(mm)	(in.)	(mm)	(sq in.)	(mm²)	(gal/ft)	(l/m)	(in.)	(mm)
4 1/2 114.3	9.50	13.9	0.205	5.2	4.090	103.9	2.77	1 786	0.68	8.48	3.965	100.7
	10.50	15.3	0.224	5.7	4.052	102.9	3.01	1 941	0.67	8.32	3.925	99.7
	11.60	16.9	0.250	6.4	4.000	101.6	3.34	2 154	0.65	8.11	3.874	98.4
	13.50	19.7	0.290	7.4	3.920	99.6	3.84	2 476	0.63	7.79	3.795	96.4
	15.10	22.0	0.337	8.6	3.826	97.2	4.41	2 844	0.60	7.42	3.701	94.0
	16.90	24.7	0.380	9.7	3.740	95.0	4.92	3 173	0.57	7.09	3.614	91.8
	17.70	25.8	0.402	10.2	3.697	93.9	5.17	3 336	0.56	6.93	3.571	90.7
	18.80	27.4	0.430	10.9	3.640	92.5	5.50	3 647	0.54	6.71	3.516	89.3
5 127	11.50	16.8	0.220	5.6	4.560	115.8	3.30	2 132	0.85	10.54	4.433	112.6
	13.00	19.0	0.253	6.4	4.494	114.1	3.78	2 436	0.82	10.23	4.370	111.0
	15.00	21.9	0.296	7.5	4.408	112.0	4.38	2 823	0.79	9.84	4.283	108.8
	18.00	26.3	0.362	9.2	4.276	108.6	5.27	3 401	0.75	9.27	4.150	105.4
	20.80	30.4	0.422	10.7	4.156	105.6	6.07	3 916	0.70	8.75	4.031	102.4
5 1/2 139.7	14.00	20.4	0.244	6.2	5.012	127.3	4.03	2 600	1.02	12.73	4.886	124.1
	15.50	22.6	0.275	7.0	4.950	125.7	4.51	2 910	1.00	12.42	4.827	122.6
	17.00	24.8	0.304	7.7	4.892	124.3	4.96	3 201	0.98	12.13	4.768	121.1
	20.00	29.2	0.361	9.2	4.778	121.4	5.83	3 760	0.93	11.57	4.654	118.2
	23.00	33.6	0.451	10.5	4.670	118.6	6.63	4 277	0.89	11.05	4.543	115.4
6 5/8 168.3	20.00	29.2	0.288	7.3	6.049	153.6	5.74	3 701	1.49	18.54	5.925	150.5
	24.00	35.0	0.352	8.9	5.921	150.4	6.94	4 475	1.43	17.76	5.795	147.2
	28.00	40.9	0.417	10.6	5.791	147.1	8.13	5 246	1.37	16.99	5.665	143.9
	32.00	46.7	0.475	12.1	5.675	144.2	9.17	5 919	1.31	16.32	5.551	141.0
7 177.8	17.00	24.8	0.231	5.9	6.538	166.1	4.91	3 171	1.74	21.66	6.413	162.9
	20.00	29.2	0.272	6.9	6.456	164.0	5.75	3 710	1.70	21.12	6.331	160.8
	23.00	33.6	0.317	8.1	6.366	161.7	6.65	4 293	1.65	20.54	6.240	158.5
	26.00	37.9	0.362	9.2	6.276	159.4	7.55	4 868	1.61	19.96	6.150	156.2
	29.00	42.3	0.408	10.4	6.184	157.1	8.45	5 450	1.56	19.38	6.059	153.9
	32.00	46.7	0.453	11.5	6.094	154.8	9.32	6 013	1.52	18.82	5.969	151.6
	35.00	51.1	0.498	12.7	6.004	152.5	10.17	6 563	1.47	18.27	5.878	149.3
	38.00	55.5	0.540	13.7	5.920	150.4	10.96	7 072	1.43	17.76	5.795	147.2
	41.00	59.8	0.590	15.0	5.820	147.8	11.88	7 662	1.38	17.17	5.693	144.6
	44.00	64.2	0.640	16.3	5.720	145.3	12.78	8 247	1.34	16.58	5.594	142.1
7 5/8 193.7	24.00	35.0	0.300	7.6	7.025	178.4	6.90	4 454	2.01	25.01	6.902	175.3
	26.40	38.5	0.328	8.3	6.969	177.0	7.52	4 850	1.98	24.61	6.843	173.8
	29.70	43.3	0.375	9.5	6.875	174.6	8.54	5 508	1.93	23.95	6.748	171.4
	33.70	49.2	0.430	10.9	6.765	171.8	9.72	6 270	1.87	23.19	6.642	168.7
	39.00	56.9	0.500	12.7	6.625	168.3	11.19	7 221	1.79	22.24	6.500	165.1

Table B2   Geometrical characteristics of casings (8 5/8 to 20 in.).
(*Source*: API Spec 5CT and ISO 11 960)

OD (in. & mm)	Nominal weight		Wall thickness		Inside diameter		Steel Cross section		Capacity		Drift diameter	
	(lbm/ft)	(kg/m)	(in.)	(mm)	(in.)	(mm)	(sq in.)	(mm²)	(gal/ft)	(l/m)	(in.)	(mm)
8 5/8 219.1	24.00	35.0	0.264	6.7	8.097	205.7	6.94	4 477	2.67	33.22	7.972	202.5
	28.00	40.9	0.304	7.7	8.017	203.6	7.95	5 126	2.62	32.57	7.894	200.5
	32.00	46.7	0.352	8.9	7.921	201.2	9.15	5 902	2.56	31.79	7.795	198.0
	36.00	52.5	0.400	10.2	7.825	198.8	10.34	6 668	2.50	31.03	7.701	195.6
	40.00	58.4	0.450	11.4	7.725	196.2	11.56	7 456	2.43	30.24	7.598	193.0
	44.00	64.2	0.500	12.7	7.625	193.7	12.76	8 234	2.37	29.46	7.500	190.5
	49.00	71.5	0.557	14.2	7.511	190.8	14.12	9 110	2.30	28.58	7.386	187.6
9 5/8 244.5	32.30	47.1	0.312	7.9	9.001	228.6	9.12	5 886	3.31	41.06	8.843	224.6
	36.00	52.5	0.352	8.9	8.921	226.6	10.25	6 615	3.25	40.33	8.764	222.6
	40.00	58.4	0.395	10.0	8.835	224.4	11.45	7 387	3.18	39.55	8.677	220.4
	43.50	63.5	0.435	11.1	8.755	222.4	12.56	8 103	3.13	38.84	8.598	218.4
	47.00	68.6	0.472	12.0	8.681	220.5	13.57	8 757	3.07	38.18	8.524	216.5
	53.50	78.1	0.545	13.8	8.535	216.8	15.54	10 028	2.97	36.91	8.378	212.8
	58.40	85.2	0.595	15.1	8.435	214.3	16.88	10 888	2.90	36.05	8.283	210.4
	61.10	89.2	0.625	15.9	8.375	212.7	17.67	11 398	2.86	35.54	8.217	208.7
	71.80	104.8	0.750	19.1	8.125	206.4	20.91	13 491	2.69	33.45	7.969	202.4
10 3/4 273.1	32.75	47.8	0.279	7.1	10.192	258.9	9.18	5 924	4.24	52.63	10.035	254.9
	40.50	59.1	0.350	8.9	10.050	255.3	11.44	7 378	4.12	51.18	9.894	251.3
	45.50	66.4	0.400	10.2	9.950	252.7	13.01	8 391	4.04	50.17	9.795	248.8
	51.00	74.4	0.450	11.4	9.850	250.2	14.56	9 394	3.96	49.16	9.693	246.2
	55.50	81.0	0.495	12.6	9.760	247.9	15.94	10 286	3.89	48.27	9.602	243.9
	60.70	88.6	0.545	13.8	9.660	245.4	17.47	11 270	3.81	47.29	9.504	241.4
	65.70	95.9	0.595	15.1	9.560	242.8	18.98	12 244	3.73	46.31	9.406	238.9
11 3/4 298.5	42.00	61.3	0.333	8.5	11.084	281.5	11.95	7 707	5.01	62.25	10.929	277.6
	47.00	68.6	0.375	9.5	11.000	279.4	13.39	8 641	4.94	61.32	10.843	275.4
	54.00	78.8	0.435	11.1	10.880	276.4	15.46	9 977	4.83	59.98	10.724	272.4
	60.00	87.6	0.489	12.4	10.772	273.6	17.30	11 160	4.73	58.80	10.614	269.6
13 3/8 339.7	48.00	70.1	0.330	8.4	12.715	323.0	13.52	8 723	7.30	90.65	12.559	319.0
	54.50	79.5	0.380	9.7	12.615	320.4	15.51	10 007	6.49	80.64	12.461	316.5
	61.00	89.0	0.430	10.9	12.515	317.9	17.48	11 280	6.39	79.37	12.358	313.9
	68.00	99.2	0.480	12.2	12.415	315.3	19.44	12 543	6.29	78.10	12.260	311.4
	72.00	105.1	0.514	13.1	12.347	313.6	20.77	13 403	6.22	77.24	12.193	309.7
16 406.4	65.00	94.9	0.375	9.5	15.250	387.4	18.40	11 870	9.49	117.85	15.063	382.6
	75.00	109.5	0.438	11.1	15.124	384.1	21.42	13 821	9.33	115.90	14.937	379.4
	84.00	122.6	0.495	12.6	15.010	381.3	24.11	15 552	9.19	114.16	14.823	376.5
18 5/8 473.1	87.50	127.7	0.435	11.1	17.755	451.0	24.86	16 039	12.86	159.73	17.571	446.3
20 508.4	94.00	137.2	0.438	11.1	19.124	485.7	26.93	17 374	14.92	185.31	18.937	481.0
	106.50	155.4	0.500	12.7	19.000	482.6	30.63	19 762	14.73	182.92	18.815	477.9
	133.00	194.1	0.635	16.1	18.730	475.7	38.63	24 925	14.31	177.76	18.543	471.0

B

Table B3 1.050 to 3 1/2 in. tubings. Dimensions and masses.
(*Source*: ISO 11960 and API 5CT)

Tubing size OD (in. & mm)	Nominal weight (lbm/ft)			Nominal weight (kg/m)			Wall thickness t		Inside dia. d		Drift diameter		Steel cross section	
	NU	EUE	IJ	NU	EUE	IJ	(in.)	(mm)	(in.)	(mm)	(in.)	(mm)	(sq in.)	(mm²)
1.050 *26.67*	1.14	1.20		1.70	1.79		0.113	2.87	0.824	20.93	0.730	18.54	0.351	227
	1.48	1.54		2.20	2.29		0.154	3.91	0.742	18.85	0.648	16.46	0.471	304
1.315 *33.40*	1.70	1.80	1.72	2.53	2.68	2.56	0.133	3.38	1.049	26.64	0.955	24.26	0.522	337
	2.19	2.24		3.26	3.33		0.179	4.55	0.957	24.31	0.863	21.92	0.689	445
1.660 *42.16*			2.10			3.13	0.125	3.18	1.410	35.81	1.316	33.43	0.627	405
	2.30	2.40	2.33	3.42	3.57	3.47	0.140	3.56	1.380	35.05	1.286	32.66	0.697	450
	3.03	3.07		4.51	4.57		0.191	4.85	1.278	32.46	1.184	30.07	0.937	605
1.900 *48.26*			2.40			3.57	0.125	3.18	1.650	41.91	1.556	39.52	0.722	466
	2.75	2.90	2.76	4.09	4.32	4.11	0.145	3.68	1.610	40.89	1.516	38.51	0.830	536
	3.65	3.73		5.43	5.55		0.200	5.08	1.500	38.10	1.406	35.71	1.130	729
	4.42			6.58			0.250	6.35	1.400	35.56	1.306	33.17	1.440	929
	5.15			7.66			0.300	7.62	1.300	33.02	1.206	30.63	1.650	1 064
2.063 *52.60*			3.25			4.84	0.156	3.96	1.751	44.48	1.657	42.09	0.975	629
	3.02			4.50			0.225	5.72	1.613	40.97	1.519	38.58	1.384	893
2 3/8 *60.33*	4.00			5.95			0.167	4.24	2.041	51.84	1.947	49.45	1.201	775
	4.60	4.70		6.85	6.99		0.190	4.83	1.995	50.67	1.901	48.28	1.361	878
	5.80	5.05		8.63	8.85		0.254	6.45	1.867	47.42	1.773	45.03	1.793	1 157
	6.60			9.82			0.295	7.49	1.785	45.34	1.691	42.95	2.063	1 331
	7.35	7.45		10.94	11.09		0.336	8.53	1.703	43.26	1.609	40.87	2.328	1 502
2 7/8 *73.03*	6.40	6.50					0.217	5.51	2.441	62.00	2.347	59.61	1.885	1 216
	7.80	7.90					0.276	7.01	2.323	59.00	2.229	56.62	2.373	1 531
	8.60	8.70					0.308	7.82	2.259	57.36	2.165	54.99	2.632	1 698
	9.35	9.45					0.340	8.64	2.195	55.75	2.101	53.37	2.890	1 865
	10.50						0.392	9.96	2.091	53.11	1.997	50.72	3.300	2 129
	11.50						0.440	11.18	1.995	50.67	1.901	48.29	3.670	2 368
3 1/2 *88.90*	7.70						0.216	5.49	3.068	77.93	2.943	74.75	2..228	1 438
	9.20	9.30					0.254	6.45	2.992	76.00	2.867	72.82	2.590	1 671
	10.20						0.289	7.34	2.922	74.22	2.797	71.04	2.915	1 881
	12.70	12.95					0.375	9.53	2.750	69.85	2.625	66.67	3.682	2 375
	14.30						0.430	10.92	2.640	67.06	2.515	63.88	4.436	2 862
	15.50						0.476	12.09	2.548	64.72	2.423	61.54	4.878	3 147
	17.00						0.530	13.46	2.440	61.98	2.315	58.80	5.384	3 474

Table B4   4 to 7-in. tubings. Dimensions and masses.
(*Source*: ISO 11960 and API 5CT)

Tubing size OD (in. & *mm*)	Nominal weight (lbm/ft)			Nominal weight (kg/m)			Wall thickness t		Inside dia. d		Drift diameter		Steel cross section	
	NU	EUE	IJ	NU	EUE	IJ	(in.)	(mm)	(in.)	(mm)	(in.)	(mm)	(sq in.)	(mm²)
4 *101.60*	9.50			14.14			0.226	5.74	3.548	90.12	3.423	86.94	2.765	1.784
		11.00			16.37		0.262	6.65	3.476	88.29	3.351	85.12	3.182	2 053
	13.20			19.64			0.330	8.38	3.340	84.84	3.215	81.66	3.974	2.564
	16.10			23.96			0.415	10.54	3.170	80.52	3.045	77.34	4.944	3 190
	18.90			28.13			0.500	12.70	3.000	76.20	2.875	73.02	5.890	3 800
	22.20			33.04			0.606	15.49	2.780	70.61	2.655	67.44	7.078	4.567
4 1/2 *114.30*	12.60	12.75		18.75	18.97		0.271	6.88	3.958	100.53	3.833	97.36	3.714	2 396
	15.20			22.62			0.337	8.56	3.826	97.18	3.701	94.01	4.585	2 958
	17.00			25.30			0.380	9.65	3.740	95.00	3.615	91.82	5.144	3 319
	18.90			28.13			0.430	10.92	3.640	92.46	3.515	89.28	5.786	3 733
	21.50			32.00			0.500	12.70	3.500	88.90	3.375	85.72	6.675	4 307
	23.70			35.27			0.560	14.22	3.380	85.85	3.255	82.68	7.421	4 788
	26.10			38.84			0.630	16.00	3.240	82.30	3.115	79.12	8.281	5 345
5 *127.00*	15			22.32			0.296	7.52	4.408	111.76	4.283	108.79	4.512	2 911
	18			26.79			0.362	9.19	4.276	108.61	4.151	105.44	5.561	3 588
	21.4			31.85			0.437	11.10	4.126	104.80	4.001	101.63	6.564	4 235
	23.2			34.53			0.478	12.14	4.044	102.72	3.919	99.54	7.148	4 612
	24.1			35.86			0.500	12.70	4.000	101.60	3.875	98.42	7.461	4 814
5 1/2 *139.70*	15.5			23.07			0.275	6.99	4.950	125.73	4.825	122.55	4.636	2 991
	17			25.30			0.304	7.72	4.892	124.26	4.767	121.08	5.105	3 294
	20			29.76			0.361	9.17	4.778	121.36	4.653	118.19	6.032	3.892
	23			34.23			0.451	10.54	4.670	118.62	4.545	115.44	6.900	4 451
7 *177.80*	23			34.23			0.317	8.05	6.366	161.70	6.241	158.52	6.810	4.394
	26			38.69			0.362	9.19	6.276	159.40	6.151	156.24	7.750	5.000
	29			43.16			0.408	10.36	6.184	157.10	6.059	153.90	8.708	5.618
	32			47.62			0.453	11.51	6.094	154.80	5.969	151.61	9.642	6.221
	35			52.09			0.498	12.65	6.004	152.50	5.879	149.33	10.561	6 814
	38			56.55			0.540	13.72	5.920	150.40	5.795	147.19	11.420	7 368

4.1 Tensile properties

Products shall conform to the tensile requirements specified in Tables B5 and B6.

4.2 Yield strength

The yield strength shall be the tensile stress required to produce the extension under load specified in Tables B5 and B6 as determined by an extensometer.

4.3 Grade color codes and tensile requirements of tubings [8]

The color and number of bands for each grade shall be as shown in Table B5. Tensile requirements are also specified in this table.

Table B5 Grade color codes and tensile requirements of tubings.
(*Source*: ISO 11 960)

Grade	Color band identification	Min. yield strength		Max. yield strength		Min. tensile strength	
		(psi)	(MPa)	(psi)	(MPa)	(psi)	(MPa)
H 40	None or black band	40 000	276	80 000	552	60 000	414
J 55	One bright green	55 000	379	80 000	552	75 000	517
K 55	Two bright green	55 000	379	80 000	552	95 000	655
N 80	One red	80 000	552	110 000	758	100 000	689
M 65	One bright green, one blue	65 000	448	85 000	586	85 000	586
L 80	One red, one brown	80 000	552	95 000	655	95 000	655
L 80 type 9CR	One red, one brown, two yellow	80 000	552	95 000	655	95 000	655
L 80 type 13CR	One red, one brown, one yellow	80 000	552	95 000	655	95 000	655
C 90 type 1	One purple	90 000	621	105 000	724	100 000	689
C 90 type 2	One purple, one yellow	90 000	621	105 000	724	100 000	689
T 95 type 1	One silver	95 000	655	110 000	758	105 000	724
T 95 type 2	One silver, one yellow	95 000	655	110 000	758	105 000	724
C 95	One brown	95 000	655	110 000	758	105 000	724
P 110	One white	110 000	758	140 000	965	125 000	862
Q 125 type 1	One orange	125 000	862	150 000	1 034	135 000	931
Q 125 type 2	One orange, one yellow	125 000	862	150 000	1 034	135 000	931
Q 125 type 3	One orange, one green	125 000	862	150 000	1 034	135 000	931
Q 125 type 4	One orange, one brown	125 000	862	150 000	1 034	135 000	931

4.4 Grade color codes and tensile requirements of tubings (non API) [2]

Special steels (non API) are used in particular conditions, shown below in Tables B6a to B6d.

Table B6a   H$_2$S resistant.

Grade	Color band identification	Min. yield strength		Max. yield strength		Min. tensile strength	
		(psi)	(MPa)	(psi)	(MPa)	(psi)	(MPa)
VM 80SS	Red + orange and orange bands	80 000	551	95 000	655	95 000	655
VM 90S SS	Purple + orange and orange bands	90 000	620	105 000	724	100 000	689
VM 95S SS	Brown + orange and orange bands	95 000	655	110 000	758	105 000	724
VM 100SS	Black + orange and orange bands	100 000	690	115 000	792	110 000	758
VM 110SS	White + orange and orange bands	110 000	758	125 000	862	120 000	828
VM 125SS	Yellow + orange and orange bands	125 000	862	140 000	965	135 000	931

Table B6b   Collapse resistant.

Grade	Color band identification	Min. yield strength		Max. yield strength		Min. tensile strength	
		(psi)	(MPa)	(psi)	(MPa)	(psi)	(MPa)
VM 80HC	Red + green band	80 000	551	110 000	758	100 000	689
VM 95HC	Brown + green band	95 000	655	125 000	862	110 000	758
VM 110HC	White + green band	110 000	758	140 000	965	125 000	862
VM 125HC	Orange + green band	125 000	862	155 000	1 069	135 000	931

Table B6c   Special deep wells.

Grade	Color band identification	Min. yield strength		Max. yield strength		Min. tensile strength	
		(psi)	(MPa)	(psi)	(MPa)	(psi)	(MPa)
VM 80 HCSS	Red + green orange and orange bands	80 000	551	95 000	655	95 000	655
VM 90HCS HCSS	Purple + green and orange bands	90 000	621	105 000	724	100 000	690
VM 95HCS HCSS	Brown + green and orange bands	95 000	655	110 000	758	105 000	724
VM 110 HCSS	White + green orange and orange bands	110 000	758	125 000	862	120 000	828

Table B6d   Special arctic (permafrost).

Grade	Color band identification	Min. yield strength		Max. yield strength		Min. tensile strength	
		(psi)	(MPa)	(psi)	(MPa)	(psi)	(MPa)
VM 55LT	Green + blue band	55 000	379	80 000	551	75 000	517
VM 80LT	Red + blue band	80 000	551	85 000	655	95 000	655
VM 95LT	Brown + blue band	95 000	655	110 000	758	105 000	724
VM 110LT	White + blue band	110 000	758	140 000	965	125 000	862
VM 125LT	Orange + blue band	125 000	862	150 000	1 034	135 000	931

Source: Vallourec & Mannesmann documentation

4.5　Performance properties [11]

The minimum performance properties, as given in Tables B7 to B9, cover the grades, sizes and weights of tubings defined from Standard API Bull. 5C2.

Table B7　Small tubing minimum performance properties.
(*Source*: API Bull. 5C2)

Tubing size OD (in. & mm)	Grade	Nominal weight (lbm/ft)			Collapse resistance		Internal yield pressure		Joint yield strength (lbf)			Joint yield strength (10³ N)		
		NU	EUE	IJ	(psi)	(MPa)	(psi)	(MPa)	NU	EUE	IJ	NU	EUE	IJ
1.050 26.7	H 40	1.14	1.20		7 680	53.0	7530	51.9	6 360	13 310		28	59	
	J 55				10 560	72.8	10 360	71.4	8 740	18 290		39	81	
	C 75				14 410	99.4	14 310	98.7	11 920	24 950		53	111	
	L/N80				15 370	106.0	15 070	103.9	12 710	26 610		57	118	
	C 90				17 290	119.2	16 950	116.9	14 000	30 000		62	133	
1.315 33.4	H 40	1.70	1.80	1.72	7 270	50.1	7080	48.8	10 960	19 760	15 970	49	88	71
	J 55				10 000	68.9	9730	67.1	15 060	27 160	21 960	67	121	98
	C 75				13 640	94.0	13 270	91.5	20 540	37 040	29 940	91	165	133
	L/N 80				14 550	100.3	14 160	97.6	21 910	39 510	31 940	97	176	142
	C 90				16 360	112.8	15 930	109.8	25 000	44 000	36 000	111	196	160
1.660 42.2	H 40	2.30	2.40	2.33	6 180	42.6	5900	40.7	15 530	26 740	22 180	69	119	99
	J 55				8 490	58.3	8120	56.0	21 360	36 770	30 500	95	164	136
	C 75				11 580	79.8	11 070	76.3	29 120	50 140	41 600	130	223	185
	L/N 80				12 360	85.2	11810	81.4	31 060	53 480	44 370	138	238	198
	C 90				13 900	95.8	13 280	91.6	35 000	60 000	50 000	156	267	222
1.900 48.3	H 40	2.75	2.90	2.76	5 640	38.9	5340	36.8	19 090	31 980	26 890	85	142	120
	J 55				7 750	53.4	7350	50.7	26 250	43 970	36 970	117	196	164
	C 75				10 570	72.9	10 020	69.1	35 800	59 960	50 420	159	267	224
	L/N 80				11 280	77.8	10 680	73.7	38 180	63 960	53 780	170	284	239
	C 90				12 620	87.0	12 020	82.9	43 000	72 000	60 000	191	320	267
2.063 52.4	H 40			3.25	5 590	39.2	5290	36.9			35 700			159
	J 55				7 690	53.0	7280	50.2			49 000			218
	C 75				10 480	72.3	9920	68.4			66 900			298
	L/N 80				11 180	77.1	10 590	73.0			71 400			318
	C 90				12 420	85.7	11 910	82.1			80 000			356

Table B8   2 3/8 and 2 7/8-in. tubing minimum performance properties.
(*Source*: API Bull. 5C2)

Tubing size OD (in. & mm)	Grade	Nominal weight (lbm/ft) NU	EUE	Collapse resistance (psi)	(MPa)	Internal yield pressure (psi)	(MPa)	Joint yield strength (lbf) NU	EUE	Joint yield strength (10^3 N) NU	EUE
2 3/8 60.3	H 40	4.00		5 230	36.1	4 920	33.9	30 100		134	
	H 40	4.60	4.70	5 890	40.6	5 600	38.6	36 000	52 200	160	232
	J 55	4.00		7 190	49.6	6 770	46.7	41 400		184	
	J 55	4.60	4.70	8 100	55.8	7 700	53.1	49 500	71 700	220	319
	C 75	4.00		9 520	65.6	9 230	63.6	56 500		251	
	C 75	4.60	4.70	11 780	81.2	10 500	72.4	67 400	97 800	300	435
	C 75	5.80	5.95	14 330	98.8	14 040	96.8	96 600	126 900	430	564
	L/N 80	4.00		9 980	68.8	9 840	67.9	60 800		270	
	L/N 80	4.60	4.70	11 780	91.2	11 200	77.2	71 900	104 300	320	464
	L/N 80	5.80	5.95	15 280	105.4	14 970	103.2	103 000	135 400	458	602
	C 90	4.00		10 940	75.4	11 070	76.3	68 000		302	
	C 90	4.60	4.70	13 250	91.4	12 600	86.9	81 000	117 000	360	520
	C 90	5.80	5.95	17 190	118.5	16 840	116.1	116 000	152 000	516	676
	P 105	4.60	4.70	15 460	106.6	14 700	101.4	94 400	136 900	420	609
	P 105	5.80	5.95	20 060	138.3	19 650	135.5	135 200	177 700	601	790
2 7/8 73.0	H 40	6.40	6.50	5 580	38.5	5 280	36.4	52 800	72 500	235	322
	J 55	6.40	6.50	7 680	53.0	7 200	49.6	72 600	99 700	323	443
	C 75	6.40	6.50	10 470	72.2	9 910	68.3	99 000	135 900	440	604
	C 75	7.80	7.90	13 020	89.8	12 600	86.9	132 100	169 000	588	752
	C 75	8.60	8.70	14 350	98.9	14 010	96.6	149 400	186 300	665	829
	L/N 80	6.40	6.50	11 160	76.9	10 570	72.9	105 600	145 000	470	645
	L/N 80	7.80	7.90	13 890	95.8	13 440	92.7	140 900	180 300	627	802
	L/N 80	8.60	8.70	15 300	105.5	14 940	103.0	159 300	198 700	709	884
	C 90	6.40	6.50	12 390	85.4	11 890	82.0	118 800	163 100	528	725
	C 90	7.80	7.90	15 620	107.7	15 120	104.2	158 500	202 800	705	902
	C 90	8.60	8.70	17 220	118.7	16 870	116.3	179 200	223 500	797	994
	P 105	6.40	6.50	14 010	96.6	13 870	95.6	138 600	190 300	616	846
	P 105	7.80	7.90	18 220	125.6	17 640	121.6	184 900	236 600	822	1 052
	P 105	8.60	8.70	20 090	138.5	19 610	135.2	209 100	260 800	930	1 160

Table B9   3 1/2 to 4 1/2-in. tubing minimum performance properties.
(*Source*: API Bull. 5C2)

Tubing size OD (in. & mm)	Grade	Nominal weight (lbm/ft)		Collapse resistance		Internal yield pressure		Joint yield strength (lbf)		Joint yield strength (10³ N)	
		NU	EUE	(psi)	(MPa)	(psi)	(MPa)	NU	EUE	NU	EUE
3 1/2 88.9	H 40	7.70		4 630	31.9	4 320	29.8	65 100		290	
	H 40	9.20	9.30	5 380	37.1	5 080	35.0	79 500	103 600	354	461
	H 40	10.20		6 060	41.8	5 780	39.9	92 600		412	
	J 55	7.70		5 940	42.0	5 940	41.0	89 500		398	
	J 55	9.20	9.30	7 400	41.2	6 990	48.2	109 400	142 500	487	634
	J 55	10.20		8 330	57.3	7 950	54.8	127 300		566	
	C 75	7.70		7 540	52.0	8 100	55.8	122 000		542	
	C 75	9.20	9.30	10 040	69.2	9 530	65.7	149 100	194 300	663	864
	C 75	10.20		11 360	78.3	10 840	74.7	173 500		772	
	C 75	12.70	12.95	14 350	98.9	14 060	96.9	231 000	276 100	1096	1 228
	L/N 80	7.70		7 870	54.3	8 640	59.6	130 100		579	
	L/N 80	9.20	9.30	10 530	72.6	10 160	70.1	159 100	207 200	708	922
	L/N 80	10.20		12 120	83.6	11 560	79.7	185 100		828	
	L/N 80	12.70	12.95	15 310	105.6	15 000	103.4	246 400	294 500	1096	1 310
	C 90	7.70									
	C 90	9.20	9.30	11 570	79.8	11 430	78.8	179 000	233 100	796	1 037
	C 90	10.20		17 220	118.7	16 880	116.4	277 200	331 300	1 233	1 474
	C 90	12.70	12.95								
	P 105	9.20	9.30	13 050	90.0	13 340	92.0	208 900	272 000	929	1 210
	P 105	12.70	12.95	20 090	138.5	19 690	135.8	323 400	386 600	1439	1 720
4 101.6	H 40	9.50		4 060	28.0	3 960	27.3	72 000		320	
	H 40		11.00	4 900	33.8	4 590	31.6		123 100		548
	J 55	9.50		5 110	35.2	5 440	37.5	99 000		440	
	J 55		11.00	6 590	45.4	6 300	43.4		169 200		753
	C 75	9.50		6 350	43.8	7 420	51.2	135 000		600	
	C 75		11.00	8 410	58.0	8 600	59.3		230 800		1 027
	L/N 80	9.50		6 590	45.4	7 910	54.5	144 000		641	
	L/N 80		11.00	8 800	60.7	9 170	63.2		246 100		1 095
	C 90	9.50		7 080	48.8	8 900	61.4	162 000		721	
	C 90		11.00	9 600	66.2	10 320	71.2		276 900		1 232
4 1/2 114.3	H 40			4 500	31.0	4 220	29.1	104 400	144 000	464	641
	J 55			5 720	39.4	5 800	40.0	143 500	198 000	638	881
	C 75	12.60	12.75	7 200	49.6	7 900	54.5	195 700	270 000	870	1 201
	L/N 80			7 500	51.7	8 430	58.1	208 700	288 000	928	1 281
	C 90			8 120	56.0	9 490	65.4	234 800	324 000	1 044	1 441

B5 TUBING CONNECTION [5]

5.1 Standard API coupling connections

Two standards API and ISO coupling connections are available:

- The **API non-upset** connection (NU) is a 10-round thread form cut on the body, wherein the joint has less strength than the pipe body (Fig. B1).
- The **API external upset** connection (EUE) is an 8-round thread form, wherein the joint has the same strength as the pipe body (Fig. B2).

For very high pressure service, the API EUE connection is available in 2 3/8, 2 7/8 and 3 1/2-in. sizes having a long thread form (EUE long threaded and coupled (T&C)), wherein the effective thread is 50% longer than standard.

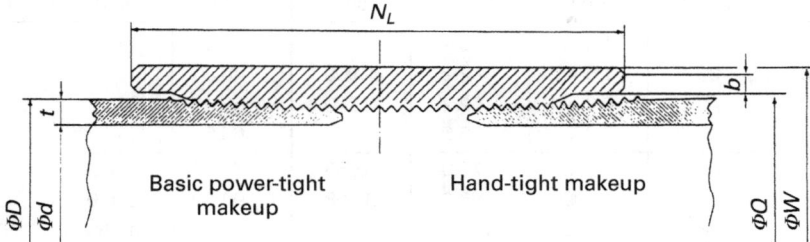

Figure B1 Non-upset tubing and coupling [8].

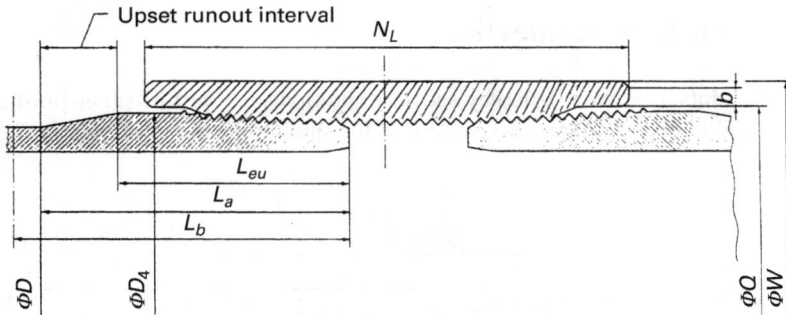

Figure B2 External-upset tubing and coupling [8].

Legend: ΦW: outside diameter of coupling; ΦQ: diameter of recess; ΦD: outside diameter of tubing; Φd: inside diameter of tubing; b: width of bearing face.

47

5.2 Special clearance couplings

Where extra clearance is needed, API couplings can be "turned down" somewhat without loss of joint tensile strength. Special clearance collars are usually marked with a black ring in the center of the color band indicating steel grade.

Special clearance (reduced outside diameter ΦWc) coupling-type thread forms have been developed for external upset tubing which have 100% joint strength.

Standard and special clearance diameters of API and ISO coupling-type connections are indicated in Table B10.

Table B10 Standard and special clearance coupling sizes.
(*Source*: ISO 11 960)

OD tubing (ΦD)		OD connection (ΦW) NU standard		OD connection (ΦW) EUE standard		OD connection (ΦWc) EUE special clearance	
(in.)	(mm)	(in.)	(mm)	(in.)	(mm)	(in.)	(mm)
1.050	26.7	1.313	33.4	1.660	42.2		
1.315	33.4	1.660	42.2	1.900	48.3		
1.660	42.2	2.054	52.2	2.200	55.9		
1.900	48.3	2.200	55.9	2.500	63.5		
2 3/8	60.3	2.875	73.0	3.063	77.8	2.910	73.9
2 7/8	73.0	3.500	88.9	3.668	93.2	3.460	87.9
3 1/2	88.9	4.250	108.0	4.500	114.3	4.180	106.2
4	101.6	4.750	120.6	5.000	127.0		
4 1/2	114.3	5.200	132.1	5.563	141.3		

5.3 Integral-joint connections

Use of these joints must be justified by special conditions. An **API integral-joint connection** (10-round thread form) has been adopted for small diameter as shown in Figure B3 and Table B11.

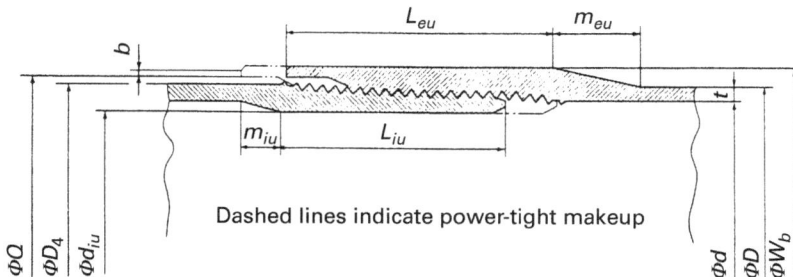

Dashed lines indicate power-tight makeup

Figure B3 Integral-joint tubing [8].

Table B11 Integral-joint connections.

OD (Φ D) tubing		OD (Φ Wb) integral-joint connection	
(in.)	(mm)	(in.)	(mm)
1.315	33.4	1.550	39.4
1.660	42.2	1.880	47.8
1.900	48.3	2.110	53.3
2.063	52.4	2.325	59.1

5.4 Special connections

Many special tubing connections, both integral joint and threaded and coupled, are available to meet needs involving very high pressures or corrosive service. The *World Oil* magazine and *Petroleum Engineer International* periodically print joint identification tables showing most special connections [6, 7].

B6 MAKEUP OF THREADED CONNECTION [2, 5, 8]

The stress induced in the connection during makeup and subsequent service determine the success of the connection as a sealing element.

Traditional field practice often dictates that API tapered thread connections are made up on one of the following bases.

6.1 Makeup position

The rules of thumb is as follows:
   (a) Last tool mark on the pin must be buried by the collar.
   (b) Add two turns past hand tight (50 ft-lb or 7 daN.m).

6.2 Field makeup

Joint life of tubing under repeated field makeup is inversely proportional to the field makeup torque applied. Therefore, in wells where leak resistance is not a great factor, minimum field makeup torque values should be used to prolong joint life.

Tables B12 to B14 contain recommended optimum makeup torque values guidelines for nonupset (NU), external upset (EUE), and integral joint tubing (IJ).

Minimum torque values are 75 percent of optimum values, and maximum torque values are 125 percent of optimum values.

A correction factor is applied for makeup torque according to type of grease [2].

B

Table B12 Small tubing makeup torque guidelines.
(*Source*: ISO 10405 and API 5CI, 17th Ed.)

Tubing size OD (in. & mm)	Grade	Threads and coupling NU			Threads and coupling EUE			Integral joint IJ		
		Nominal weight (lbm/ft)	Torque (lbf-ft)	Torque (N.m)	Nominal weight (lbm/ft)	Torque (lbf-ft)	Torque (N.m)	Nominal weight (lbm/ft)	Torque (lbf-ft)	Torque (N.m)
1.050 *26.7*	H 40	1.14	140	190	1.20	460	630			
	J 55		180	240		600	810			
	C 75		230	320		780	1 060			
	L 80		240	330		810	1 090			
	N 80		250	340		830	1 130			
	C 90		260	350		880	1 190			
1.315 *33.4*	H 40	1.70	210	280	1.80	440	590	1.72	310	410
	J 55		270	370		570	770		400	540
	C 75		360	480		740	1 010		520	700
	L 80		370	500		760	1 040		530	720
	N 80		380	510		790	1 070		550	740
	C 90		400	540		830	1 130		580	780
1.660 *42.2*	H 40	2.30	270	360	2.40	530	720	2.33	380	520
	J 55		350	470		690	940		500	680
	C 75		460	620		910	1 230		650	890
	L 80		470	640		940	1 270		680	920
	N 80		490	660		960	1 300		690	940
	C 90		510	700		1 020	1 380		730	1 000
1.900 *48.3*	H 40	2.75	320	430	2.90	670	910	2.76	450	600
	J 55		410	560		880	1 190		580	790
	C 75		540	730		1 150	1 560		760	1 030
	L 80		560	760		1 190	1 610		790	1 070
	N 80		570	780		1 220	1 650		810	1 100
	C 90		610	830		1 300	1 760		860	1 160
2.063 *52.4*	H 40							3.25	570	770
	J 55								740	1 010
	C 75								970	1 320
	L 80								1 010	1 370
	N 80								1 030	1 400
	C 90								1 100	1 490

Table B13 2 3/8 and 2 7/8-in. tubing makeup torque guidelines.
(*Source*: ISO 10405 and API 5 CI, 17th Ed.)

Tubing size OD (in. & mm)	Grade	Threads and coupling NU			Threads and coupling EUE			Integral joint IJ		
		Nominal weight	Torque		Nominal weight	Torque		Nominal weight	Torque	
		(lbm/ft)	(lbf-ft)	(N.m)	(lbm/ft)	(lbf-ft)	(N.m)	(lbm/ft)	(lbf-ft)	(N.m)
2 3/8 *60.3*	H 40	4.00	470	630						
	H 40	4.60	560	760	4.70	990	1 340			
	J 55	4.00	610	830						
	J 55	4.60	730	990	4.70	1 290	1 750			
	C 75	4.00	800	1 090						
	C 75	4.60	960	1 300	4.70	1 700	2 310			
	C 75	5.80	1 380	1 860	5.95	2 120	2 870			
	L 80	4.00	830	1 130						
	L 80	4.60	990	1 350	4.70	1 760	2 390			
	L 80	5.80	1 420	1 930	5.95	2 190	2 970			
	N 80	4.00	850	1 160						
	N 80	4.60	1 020	1 380	4.70	1 800	2 450			
	N 80	5.80	1 460	1 980	5.95	2 240	3 040			
	C 90	4.00	910	1 230						
	C 90	4.60	1 080	1 470	4.70	1 920	2 610			
	C 90	5.80	1 550	2 110	5.95	2 390	3 250			
	P 105	4.60	1 280	1 740	4.70	2 270	3 080			
	P 105	5.80	1 840	2 490	5.95	2 830	3 830			
2 7/8 *73.0*	H 40	6.40	900	1 080	6.50	1 250	1 700			
	J 55	6.40	1 050	1 420	6.50	1 650	2 230			
	C 75	6.40	1 380	1 880	6.50	2 170	2 940			
	C 75	7.80	1 850	2 500	7.90	2 610	3 540			
	C 75	8.60	2 090	2 830	8.70	2 850	3 860			
	L 80	6.40	1 430	1 940	6.50	2 250	3 050			
	L 80	7.80	1 910	2 590	7.90	2 710	3 680			
	L 80	8.60	2 160	2 930	8.70	2 950	4 000			
	N 80	6.40	1 470	1 990	6.50	2 300	3 120			
	N 80	7.80	1 960	2 650	7.90	2 770	3 760			
	N 80	8.60	2 210	3 000	8.70	3 020	4 090			
	C 90	6.40	1 570	2 130	6.50	2 460	3 340			
	C 90	7.80	2 090	2 840	7.90	2 970	4 020			
	C 90	8.60	2 370	3 210	8.70	3 230	4 380			
	P 105	6.40	1 850	2 510	6.50	2 910	3 940			
	P 105	7.80	2 470	3 350	7.90	3 500	4 750			
	P 105	8.60	2 790	3 790	8.70	3 810	5 170			

Table B14   3 1/2 to 4 1/2-in. tubing makeup torque guidelines.
(*Source*: ISO 10405 and API 5CI, 17th Ed.)

Tubing size OD (in. & *mm*)	Grade	Threads and coupling NU			Threads and coupling EUE			Integral joint IJ		
		Nominal weight	Torque		Nominal weight	Torque		Nominal weight	Torque	
		(lbm/ft)	(lbf-ft)	(N.m)	(lbm/ft)	(lbf-ft)	(N.m)	(lbm/ft)	(lbf-ft)	(N.m)
3 1/2 *88.9*	H 40	7.70	920	1 250						
	H.40	9.20	1 120	1 520	9.30	1 730	2 340			
	H 40	10.20	1 310	1 770						
	J 55	7.70	1 210	1 640						
	J 55	9.20	1 480	2 010	9.30	2 280	3 090			
	J 55	10.20	1 720	2 330						
	C 75	7.70	1 600	2 170						
	C 75	9.20	1 950	2 650	9.30	3 010	4 080			
	C 75	10.20	2 270	3 080						
	C 75	12.70	3 030	4 100	12.95	4 040	5 480			
	L 80	7.70	1 660	2 250						
	L 80	9.20	2 030	2 750	9.30	3 030	4 240			
	L 80	10.20	2 360	3 200						
	L 80	12.70	3 140	4 260	12.95	4 200	5 700			
	N 80	7.70	1 700	2 300						
	N 80	9.20	2 070	2 810	9.30	3 200	4 330			
	N 80	10.20	2 410	3 270						
	N 80	12.70	3 210	4 350	12.95	4 290	5 820			
	C 90	7.70	1 820	2 460						
	C 90	9.20	2 220	3 010	9.30	3 430	4 650			
	C 90	10.20	2 590	3 510						
	C 90	12.70	3 440	4 670	12.95	4 610	6 250			
	P 105	9.20	2 620	3 550	9.30	4 050	5 490			
	P 105	12.70	4 060	5 510	12.95	5 430	7 370			
4 *101.6*	H 40		930	1 260		1 940	2 630			
	J 55		1 220	1 660		2 560	3 470			
	C 75	9.50	1 620	2 200	11.00	3 390	4 600			
	L 80		1 680	2 280		3 530	4 780			
	N 80		1 720	2 330		3 600	4 880			
	C 90		1 940	2 630		3 870	5 250			
4 1/2 *114.3*	H 40		1 320	1 780		2 160	2 930			
	J 55		1 740	2 360		2 860	3 870			
	C 75	12.60	2 300	3 120	12.75	3 780	5 130			
	L 80		2 400	3 250		3 940	5 340			
	N 80		2 440	3 310		4 020	5 450			
	C 90		2 630	3 570		4 330	5 870			

B7   API AND BUTTRESS TUBING THREAD FORM

7.1   API tubing thread form [2]

The tubing thread form recommended by API standard is shown in Fig. B4. The geometrical characteristics of these threads are given in Table B15.

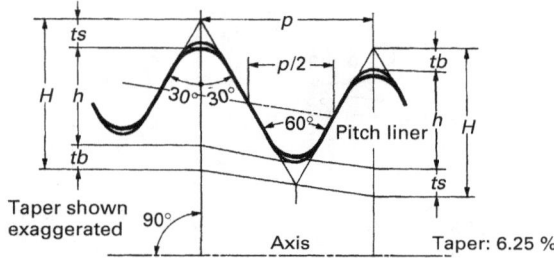

Figure B4

Table B15   Geometrical characteristics of threads.

Thread element	10 threads per inch p = 2.540 mm (1/10 in.)	8 threads per inch p = 3.175 mm (1/8 in.)
$H = 0.866\,p$	2.200 mm	2.750 mm
$h = 0.626\,p - 0.178$	1.412 mm	1.810 mm
$tb = 0.120\,p + 0.051$	0.356 mm	0.432 mm
$ts = 0.120\,p + 0.127$	0.432 mm	0.508 mm

OD (in.)	Threads per inch		
	Tubing without upset	Tubing with external upset	Tubing with integral joint
1.050	10	10	–
1.315	10	10	10
1.660	10	10	10
1.900	10	10	10
2.063	–	–	10
2 3/8	10	8	–
2 7/8	10	8	–
3 1/2	10	8	–
4	8	8	–
4 1/2	8	8	–

mm × 0.0394 = in.

7.2 Buttress thread form [2]

Buttress connections are similar to API round thread connections in that adequate bearing pressure between mating thread surfaces must be established, and voids must be filled with thread compound solids to transmit bearing loads across void spaces (Fig. B5).

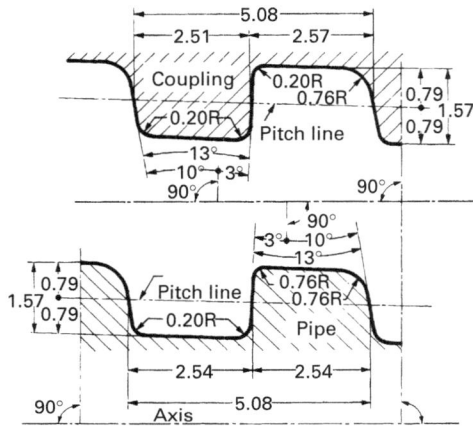

Figure B5 Taper: 6.25%. 5 threads per in. Thread crests and roots are parallel to cone. Dimensions in millimeters unless otherwise indicated.

B8 ELONGATION [8]

The stretch or elongation of oil well tubular material resulting from an applied pulling force is a commonly required determination.

The amount of stretch that will occur when a pull force is applied varies with the amount of pull, the length of material being stretched, the elasticity of the material, and its cross-sectional area.

The minimum elongation in 2 in. or 50.8 mm (gauge length of the tensile specimen) shall be that determined by the following formula:

$$e = 1944 \frac{A^{0.2}}{U^{0.9}} \tag{B1}$$

where

  e minimum elongation in 50.8 mm (2 in.) in per cent rounded to nearest 0.5%
  A cross sectional area of the tensile test specimen in mm^2, based on specified outside diameter or nominal specimen width and specified wall thickness, rounded to the nearest 10 mm^2, or 490 mm^2, whichever is smaller
  U specified tensile strength (MPa).

The minimum elongations for both round bar tensile specimens (the 8.9 mm diameter with 35.6 mm gauge length, and the 12.7 mm diameter with 50.8 mm gauge length) shall be determined with an area A of 130 mm^2.

B9 TUBING MOVEMENT FORMULAS [4]

Changes in temperature and pressure cause contraction or expansion of a tubing string.

The formulas for calculating the forces developed by this contraction/expansion are given below.

F_1	piston effect	$F_1 = (A_p - A_i)\Delta P_i - (A_p - A_o)\Delta P_o$	(B2)
F_2	buckling effect	$F_2 =$ usually negligible	
F_3	ballooning effect	$F_3 = 0.6(\Delta P_{ia}A_i - \Delta P_{oa}A_o)$	(B3)
F_4	temperature effect	$F_4 = 207A_s\Delta t$	(B4)

where

A_s	cross section area of tubing (sq in.) $= A_o - A_i$
A_p	area of packer bore (sq in.)
A_o	area of tubing OD (sq in.)
A_i	area of tubing ID (sq in.)
ΔP_i	change in tubing pressure at packer (psi)
ΔP_o	change in annulus pressure at packer (psi)
ΔP_{ia}	change in average tubing pressure (psi)
ΔP_{oa}	change in average annulus pressure (psi)
Δt	change in average tubing temperature (°F).

The equivalent tubing movement can be calculated using the formula:

$$\Delta L = \frac{FL}{EA_s} \qquad \text{(B5)}$$

where

ΔL	stretch or contraction (ft or cm)
F	force (lbf or daN)
L	tubing length (ft or m)
E	elastic factor of steel
A_s	cross section area of tubing (sq in. or cm^2).

B10 TUBING CAPACITY

Table B16 and B17 introduce capacities of tubings in relation with nominal size and nominal weight.

Values are given in U.S. and metric units.

B

Table B16 1.050 to 3 1/2-in. tubings. Capacity.

Tubing size OD (in. & mm)	Nominal weight (lbm/ft)			Nominal weight (kg/m)			Wall thickness t		Inside dia. d		Capacity	
	NU	EUE	IJ	NU	EUE	IJ	(in.)	(mm)	(in.)	(mm)	(gal/ft)	l/m
1.050 26.67	1.14	1.20		1.70	1.79		0.113	2.87	0.824	20.93	0.027	0.344
	1.48	1.54		2.20	2.29		0.154	3.91	0.742	18.85	0.022	0.279
1.315 33.40	1.70	1.80	1.72	2.53	2.68	2.56	0.133	3.38	1.049	26.64	0.045	0.557
	2.19	2.24		3.26	3.33		0.179	4.55	0.957	24.31	0.037	0.464
1.660 42.16			2.10			3.13	0.125	3.18	1.410	35.81	0.082	1.007
	2.30	2.40	2.33	3.42	3.57	3.47	0.140	3.56	1.380	35.05	0.077	0.965
	3.03	3.07		4.51	4.57		0.191	4.85	1.278	32.46	0.066	0.827
1.900 48.26			2.40			3.57	0.125	3.18	1.650	41.91	0.111	1.380
	2.75	2.90	2.76	4.09	4.32	4.11	0.145	3.68	1.610	40.89	0.105	1.313
	3.65	3.73		5.43	5.55		0.200	5.08	1.500	38.10	0.092	1.140
	4.42			6.58			0.250	6.35	1.400	35.56	0.080	0.993
	5.15			7.66			0.300	7.62	1.300	33.02	0.069	0.856
2.063 52.60			3.25			4.84	0.156	3.96	1.751	44.48	0.125	1.554
	3.02			4.50			0.225	5.72	1.613	40.97	0.106	1.318
2 3/8 60.33	4.00			5.95			0.167	4.24	2.041	51.84	0.170	2.110
	4.60	4.70		6.85	6.99		0.190	4.83	1.995	50.67	0.162	2.016
	5.80	5.05		8.63	8.85		0.254	6.45	1.867	47.42	0.143	1.766
	6.60			9.82			0.295	7.49	1.785	45.34	0.130	1.614
	7.35	7.45		10.94	11.09		0.336	8.53	1.703	43.26	0.118	1.470
2 7/8 73.03	6.40	6.50					0.217	5.51	2.441	62.00	0.243	3.019
	7.80	7.90					0.276	7.01	2.323	59.00	0.220	2.734
	8.60	8.70					0.308	7.82	2.259	57.36	0.208	2.584
	9.35	9.45					0.340	8.64	2.195	55.75	0.196	2.441
	10.50						0.392	9.96	2.091	53.11	0.178	2.251
	11.50						0.440	11.18	1.995	50.67	0.162	2.016
3 1/2 88.90	7.70						0.216	5.49	3.068	77.93	0.384	4.770
	9.20	9.30					0.254	6.45	2.992	76.00	0.365	4.536
	10.20						0.289	7.34	2.922	74.22	0.348	4.326
	12.70	12.95					0.375	9.53	2.750	69.85	0.308	3.832
	14.30						0.430	10.92	2.640	67.06	0.265	3.532
	15.50						0.476	12.09	2.548	64.72	0.265	3.290
	17.00						0.530	13.46	2.440	61.98	0.247	3.017

Table B17   4 to 7-in. tubings. Capacity.

Tubing size OD (in. & mm)	Nominal weight (lbm/ft)			Nominal weight (kg/m)			Wall thickness t		Inside dia. d		Capacity	
	NU	EUE	IJ	NU	EUE	IJ	(in.)	(mm)	(in.)	(mm)	(gal/ft)	(l/m)
4 101.60	9.50			14.14			0.226	5.74	3.548	90.12	0.513	6.379
		11.00			16.37		0.262	6.65	3.476	88.29	0.493	6.122
	13.20			19.64			0.330	8.38	3.340	84.84	0.455	5.653
	16.10			23.96			0.415	10.54	3.170	80.52	0.410	5.092
	18.90			28.13			0.500	12.70	3.000	76.20	0.322	4.560
	22.20			33.04			0.606	15.49	2.780	70.61	0.315	3.916
4 1/2 114.30	12.60	12.75		18.75	18.97		0.271	6.88	3.958	100.53	0.639	7.937
	15.20			22.62			0.337	8.56	3.826	97.18	0.597	7.417
	17.00			25.30			0.380	9.65	3.740	95.00	0.570	7.088
	18.90			28.13			0.430	10.92	3.640	92.46	0.541	6.714
	21.50			32.00			0.500	12.70	3.500	88.90	0.500	6.207
	23.70			35.27			0.560	14.22	3.380	85.85	0.466	5.788
	26.10			38.84			0.630	16.00	3.240	82.30	0.428	5.320
5 127.00	15			22.32			0.296	7.52	4.408	111.76	0.790	9.810
	18			26.79			0.362	9.19	4.276	108.61	0.746	9.264
	21.4			31.85			0.437	11.10	4.126	104.80	0.694	8.626
	23.2			34.53			0.478	12.14	4.044	102.72	0.667	8.287
	24.1			35.86			0.500	12.70	4.000	101.60	0.653	8.107
5 1/2 139.70	15.5			23.07			0.275	6.99	4.950	125.73	1.000	12.415
	17			25.30			0.304	7.72	4.892	124.26	0.976	12.127
	20			29.76			0.361	9.17	4.778	121.36	0.931	11.567
	23			34.23			0.451	10.54	4.670	118.62	0.890	11.051
7 177.80	23			34.23			0.317	8.05	6.366	161.70	1.653	20.535
	26			38.69			0.362	9.19	6.276	159.40	1.606	19.955
	29			43.16			0.408	10.36	6.184	157.10	1.561	19.384
	32			47.62			0.453	11.51	6.094	154.80	1.515	18.820
	35			52.09			0.498	12.65	6.004	152.50	1.470	18.265
	38			56.55			0.540	13.72	5.920	150.40	1.430	17.765

B11 ANNULAR VOLUME BETWEEN CASING AND TUBING

$$V = 0.0007854\left(D_o^2 - D_i^2\right) \tag{B6}$$

where

　V　annular volume (l/m)
　D_o　outside diameter of tubing (mm)
　D_i　inside diameter of casing (mm).

Tables B18 and B19 indicate volumes from 4 1/2 to 9 5/8-in. casings.

- *Conversion in U.S. units:*

l/m × 0.0805 = gal/ft　　　l/m × 0.00192 = bbl/ft

Table B18　Annular volume between 41/2 to 6 5/8-in. casing and tubing [2].

			Nominal size of inner string (in.)									
		0*	1.050	1.315	1.660	1.900	2.063	2 3/8	2 7/8	3 1/2	4	4 1/2
		0** (l/m)	0.56	0.88	1.40	1.84	2.16	2.88	4.22	6.26	8.17	10.33
4 1/2	9.50	8.48	7.92	7.60	7.08	6.64	6.32	5.60	4.26			
	10.50	8.32	7.76	7.44	6.92	6.48	6.16	5.44	4.10			
	11.60	8.11	7.55	7.23	6.71	6.27	5.95	5.23	3.89			
	13.50	7.79	7.23	6.91	6.39	5.95	5.63	4.91	3.57			
	15.10	7.42	6.86	6.54	6.02	5.58	5.26	4.54	3.20			
	16.90	7.09	6.53	6.21	5.69	5.25	4.93	4.21				
	17.70	6.93	6.37	6.05	5.53	5.09	4.77	4.05				
	18.80	6.71	6.15	5.83	5.31	4.87	4.55	3.83				
5	11.50	10.54	9.98	9.66	9.14	8.70	8.38	7.66	6.32			
	13.00	10.23	9.67	9.35	8.83	8.39	8.07	7.35	6.01			
	15.00	9.84	9.28	8.96	8.44	8.00	7.68	6.96	5.62			
	18.00	9.27	8.71	8.39	7.87	7.43	7.11	6.39	5.05			
	20.80	8.75	8.19	7.87	7.35	6.91	6.59	5.87	4.53			
5 1/2	14.00	12.73	12.17	11.85	11.33	10.89	10.57	9.85	8.51	6.47		
	15.50	12.42	11.86	11.54	11.02	10.58	10.26	9.54	8.20	6.16		
	17.00	12.13	11.57	11.25	10.73	10.29	9.97	9.25	7.91	5.87		
	20.00	11.57	11.01	10.69	10.17	9.73	9.41	8.69	7.35	5.31		
	23.00	11.05	10.49	10.17	9.65	9.21	8.89	8.17	6.83	4.79		
6 5/8	20.00	18.54	17.98	17.66	17.14	16.70	16.38	15.66	14.32	12.28	10.37	8.21
	24.00	17.76	17.20	16.88	16.36	15.92	15.60	14.88	13.54	11.50	9.59	7.43
	28.00	16.99	16.43	16.11	15.59	15.15	14.83	14.11	12.77	10.73	8.82	6.66
	32.00	16.32	15.76	15.44	14.92	14.48	14.16	13.44	12.10	10.06	8.15	

Left axis: Nominal size and weight of outer string (in. and lb/ft)

* The zero vertical column gives the capacity of the casing in liters per meter.
** The zero horizontal line gives the total displacement of tubing with coupling in liters per meter.
　l/m × 0.0805 = gal/ft　　l/m × 0.00192 = bbl/ft.

Table B19 Annular volume between 7 to 9 5/8-in. casing and tubing [2].

Outer size	Outer weight	0*	1.050	1.315	1.660	1.900	2.063	2 3/8	2 7/8	3 1/2	4	4 1/2
	0**	(l/m)	0.56	0.88	1.40	1.84	2.16	2.88	4.22	6.26	8.17	10.33
7	17.00	21.66	21.10	20.78	20.26	19.82	19.50	18.78	17.44	15.40	13.49	11.33
	20.00	21.66	21.10	20.78	20.26	19.82	19.50	18.78	17.44	15.40	13.49	11.33
	23.00	20.54	19.98	19.66	19.14	18.70	18.38	17.66	16.32	14.28	12.37	10.21
	26.00	19.96	19.40	19.08	18.56	18.12	17.80	17.08	15.74	13.70	11.79	9.63
	29.00	19.38	18.82	18.50	17.98	17.54	17.22	16.50	15.16	13.12	11.21	9.05
	32.00	18.82	18.26	17.94	17.42	16.98	16.66	15.94	14.60	12.56	10.65	8.49
	35.00	18.27	17.71	17.39	16.87	16.43	16.11	15.39	14.05	12.01	10.10	7.94
	38.00	17.76	17.20	16.88	16.36	15.92	15.60	14.88	13.54	11.50	9.59	7.43
	41.00	17.17	16.61	16.29	15.77	15.33	15.01	14.29	12.95	10.91	9.00	6.84
	44.00	16.58	16.02	15.70	15.18	14.74	14.42	13.70	12.36	10.32	8.41	6.25
7 5/8	24.00	25.01	24.45	24.13	23/61	23.17	22.85	22.13	20.79	18.75	16.84	16.68
	26.40	24.61	24.05	23.73	23.21	22.77	22.45	21.73	20.39	18.35	16.44	14.28
	29.70	23.95	23.39	23.07	22.55	22.11	21.79	21.07	19.73	17.69	15.78	12.86
	33.70	23.19	22.63	22.31	21.79	21.35	21.03	20.31	18.97	16.96	15.02	12.86
	39.00	22.24	21.68	21.36	20.84	20.40	20.08	19.36	18.02	15.95	14.07	11.91
8 5/8	24.00	33.22	32.66	32.34	31.82	31.38	31.06	30.34	29.00	26.96	25.05	22.89
	28.00	32.57	32.01	31.69	31.17	30.73	30.41	29.69	28.35	26.31	24.40	22.24
	32.00	31.79	31.23	30.91	30.39	29.95	29.63	28.91	28.57	25.53	23.62	21.46
	36.00	31.03	30.47	30.15	29.63	29.19	28.87	28.15	26.81	24.77	22.86	20.70
	40.00	30.24	29.68	29.36	28.84	28.40	28.08	27.36	26.02	23.98	22.07	19.91
	44.00	29.46	28.90	28.58	28.06	27.62	27.30	26.58	25.24	23.20	21.29	19.13
	49.00	28.58	28.02	27.70	27.18	26.74	26.42	25.70	24.36	22.32	20.41	18.25
9 5/8	32.30	10.06	40.50	10.18	39.66	39.22	38.90	38.18	36.84	34.80	32.89	30.73
	36.00	40.33	39.77	39.45	38.93	38.49	38.17	37.45	36.11	34.07	32.16	30.00
	40.00	39.55	38.99	38.67	38.15	37.71	37.39	36.67	35.33	33.29	31.38	29.22
	43.50	38.84	38.28	37.96	37.44	37.00	36.68	35.96	34.62	32.58	30.67	28.51
	47.00	38.18	37.62	37.30	36.78	36.34	36.02	35.30	33.96	31.92	30.01	27.85
	53.50	36.91	36.35	36.03	35.51	35.07	34.75	34.03	32.69	30.65	28.74	26.58
	58.40	36.05	35.49	35.17	34.65	34.21	33.89	33.17	31.83	29.79	27.88	25.72
	61.10	35.54	34.98	34.66	34.14	33.70	33.38	32.66	31.32	29.28	27.37	25.21
	71.80	33.45	32.89	32.57	32.05	31.61	31.29	30.57	29.23	27.19	25.28	23.12

Column header: Nominal size of inner string (in.). Left axis: Nominal size and weight of outer string (in. and lb/ft).

* The zero vertical column gives the capacity of the casing in liters per meter.
** The zero horizontal line gives the total displacement of tubing with coupling in liters per meter.
l/m × 0.0805 = gal/ft   l/m × 0.00192 = bbl/ft.

B

REFERENCES

1   Publication de la Chambre Syndicale de la Recherche et de la Production du Pétrole et du Gaz Naturel (1970) *Formulaire du Producteur*. Editions Technip, Paris
2   Gabolde G, Nguyen JP (1999) *Drilling Data Handbook*. Editions Technip, Paris
3   Dowell Schlumberger (1982) *Field Data Handbook*
4   Baker Packers (1984) *Baker Tech Facts*
5   Allen TO, Roberts AP (1994) *Production Operations*. OGCI, Tulsa, OK
6   1999 Tubing Reference Tables. *World Oil,* January 1999
7   1997 Tubing Guide. Hart's Petroleum Engineer International
8   Standard ISO 11960 (1998) *Petroleum and natural gas industries. Steel pipes for use as casing or tubing for wells.* International Organization for standardization, ISO, Geneva, Switzerland
9   Standard ISO 10 405 (1998) *Petroleum and natural gas industries. Care and use of casing and tubing.* Geneva, Switzerland
10 Standard API RP 5C1 (1994*) Recommended Practice for Care and Use of Casing and Tubing*. 17th edition. API, Washington DC
11 Standard API Bulletin 5C2 (1987) *Bulletin on Performance Properties of Casing, Tubing and Drill Pipe*, 20th edition. API, Washington DC.

C

Coiled Tubing

C

Coiled Tubing

Although coiled tubing has been in use for some time in oil and gas well operations, it is a "relatively" new type of well servicing equipment.

Today, coiled tubing is being used to cleanout wellbores at depths much greater than 3 000 m (10 000 ft). It has been used, not just as concentric, velocity or syphon string, but as the primary production tubing during initial well completions. Coiled tubing is being used to assist wireline logging operations, and to operate in horizontal drain holes.

This chapter will provide readers with basic knowledge concerning the fundamentals of coiled tubing technology as well as what the latest techniques and developments will mean to industry.

C1 COILED TUBING UNIT EQUIPMENT DESIGN [1]

The coiled tubing unit is a portable, hydraulically-powered service system designed to run and retrieve a continuous string of tubing concentric to larger ID production pipe or casing strings. The basic coiled tubing unit components are as follows:

- Tubing injector head
- Coiled tubing reel
- Wellhead blowout preventer stack
- Hydraulic power-drive unit
- Control console.

A simplified coiled tubing unit is shown in Fig. C1.

1.1 Tubing injector heads

They are designed to perform three basic functions:

- Provide the thrust to snub tubing into the well against pressure or to overcome wellbore friction.
- Control the rate of tubing entry into the well under various well conditions.
- Support the full suspended tubing weight and accelerate it to operating speed when extracting it from the well.

Figure C2 illustrates a simplified coiled tubing injector head rig-up and wellhead blowout preventer stack.

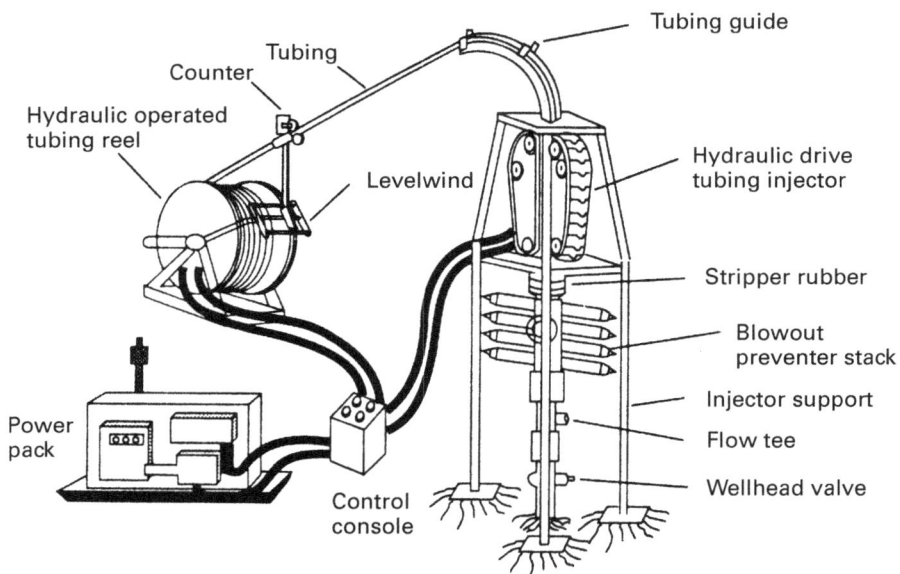

Figure C1 Mechanical components of a hydraulic coiled tubing unit [1].

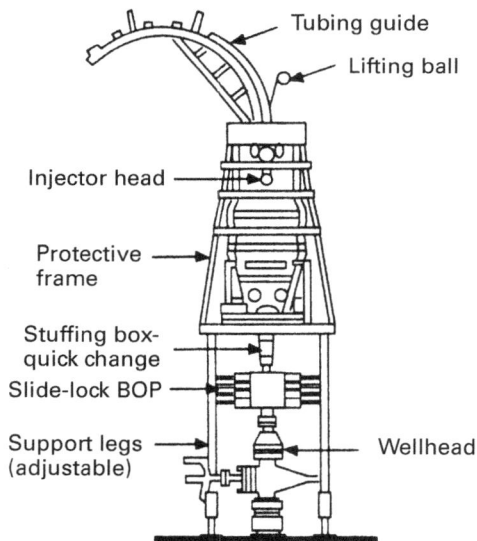

Figure C2 Tubing injector head on adjustable support legs and BOP stack [1].

1.2 Coiled tubing reel

It is a manufactured steel spool. Spooled pipe capacities are contingent on the core diameter (Fig. C1).

Reel rotation is controlled by a hydraulic motor that is mounted for direct drive on the reel shaft or operated by a chain-and-sprocket drive assembly. This motor is used to maintain a constant pull on the tubing and keep the pipe wrapped tightly on the reel.

The tubing is guided onto the spool through a mechanism called the "level-wind assembly" to properly align the pipe as it is wrapped on or spooled off the reel.

1.3 Wellhead blowout preventer stack

The BOP stack is composed of four hydraulically-operated rams, generally rated for a minimum working pressure of 70 MPa (10 000 psi). The four BOP compartments are equipped (from top down) with:
  (a) Blind rams
  (b) Tubing shear rams
  (c) Slip rams
  (d) Pipe rams.

1.4 Hydraulic power-drive units

They are sized to operate all of the coiled tubing unit components. The prime mover assembly size will vary with hydraulic-drive unit needs.

1.5 Control console

The console includes all of the controls and gauges required to operate and monitor the coiled tubing unit components:
  (a) Reel and injector heads are activated through valves that determine tubing motion direction at operating speed.
  (b) Control system regulate the drive chain, stripper rubber, and blowout preventers.

C2 WORKOVER SAFETY [2]

2.1 Pre-job requirements

Where rig operations are involved, site inspections must be performed before moving coiled tubing equipment to location.

2.1.1 Location

- Inspect roads, bridges, overhead lines, and locations prior to moving coiled tubing unit and identify any problems or limitations.
- Inspect location for hazards (e.g., electrical, fire, environmental, etc.).
- Onsite supervisor should locate flowlines, power cables, injection lines, and ground wires prior to setting anchors.

2.1.2 Coiled tubing

- A permanent "depth flag" of some type should be used of the tube OD about 90 m (300 ft) from the end of tubing. This is used to verify the depth counter when pulling out of the hole.
- Coiled tubing should be "pickled" with a sufficient volume of properly inhibited hydrochloric (HCl) acid to remove rust, scale, and foreign debris. The acid should then be displaced with a neutralizing soda ash solution.
- Coiled tubing reels should be pressure tested to 35 MPa (5 000 psi) with liquid after "pickling" is completed and held for a minimum of 5 min.
- A 1-in. ball valve should be threaded onto the end of the coiled tubing. With the valve open, displace liquid in the tubing with nitrogen, allow pressure to bleed down to about 70 kPa (10 psi) and then close the valve to maintain a nitrogen blanket in the tubing.

2.1.3 Power packs

- Ensure that all exhaust manifolds and mufflers are wrapped and insulated to comply with personnel protection requirements.
- All power packs must be equipped with spark arresters and pollution pan skids to contain pollutants and prevent accidental discharge to the environment.
- All diesel engines must be equipped with remote-operated or automatic shut-down devices.

2.1.4 Blowout preventers

- Blowout preventers (BOPs) must be hydraulically operated by controls located at the operator console.
- BOP stack order must be equipped from top down as follows:
  - Blind rams
  - Cutter rams
  - Kill/return spool with isolation valve
  - Uni-directional slip rams
  - Tubing rams.

2.1.5 Tubing injector

Injector heads must be equipped with 4 telescoping legs to stabilize and properly support the injector.

2.2 Operations safety

Several safety items which pertain to coiled tubing operations in general are offered below:

- At no time should produced hydrocarbons be reverse circulated up the coiled tubing string.
- At no time should natural gas be injected down coiled tubing for jet lifting, foam washing, etc.
- Make all necessary safety provisions for handling caustic workover materials available to operations personnel.
- When energized fluids are used in workovers (nitrogen, CO_2, etc.), it is recommended that use of certified high pressure hoses be limited.
- Place boards or plywood sheets beneath nitrogen or CO_2 transfer hose connections to prevent damage to steel structures in the event of a leak.
- Liquid nitrogen and CO_2 can cause severe flesh burns on contact.
- Be prepared for hydraulic oil spills which may occur on location during rigging up and rigging down. Also, be prepared for possible additional hydraulic oil discharge from leaking connections during the workover. Placement of grass mats soaked with soap around the worksite will allow for oil removal from the soles of boots and minimize potential for slips and falls.
- During workover, the well must be continually monitored and should not be left unattended at any time unless it is shut in and secured.
- Acid pumped through coiled tubing must be properly inhibited to protect the coiled tube material and should be only be handled and pumped by personnel specifically trained for acid service. Coiled tubing unit wash pumps are not designed or maintained for acid service and should not be used unless specifically prepared for corrosive service.
- When coiled tubing work is being performed in a well which is underbalanced, all active wells requiring production processing on the platform should be shut in.

C3 TUBE TECHNOLOGY AND CAPABILITIES

Factors affecting coiled tubing performance are presented in this paragraph. The information presented here is considered proven technology (API RP 5C7) [11].

3.1 Mechanical testing properties [11]

Tensile and hardness requirements for manufactured coiled tubing are given in Table C1.

Note: The mechanical properties do not necessarily remain the same after spooling.

Table C1 Tensile and hardness requirements for manufactured coiled tubing.

Grade	Minimum yield strength		Minimum tensile strength		Maximum hardness
	(psi)	(MPa)	(psi)	(MPa)	(HRC)
CT 55	55 000	380	70 000	480	22
CT 70	70 000	480	80 000	550	22
CT 80	80 000	550	90 000	620	22
CT 90	90 000	620	100 000	690	22

3.2 Dimension and weight characteristics [11]

Dimension and weight characteristics are given in Tables C2a and C2b.

Table C2a Coiled tubing dimensions and weights [11].

Specified diameter D		Plain end weight		Specified wall thickness t		Minimum wall thickness t_{min}		Inside diameter d	
(in.)	(mm)	(lbm/ft)	(kg/m)	(in.)	(mm)	(in.)	(mm)	(in.)	(mm)
0.750	19.05	0.59	0.878	0.083	2.10	0.078	1.98	0.584	14.83
1.000	25.40	0.74	1.101	0.075	1.91	0.070	1.78	0.850	21.59
1.000	25.40	0.79	1.176	0.080	2.03	0.075	1.91	0.840	21.34
1.000	25.40	0.85	1.265	0.087	2.21	0.082	2.08	0.826	20.98
1.000	25.40	0.92	1.369	0.095	2.41	0.090	2.29	0.810	20.57
1.000	25.40	0.98	1.458	0.102	2.59	0.097	2.46	0.796	20.22
1.000	25.40	1.04	1.548	0.109	2.77	0.104	2.64	0.782	19.86
1.000	25.40	1.17	1.741	0.125	3.17	0.117	2.97	0.750	19.05
1.250	31.75	0.94	1.399	0.075	1.91	0.070	1.78	1.100	27.94
1.250	31.75	1.00	1.488	0.080	2.03	0.075	1.91	1.090	27.69
1.250	31.75	1.08	1.607	0.087	2.21	0.082	2.08	1.076	27.33
1.250	31.75	1.17	1.741	0.095	2.41	0.090	2.29	1.060	26.92
1.250	31.75	1.25	1.860	0.102	2.59	0.097	2.46	1.046	26.57
1.250	31.75	1.33	1.979	0.109	2.77	0.104	2.64	1.032	26.21
1.250	31.75	1.50	2.232	0.125	3.17	0.117	2.97	1.000	25.40
1.250	31.75	1.60	2.381	0.134	3.40	0.126	3.20	0.982	24.94
1.250	31.75	1.82	2.708	0.156	3.56	0.148	3.76	0.938	23.83
1.250	31.75	2.01	2.976	0.175	4.44	0.167	4.24	0.900	22.86
1.500	38.10	1.43	2.128	0.095	2.41	0.090	2.29	1.310	33.27
1.500	38.10	1.52	2.262	0.102	2.59	0.097	2.46	1.296	32.92
1.500	38.10	1.62	2.411	0.109	2.77	0.104	2.64	1.282	32.56
1.500	38.10	1.84	2.738	0.125	3.17	0.117	2.97	1.250	31.75
1.500	38.10	1.95	2.902	0.134	3.40	0.126	3.20	1.232	31.29
1.500	38.10	2.24	3.333	0.156	3.56	0.148	3.76	1.188	30.17
1.500	38.10	2.48	3.691	0.175	4.44	0.167	4.24	1.150	29.21

Table C2b Coiled tubing dimensions and weights [11].

Specified diameter D		Plain end weight		Specified wall thickness t		Minimum wall thickness t_{min}		Inside diameter d	
(in.)	(mm)	(lbm/ft)	(kg/m)	(in.)	(mm)	(in.)	(mm)	(in.)	(mm)
1.750	44.45	1.80	2.678	0.102	2.59	0.097	2.46	1.546	39.27
1.750	44.45	1.91	2.842	0.109	2.77	0.104	2.64	1.532	38.91
1.750	44.45	2.17	3.229	0.125	3.17	0.117	2.97	1.500	38.10
1.750	44.45	2.31	3.438	0.134	3.40	0.126	3.20	1.482	37.64
1.750	44.45	2.66	3.956	0.156	3.56	0.148	3.76	1.438	36.53
1.750	44.45	2.94	4.375	0.175	4.44	0.167	4.24	1.400	35.56
1.750	44.45	3.14	4.673	0.188	4.77	0.180	4.57	1.374	34.90
2.375	60.33	2.64	3.929	0.109	2.77	0.104	2.64	2.157	54.79
2.375	60.33	3.00	4.465	0.125	3.17	0.117	2.97	2.125	53.98
2.375	60.33	3.21	4.777	0.134	3.40	0.126	3.20	2.107	53.18
2.375	60.33	3.70	5.506	0.156	3.56	0.148	3.76	2.063	52.40
2.375	60.33	4.11	6.116	0.175	4.44	0.167	4.24	2.025	51.44
2.375	60.33	4.39	6.533	0.188	4.77	0.180	4.57	1.999	50.77
2.875	73.03	3.67	5.462	0.125	3.17	0.117	2.97	2.625	66.68
2.875	73.03	3.93	5.848	0.134	3.40	0.126	3.20	2.607	66.22
2.875	73.03	4.53	6.741	0.156	3.56	0.148	3.76	2.563	65.10
2.875	73.03	5.05	7.516	0.175	4.44	0.167	4.24	2.525	64.14
2.875	73.03	5.40	8.036	0.188	4.77	0.180	4.57	2.499	63.47
2.875	73.03	5.79	8.616	0.203	5.16	0.195	4.95	2.469	62.71
3.500	88.90	4.82	7.173	0.134	3.40	0.126	3.20	3.232	82.09
3.500	88.90	5.57	8.289	0.156	3.56	0.148	3.76	3.188	80.98
3.500	88.90	6.21	9.241	0.175	4.44	0.167	4.24	3.150	80.01
3.500	88.90	6.65	9.896	0.188	4.77	0.180	4.57	3.124	79.35
3.500	88.90	7.15	10.640	0.203	5.16	0.195	4.95	3.094	78.59

3.3 Hydrostatic pressure test [11]

Hydrostatic pressure tests are performed by the manufacturer on spooled coiled tubing. The test pressures specified herein are based on the following formula; they should not exceed 70 MPa (10 000 psi). The minimum hold time at the hydrostatic test pressure shall be 15 minutes. Failure will be defined as pressure loss greater than 350 kPa (50 psi) during the hold period or any visible fluid loss.

The test pressure for a tapered coiled tubing string shall be based on the thinnest-wall segment of the string:

$$P = \frac{2 \times f \times Y \times t_{min}}{D} \tag{C1}$$

where

P hydrostatic test pressure (MPa or psi)

f test factor $= 0.80$

Y specified minimum yield strength (MPa or psi) (see Table C1)

t_{min} minimum specified wall thickness of the thinnest wall segment of tubing on the spool (mm or in.). $t_{min} \approx 0.95\, t$ (see Tables C2a and b)

D specified outside diameter (mm or in.).

3.4 Calculated performance properties of new coiled tubing [11]

3.4.1 Pipe body yield load

The pipe body yield load is defined as the axial tension load (in the absence of pressure or torque) which produces a stress in the tube equal to the specified minimum yield strength Y in tension:

$$L_Y = 3.1416(D - t_{min})t_{min}Y \qquad (C2)$$

where

L_Y pipe body yield load (daN or pounds)

Y specified minimum yield strength (bar or psi)

D specified outside diameter (cm or in.)

t_{min} minimum wall thickness (cm or in.).

3.4.2 Internal yield pressure

The internal yield pressure is defined as the internal pressure which produces a stress in the tubing equal to the specified minimum yield strength Y, based on the specified outside diameter and the minimum wall thickness, using Eq. 31 from API Bulletin 5C3 [12]:

$$P = \frac{2 \times Y \times t_{min}}{D} \qquad (C3)$$

where

P hydrostatic test pressure (MPa or psi)

Y specified minimum yield strength (MPa or psi) (see Table C1)

t_{min} minimum specified wall thickness of the thinnest wall segment of tubing on the spool (mm or in.). $t_{min} \approx 0.95\, t$ (see Tables C2)

D specified outside diameter (mm or in.).

3.4.3 Torsional yield strength

Torsional yield strength is defined as the torque required to yield the coiled tubing (in the absence of pressures or axial stress) and is calculated as shown in Eqs. C4 and C5.

- *In metric units*

$$T_f = \frac{Y \times \left[D^4 - (D - 2t_{min})^4 \right]}{7.113 \times D} \qquad (C4)$$

where

T_f torsional yield strength (daN per meter)
Y specified minimum yield strength (MPa) (see Table C1)
t_{min} minimum specified wall thickness of the thinnest wall segment of tubing on the spool (cm). $t_{min} \approx 0.95\ t$ (see Tables C2)
D specified outside diameter (cm).

- *In U.S. units*

$$T_f = \frac{Y \times \left[D^4 - \left(D - 2t_{min}\right)^4\right]}{105.86 \times D} \tag{C5}$$

where

T_f torsional yield strength (pounds per foot)
Y specified minimum yield strength (psi) (see Table C1)
t_{min} minimum specified wall thickness of the thinnest wall segment of tubing on the spool (in.). $t_{min} \approx 0.95\ t$ (see Tables C2)
D specified outside diameter (in.).

3.5 Coiled tubing string design and working life [11]

Topics on coiled tubing string design considerations, ultra-low cycle fatigue prediction methods, diametral growth, other OD anomalies, collapse derating, discussion on corrosion effects, and common weld survivability, are developed in API RP 5C7 (Paragraph 5).

The useful working life of coiled tubing is limited by several factors, including the following:
- Fatigue
- Diameter growth and ovality
- Mechanical damage (kinks, surface anomalies)
- Corrosion
- Welds.

C4 SAND AND SOLIDS WASHING [4]

Operations involving sand or solids washing are the most common of today's coiled tubing workover services.

Some recommendations

Do:
- Require a flow tee to direct returns out of the well. Place the tee directly below the BOP.
- Install an adjustable choke on the return line and have a replacement stern on location.
- Plan for wash fluid loss and have additional fluid on location.

- Keep adequate tankage on location to capture all returns and solids from the well. Plan to have liquids treated through production facilities or sent to an approved disposal site. Solids should be cleaned and dumped, or sent to an appropriate disposal site.
- Inject coiled tubing into the well no faster than 10 m/min (30 to 40 ft/min) if top of fill is unknown. If top of fill has been located, insertion rate should not exceed 20 m/min (60 ft/min).
- Maintain returns throughout the wash program. If observed returns decrease or stop, pull coiled tubing up the hole until returns are reestablished.
- **Wash solids slowly.** When breaking through bridges, allow sufficient time to circulate solids from the well before continuing downhole.
- Check tubing drag every 300 to 500 m (1 000 to 1 500 ft). Have coiled tubing representatives identify tubing sections that have been cycled extensively and avoid conducting periodic drag tests in these interval.
- Monitor surface pump pressure and return choke pressures while circulating large slugs of solids laden fluids.

Do not:

- Allow coiled tubing to stay stationary for longer than half of the time required to circulate bottoms-up.
- Shut down pumps for any reason, until coiled tubing is out of the well.
- Exceed a design fluid circulation pressure of 25 MPa (3 500 psi).
- Wash out of production tubing into casing without circulating at least one tubing volume up the annulus.

C5 UNLOADING WELLS WITH LIGHTER FLUIDS [5]

Techniques are used for initiating production from overbalanced or "logged-up" wells using coiled tubing.

Some recommendations

Do:

- Determine reservoir performance parameters, including static BHP, desired pressure drawdown, fluid type, solution GLR and PI prior to designing wellbore unloading programs.
- Obtain information on all downhole tools and completion equipment that could cause flow restrictions.
- Determine the most appropriate method for unloading the wellbore based on the "soft-start" pressure drawdown concept.
- If an N_2 unloading method is selected, design for the lowest possible N_2 circulation rates to minimize frictional pressure losses within the system.
- Rig up high-pressure piping for pump and return lines and secure to location anchors.

- Install an adjustable choke on the return line and have a replacement stem available on location. Verify choke calibration with provided documentation.
- Request that service companies provide a flow tee to direct return flow out of the well. Place the flow tee directly beneath the BOP stack.

Do not:
- **Get in a hurry when unloading wells to initiate production!**
- Pump natural gas through a coiled tubing string.
- Circulate N_2 below the predetermined maximum depth or attempt to increase fluid lift rates without compensating for surface choke pressure.
- Increase N_2 circulation rates to increase fluid production rates without evaluating all possible causes for reduced flow.
- Discontinue pumping of N_2 down coiled tubing unless pulling out of the hole. Leave N_2 unit hooked-up until the coiled tubing is out of the wellbore.

C6 COILED TUBING ASSISTED LOGGING AND PERFORATING [6]

Logging with coiled tubing is outwardly simple, offering advantages that may, in some applications, not be available with other methods. There are, however, costly and potentially hazardous pitfalls if the technology is not used properly.

6.1 Advantages of coiled tubing conveyed wireline operations

The advantages of coiled tubing for wireline operations can be listed as follows:
- Convey tools over long distances in high-angle extended-reach and horizontal wells.
- Allow for continuous movement.
- Convey tools through short sections of corkscrewed or twisted pipe.
- Introduce or reverse circulating fluid downhole.
- Provide constant pressure control.
- Minimize the danger of being "blown up hole".
- Record data while drilling, stimulating or performing other tasks.
- Allow electric line to remain inside the coiled tubing for higher reliability.
- Assist specialized applications, like borehole seismic.

6.2 Operational guidelines

Contacting prospective service companies and providing them with a detailed explanation of required CT logging services is necessary. Check to see that they provide or can subcontract basic equipment and information, including:

- **For CT unit and all peripherals:**
  - Unit size for operations
  - Proper tube weight and size
  - Proper reel diameter
  - Proper tubing guide radius
  - Adequate injector pull and speed (high and low)
  - Necessary instrumentation (weight, pressure, running speed, etc.)
  - Computers as needed.
- **For CT electric line reel:**
  - Coiled tubing (OD and ID)
  - Condition and history of CT reel
  - Wireline (size, conductors, and condition)
  - Type and condition of collector.
- **For crane or deployment system:**
  - Required risers or lubricator
  - Adequate support to deploy tools into high pressure wells if a deployment system is not used
  - Cross-overs to wellhead and CT
  - A means to shear tools.
- **For BOP stack:**
  - Large enough ID to pass tools
  - Properly positioned BOPs close to the top of the tree, not just below the injector when a long riser is used
  - Pressure and H_2S service ratings
  - Blind, cutter, slip and pipe rams.
- **For downhole equipment:**
  - Bump-up sub
  - CT connector
  - Safety release system
  - Back-pressure (check) valves
  - Wireline adapter for logging company connector.

C7 CEMENTING [7]

Basic coiled tubing workover (CTWO) squeeze procedures have been refined over the years and a variety of special cement blends have been developed that reflect the changing requirements of dynamic field production.

The advantages of coiled tubing cementing operations are:

- Wells can be safely, efficiently and economically squeezed.
- Squeeze operations can be completed in 12 hours or less.
- Uncontaminated cement can be safely reversed out up the coiled tubing provided the CT is not damaged.

- Cement that is to be circulated out can be contaminated with a bio-polymer system.
- Wells can be squeezed, tested, perforated and returned to production in three days to minimize downtime and lost production.

C8 FISHING [8]

In areas where pulling tubing is expensive, coiled tubing offers a viable alternative to conventional rig work.

To properly evaluate a well as a candidate for coiled tubing fishing and make proper decisions during the operation, supervisors must fully understand the advantages, disadvantages, strengths and limitations of coiled tubing. They must also understand the many available tools and their appropriate applications.

8.1 Advantages

- It offers additional tensile strength above that of braided line and the ability to use heavier tools is helpful in most applications.
- The capacity to circulate fluid through the system can also be helpful in some situations.
- Relatively low cost, quick rig up and fast time are advantages in certain applications.

8.2 Disadvantages

- Relatively low tensile strength capacity restricts overpull and inability to rotate limits the use of bent subs, wall hooks and some types of releasing mechanisms that are incorporated into conventional overshots and spears.
- Coiled tubing is more expensive than braided line operations and cannot use spang jars as effectively due to limited running speed.

C9 VELOCITY STRINGS [9]

Installation (hanging off) of a concentric string of coiled tubing inside existing production tubing (Fig. C3) is an economically viable, safe, convenient and effective alternative for returning some of these liquid loaded (logged-up) wells to flowing status.

In Fig. C3, the well was originally configured with 2 7/8-in. production tubing from surface to 9 702 ft (2 950 m). Prior to velocity string installation, fluid flowed up 5-in. casing from mid-perforations at 11 381 ft (3 470 m) to 9 702 ft (2 950 m).

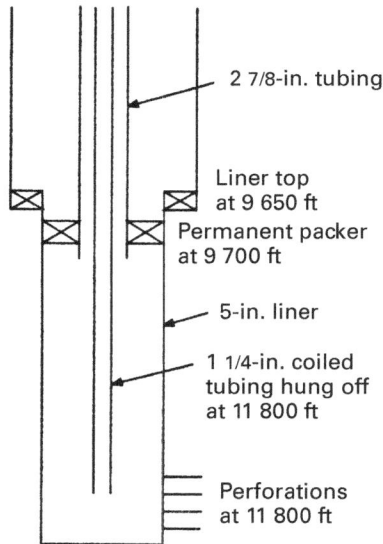

Figure C3 Schematics of concentric coiled tubing [9].

Detailed installation procedure of concentric coiled tubing velocity strings

1. Cut paraffin if necessary.
2. Swab down to packer.
3. Close lower master valve and bleed off pressure.
4. Remove tree above lower master valve.
5. Install tubing hanger on lower master valve.
6. Place packoff assembly in tubing hanger.
7. Rotate lock pins in until tips are touching the top plate of the packoff assembly. This prevents pressure from displacing the packoff assembly when the master valve is opened.
8. Connect blowout preventers.
9. Connect coiled tubing unit to BOP stack.
10. Connect access window to tubing hanger and BOP stack.
11. Run tubing slowly through BOP stack, access window and packoff assembly until it reaches the lower master valve. Tubing must be sealed on the end with a pumpout plug
12. Close access window.
13. Energize stripper rubber at the top of the coiled tubing unit.
14. Open lower master valve.
15. Run tubing to desired depth.

16. Rotate lock pins in tightly to energize packoff.
17. Tighten packing gland nuts.
18. Bleed off pressure from BOP stack to verify that packoff is properly energized.
19. Raise access window.
20. Secure segmented wraparound slips around tubing.
21. Lower tubing into well until weight indicator reads "0".
22. Rough cut tubing a minimum of six inches above tubing hanger top flange. Remove access window, BOP stack and coiled tubing unit.
23. Make a final cut on the tubing to fit into wireline guide and place guide over tubing.
24. Install tree over coiled tubing hanger.
25. Rigup nitrogen truck and pump 10 000 std cu ft (≈ 300 m^3) or more nitrogen (N$_2$) down the coiled tubing to displace end plug.
26. Pump N$_2$ down the existing tubing-coiled tubing annulus to unload fluid.
27. Secure lower master valve in full open position.
28. Return well to production.

C10 PRODUCTION APPLICATIONS [10]

Coiled tubing is being used with increasing frequency in conventional or traditional production operations.

10.1 Advantages of coiled tubing as production tubulars

The advantages of using coiled tubing as production tubulars are as follows:
- It can be run in underbalanced well conditions to minimize formation damage from completion or workover operations.
- Installation and removal is generally faster than jointed pipe.
- Joint connections are reduced or eliminated, minimizing potential for leaks and the need for testing connections.
- Costs are competitive with jointed pipe in most sizes.
- It is compatible with most artificial lift methods.

10.2 Coiled tubing installations

10.2.1 General procedure for hanging coiled tubing from surface as a production or injection string

1. Rig up coiled tubing unit and kill well if necessary.
2. Install coiled tubing tubing head. This may already be in place or may be an addition to existing wellhead equipment. Many times, the tubing head will be installed on the lower master valve.

3. Nipple up blowout preventers (BOPs) with window on tubing head (Fig. C4).

4. Run coiled tubing with shear-out or pump out plug on the end to prevent possible well flowback through the coiled tubing, accessories such as seals for a packer installation, and landing nipples or gas lift mandrels as needed. Use BOPs or tubing stripper for annular well control.

5. When end of coiled tubing is at desired depth, close lower set of BOPs and check for leaks.

6. Carefully measure distance from bottom flange of access window to tubing head lock screws to insure that, while landing hanger, the assembly sets completely in the hanger profile (Fig. C5).

7. Attach hanger and slips to coiled tubing (both are wraparound style) and slowly lower assembly to top of the lower set of BOP rams.

8. Close upper BOPs, open lower BOPs and allow pressure to equalize across the spool.

9. Lower hanger to depth of bowl and land tubing with weight on hanger. Carefully engage lock-down screws. Pressure test hanger.

10. Rough cut coiled tubing at the window, and nipple down BOPs and window assembly.

11. Make a final (smooth) cut on the coiled tubing, and bevel to fit adapter and avoid damaging adapter seals, install remaining wellhead equipment (Fig. C6) and connect flowline (see figure on page 80).

12. Pressure up on coiled tubing to shear out bottom plug.

13. Place well in service.

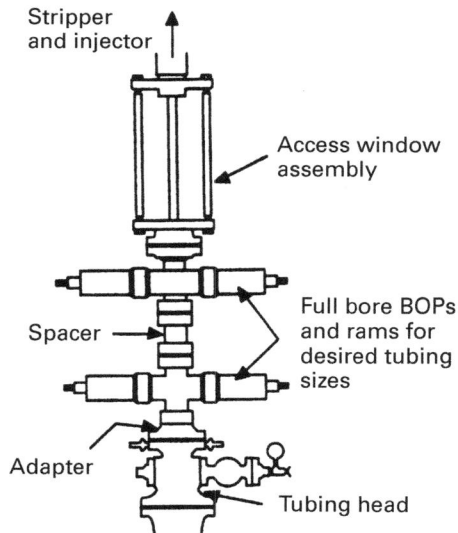

Figure C4 Coiled tubing rig-up and blowout preventer configuration for "live well" completion work [10].

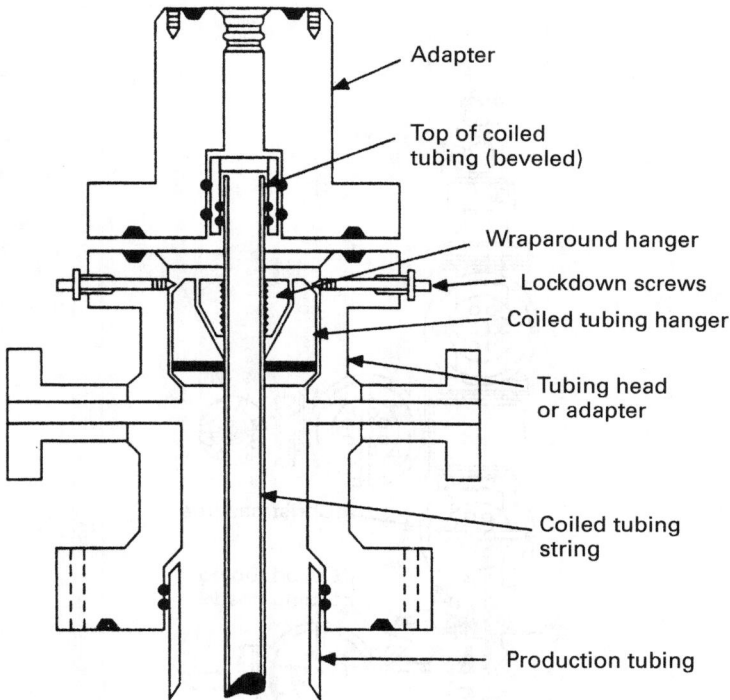

Figure C5 Typical tubing head and adapter for hanging coiled tubing off from surface [10].

10.2.2 Example procedure for hanging a partial length of coiled tubing (modified velocity string or tubing patch) from a packer and stinging into a lower sealbore

1. Set special wireline mandrel with sealbore extension in a lower landing nipple. A flapper valve on the end of the mandrel prevents well flow during installation. Kill well if necessary. A valve can be set in an upper gas lift mandrel to help unload the well after workover and deeper mandrels can be left open to provide a means to check for communication below the top coiled tubing packer. The tubing by casing annulus should be filled with inhibited water.

2. Rig up a unit with the reel of coiled tubing to be installed. Nipple up two sets of BOPs, one with rams to fit the installation string and the other with rams for the workstring and a lubricator. Pressure test surface equipment.

3. Run design coiled tubing string to be installed with a pump-out plug (for well control), seal assembly, locator sub, and landing nipple to sting into the mandrel sealbore. A telescoping swivel tool below the packer minimizes packer hanging weight during setting.

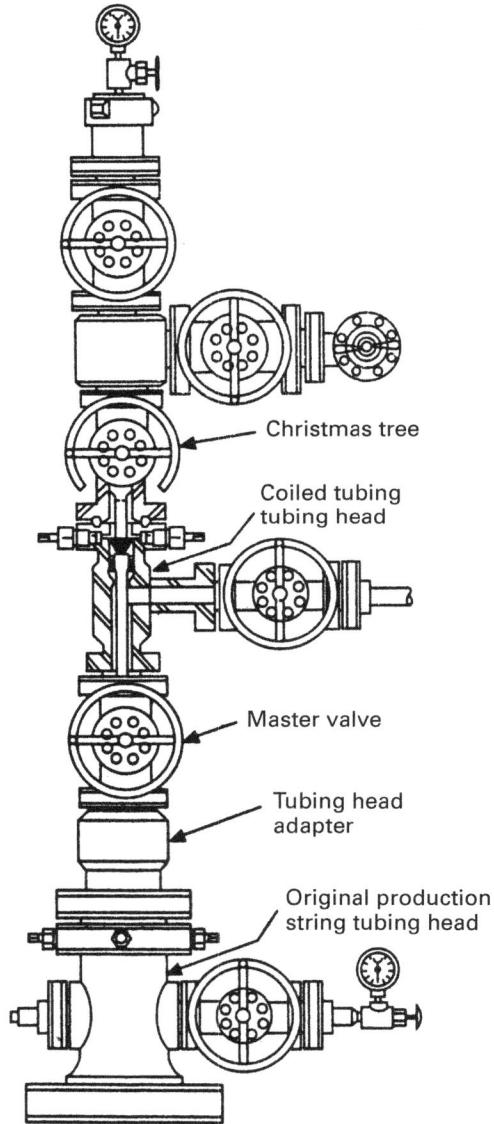

Christmas tree

Coiled tubing
tubing head

Master valve

Tubing head
adapter

Original production
string tubing head

Figure C6 Typical wellhead configuration for hanging coiled tubing
off from surface [10].

Once in place, the seal assembly holds flapper valve open. Run coiled tubing installation string and hang it off on slips in the BOPs. Bleed down and remove lubricator. Cut coiled tubing using manual pipe cutters. Install a short lubricator during pipe switch. Rig down coiled tubing unit with installation string and pick up unit with coiled tubing workstring, chains and packoff as needed.

4. Make up a retrievable hydraulic-set packer and setting tool on top of the coiled tubing installation string using a slip-type connector. Use a similar connector to attach setting tool to the coiled tubing workstring. Pull test connections against BOPs and run assembly in at about 30 m/min (100 ft/min).

5. Sting seal assembly into lower sealbore and slack off weight on mandrel to verify location. Pick up coiled to a neutral position and place a properly sized ball in the coiled tubing workstring to facilitate pressure setting of the retrievable packer.

6. Pump the ball out of the reel coil and allow it to fall to the packer setting tool. Apply design surface setting pressure.

7. In the event of early release from the packer or if it fails to set and pressure test, packer and coiled tubing assembly can be pulled to surface, leaving coiled tubing installation string suspended in the well. The bottom seal assembly may not need to be redressed if it has been cycled only once and exposed only to non-damaged fluids. Packer and setting tools can be redressed, and the assembly rerun and stung into the sealbore in the tubing tailpipe.

8. Slack off partial string weight on the lower mandrel and set the packer with design surface pressure. After pressure testing the annulus, pull the coiled tubing workstring out leaving coiled tubing installation string and assembly set inside the production tubing.

9. Shear the seal assembly bottom pump-out plug and pressure test the tubing by casing annulus again. Rig down the coiled tubing unit and lift the well in a gas lift hookup, or coiled tubing and lighter fluids.

C11 ADVANCED-COMPOSITE SPOOLABLE TUBING [13]

Development of advanced-composite spoolable tubing offers several new solutions to many challenging oilfield operations.

Such attributes as excellent corrosion resistance and low material density and weight, coupled with high working-pressure rating and extensive fatigue resistance, make these products attractive for a number of oilfield tubular applications, including well-servicing strings and corrosion-resistant completion strings.

11.1 Production-tubing design criteria

Design criteria for a standard advanced-composite production-tubing product include the following:

- Minimum working pressure of 5 000 psi (or 35 MPa)
- Minimum collapse resistance of 3 000 psi (or 21 MPa)
- Minimum depth rating to 12 500 ft (or 3 800 m)

C

- Temperature rating to 250°F (or 120°C)
- Compatibility with surface installation.

11.2 Standardized coiled tubing products

Work currently is proceeding on four standard spoolable composite products for general workover applications. Product specifications are given in Table C3.

Table C3 Standardized coiled tubing products [13].

Coiled tubing size	Working pressure		Maximum snubbing pressure	
(in.)	(psi)	(MPa)	(psi)	(MPa)
1 1/2	6 000	420	3 000	210
1 1/2	7 500	520	3 000	210
2 3/8	5 000	350	2 500	175
2 7/8	5 000	350	2 500	175

REFERENCES

1   Sas-Jaworsky II A (1991) Coiled tubing… operations and services. Part 1. *World Oil,* November 1991, 41–47
2   Sas-Jaworsky II A (1991) Coiled tubing… operations and services. Part 2. *World Oil,* December 1991, 71–78
3   Sas-Jaworsky II A (1992) Coiled tubing… operations and services. Part 3. *World Oil,* January 1992, 95–101
4   Sas-Jaworsky II A (1992) Coiled tubing… operations and services. Part 4. *World Oil,* March 1992, 71–79
5   Sas-Jaworsky II A (1991) Coiled tubing… operations and services. Part 5. *World Oil,* April 1992, 59–66
6   Blount CG, Walker EJ (1992) Coiled tubing… operations and services. Part 6. *World Oil,* May 1992, 89–96
7   Walker EJ, Gantt L, Crow W (1992) Coiled tubing… operations and services. Part 7. Cementing. *World Oil,* June 1992, 69–76
8   Welch JL, Stephens RK (1992) Coiled tubing… operations and services. Part 9. Fishing. *World Oil,* September 1992, 81–85
9   Brown PT, Wimberly RD (1992) Coiled tubing… operations and services. Part 10. Velocity strings. *World Oil,* October 1992, 75–85
10 Hightower CM (1992) Coiled tubing… operations and services. Part 11. Production applications. *World Oil,* November 1992, 49–56
11 API Recommended Practice 5C7 (1996) *Recommended Practice for Coiled Tubing Operations in Oil and Gas Well Services,* First Edition
12 API Bulletin 5C3 (1989) *Bulletin on Formulas and Calculations for Casing, Tubing, Drill Pipe, and Line Pipe Properties*, Fifth Edition
13 Hampton S, Feechan M, Berning SA (1999) Advanced-composite spoolable tubing. *Journal of Petroleum Technology,* Vol. 51, No. 5, May 1999, 58–60.

D

Packers

Packers

D1 INTRODUCTION [1]

Once a tubing-packer system has been selected, designed, and installed in a well there are four modes of operation:
- (a) Shut-in
- (b) Producing
- (c) Injection
- (d) Treating.

These operational modes with their respective temperature and pressure profiles have considerable impact on the length and force changes on the tubing-to-packer connections.

There are two principal types of packers:
- Retrievable packers are run and pulled on the tubing string on which they are installed.
- Permanent and semipermanent packers can be run on wireline and tubing.

D2 TUBING-TO-PACKER CONNECTIONS [1]

There are three methods for connecting a packer and a tubing string:
- Tubing is latched or fixed on the packer, allowing no movement (retrievable packers). Tubing can be set either in tension, compression, or neutral.
- Tubing is landed with a seal assembly and locator sub that allows limited movement (permanent or semipermanent packers only). The tubing can be set only in compression or neutral.
- Tubing is stung into the packer with a long seal assembly that allows essentially unlimited movement (permanent packers only). The tubing is left in neutral and it cannot be set in tension or compression.

D3 DIFFERENT TYPES OF PACKERS [1]

3.1 Representation

Different types of packers are represented on Figs. D1a and D1b.

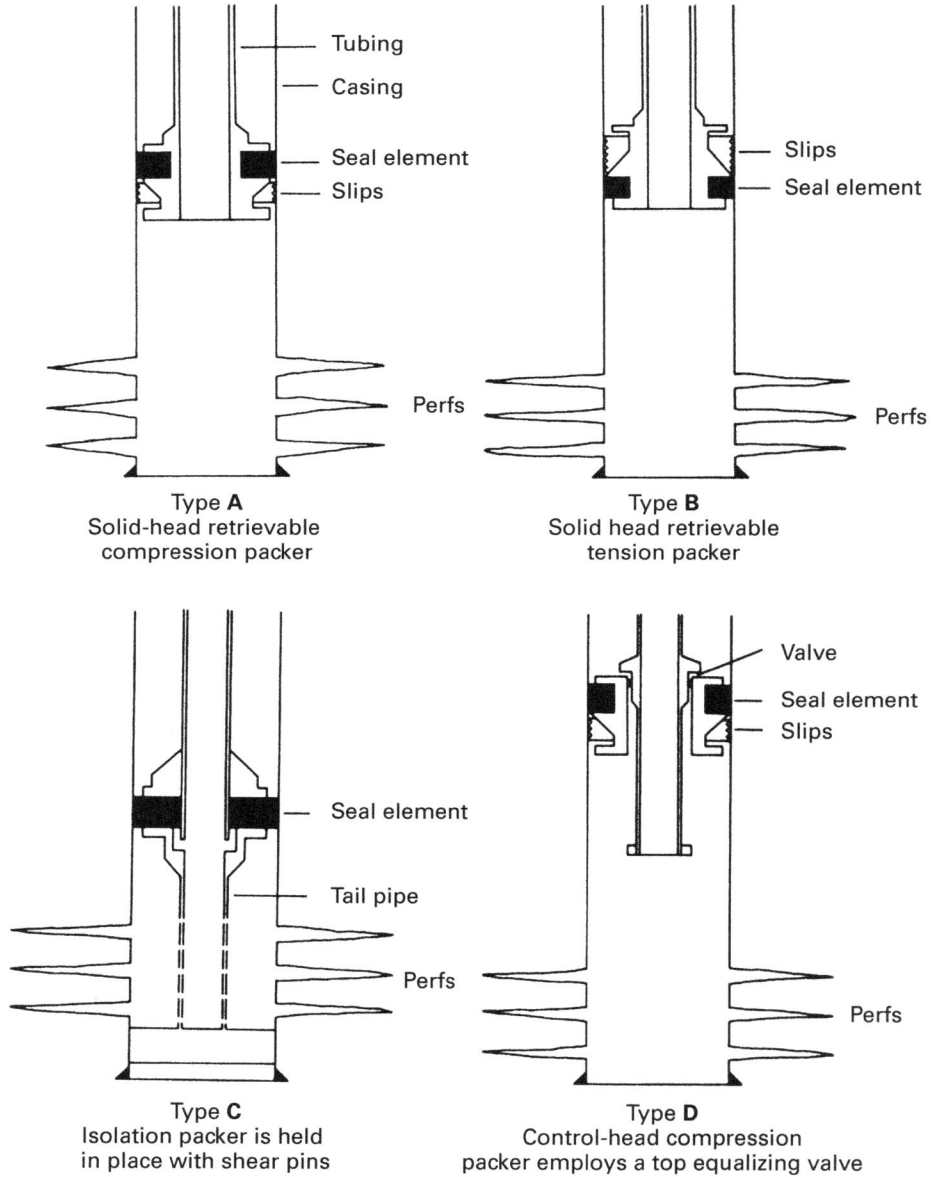

Type **A**
Solid-head retrievable
compression packer

Type **B**
Solid head retrievable
tension packer

Type **C**
Isolation packer is held
in place with shear pins

Type **D**
Control-head compression
packer employs a top equalizing valve

Figure D1a Retrievable packers [1].

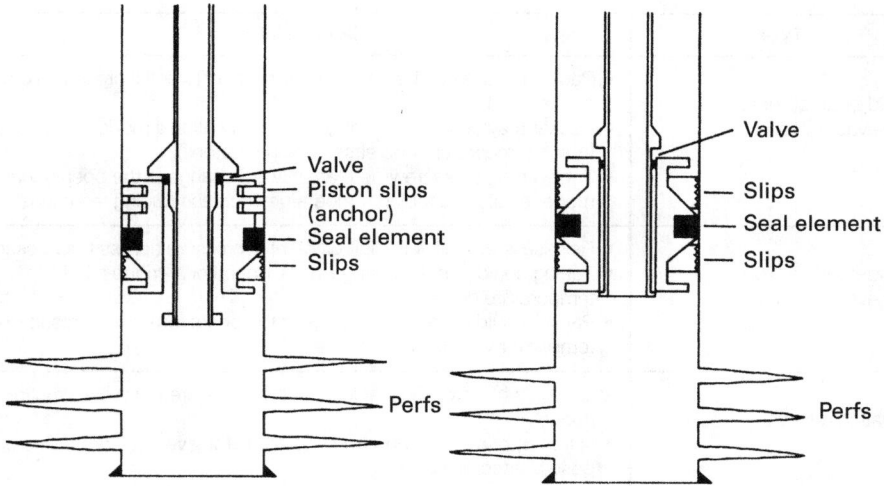

Type **E**
Solid head retrievable tension
packer is held by an anchor
containing piston slips

Type **F**
Mechanically set dual-slip packer
has slips above and
below rubber element

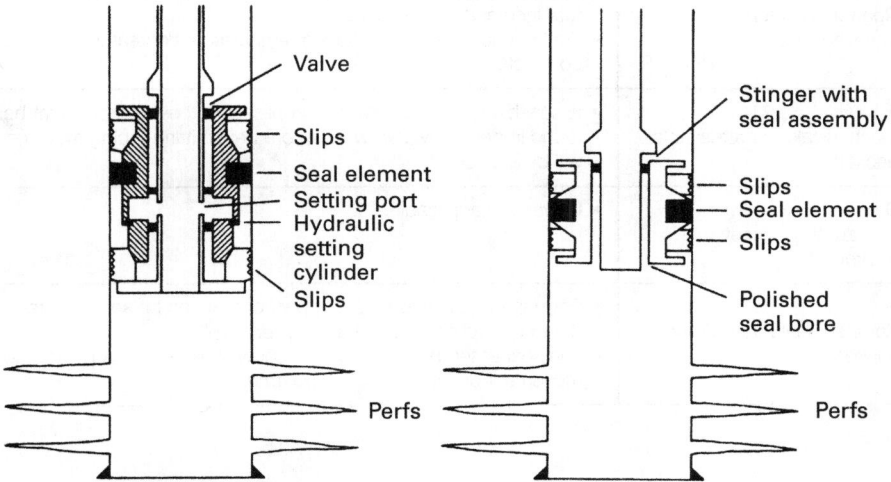

Type **G**
Hydraulic packer
is set by tubing pressure

Type **H**
Retrievable, permanent-type packer
is made with polished sealbore

Figure D1b Retrievable and permanent packers [1].

Table D1 Retrievable and permanent packer utilization and constraints [1].

Type	Constraints
A Solid-head compression retrievable packer	• Packer release can be hampered by high differential pressure across packer. • Packer may unseat if a change in the operational mode results in a tubing temperature decrease (tubing shortens). • Tubing may corkscrew permanently if a change in the operational mode results in a tubing temperature increase (tubing lengthens).
B Solid-head tension retrievable packer	• Release is difficult with high differential pressure across the packer. • Tubing could part if a change in the operational mode results in a temperature decrease. • Packer could release if a change in the operational mode results in a temperature increase.
C Isolation retrievable packer	• Is used when two mechanically set packers are to be set simultaneously. • Is for temporary use only and should be retrieved as soon as its purpose is accomplished.
D Control-head compression retrievable packer	• The bypass or equalizing valve could open if an operational mode change results in a tubing temperature decrease. • Tubing could corkscrew permanently, if an operational mode change results in a tubing temperature increase.
E Control-head tension retrievable packer	• Premature bypass valve opening could occur with a tubing temperature increase as the tubing elongates. • Tubing could part with a tubing temperature decrease as the tubing contracts.
F Mechanically set retrievable packer	• Is suitable for almost universal application, the only constraint being found in deep deviated wells where transmitting tubing movement will be a problem.
G Hydraulic-set retrievable packer	• Universally applicable.
H Polished sealbore permanent packer	• Permanent or semipermanent packer that can be set with precision depth control on conductor wireline. • The seal assembly length should allow sufficient free upward tubing movement during stimulation treatments.

3.2 Constraints

The constraints of retrievable and permanent packers are reported on Table D1.

3.3 Purchase price

Table D2 presents a range of packer cost indices.

Table D2   Cost comparison of production packers [1].

Packer type	Tubing-casing size (in.)	Typical cost index
Compression	2 × 5 1/2	1.00
Tension set	2 × 5 1/2	0.925
Mechanical set	2 × 5 1/2	1.54
Hydraulic set	2 × 5 1/2	2.30
Dual	2 × 2 × 7	5.85
Permanent*	2 × 5 1/2	1.85
Semipermanent*	2 × 5 1/2	2.30

* Electric-line setting charge not included.

D4   TUBING RESPONSE CHARACTERISTICS [1]

Depending on:
- (a) how the tubing is connected to the packer
- (b) the type of packer
- (c) how the packer is set,

temperature and pressure changes will affect the following:

- Length variation in the tubing string will result if the seals are permitted to move inside a permanent polished seal-bore packer.
- Tensile or compressive forces will be induced in the tubing and packer system if tubing motion is not permitted (latched connection).
- A permanent packer will be unsealed if motion is permitted (tubing contraction) and if the seal assembly section is not long enough.
- Unseating of a solid-head tension (or compression) packer will occur if it is not set with sufficient strain (or weight) to compensate for tubing movement.
- The equalizing valve will open prematurely on control-head packers (tension or compression).

4.1   Temperature effect

4.1.1   Influence of thermal expansion

Thermal expansion or contraction causes major length changes in the tubing as shown in the following:

- *In metric units*

$$\Delta L_t = 1.4935 \times 10^{-5} \times L_t \times \Delta T \qquad \text{(D1)}$$

where
- ΔL_t change in tubing length (m)
- L_t tubing length (m)
- ΔT change in average temperature (°C).

- *In U.S. units*

$$\Delta L_t = 8.28 \times 10^{-5} \times L_t \times \Delta T \tag{D2}$$

where
  ΔL_t change in tubing length (ft)
  L_t tubing length (ft)
  ΔT change in average temperature (°F).

4.1.2 Temperature-induced force

If the motion is constrained, forces will be induced as a result of the temperature change. The temperature-induced force is:

- *In metric units*

$$F = 742 \times A_{tw} \times \Delta T \tag{D3}$$

where
  F force (tensile or compressive, depending on direction of T) (N)
  A_{tw} cross-sectional area of the tubing wall (m²).

- *In U.S. units*

$$F = 207 \times A_{tw} \times \Delta T \tag{D4}$$

where
  F force (tensile or compressive, depending on direction of T) (lbf)
  A_{tw} cross-sectional area of the tubing wall (sq in.).

4.2 Piston effect

The length change or force induced by the piston effect is caused by pressure changes inside the annulus and tubing at the packer, acting on different areas (Fig. D2).
  The force and length changes can be calculated as follows:

- *In metric units*

$$F = \Delta p_t (A_{pi} - A_{ti}) - \Delta p_{an}(A_{pi} - A_{to}) \tag{D5}$$

and

$$\Delta L_t = \frac{3.6576 L_t}{EA_{tw}}\left[\Delta p_t(A_{pi} - A_{ti}) - \Delta p_{an}(A_{pi} - A_{to})\right] \tag{D6}$$

where
  E modulus of elasticity (kPa)
  A_{pi} area of packer ID (m²)
  A_{ti} area of tubing ID (m²)
  A_{to} area of tubing OD (m²)
  Δp_t change in tubing pressure at packer (kPa)
  Δp_{an} change in annulus pressure at packer (kPa).

D   *Packers*

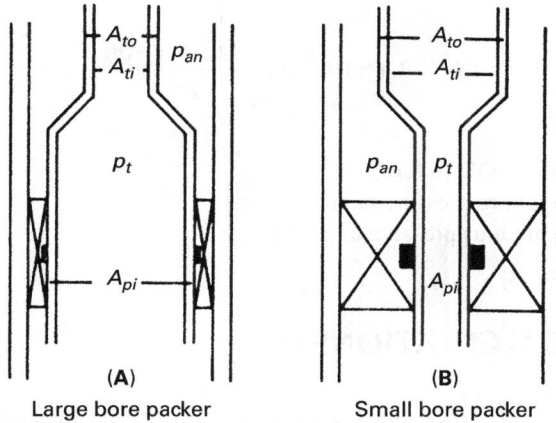

Figure D2 Tubing and packer systems, illustrating different areas
and pressures necessary for movement or force calculations [1].

- *In U.S. units*

$$F = \Delta p_t (A_{pi} - A_{ti}) - \Delta p_{an}(A_{pi} - A_{to}) \qquad \text{(D7)}$$

and

$$\Delta L_t = \frac{12 L_t}{E A_{tw}}\left[\Delta p_t(A_{pi} - A_{ti}) - \Delta p_{an}(A_{pi} - A_{to})\right] \qquad \text{(D6)}$$

where

E modulus of elasticity (psi)
A_{pi} area of packer ID (sq in.)
A_{ti} area of tubing ID (sq in.)
A_{to} area of tubing OD (sq in.)
Δp_t change in tubing pressure at packer (psi)
Δp_{an} change in annulus pressure at packer (psi).

4.3 Ballooning and reverse ballooning

Internal pressure swells or balloons the tubing and causes it to shorten. Likewise, pressure
in the annulus squeezes the tubing, causing it to elongate. The ballooning and reverse ballooning length change and force are given by:

- *In metric units*

$$\Delta L_t = 7.3 \times 10^{-8} \times L_t \frac{\Delta p_t - F_{oi}^2 \Delta p_{an}}{F_{oi}^2 - 1} \qquad \text{(D9)}$$

where

F_{oi} ratio of tubing OD to ID
Δp_t change in tubing pressure at packer (kPa)
Δp_{an} change in annulus pressure at packer (kPa).

- *In U.S. units*

$$\Delta L_t = 2.4 \times 10^{-7} \times L_t \frac{\Delta p_t - F_{oi}^2 \, \Delta p_{an}}{F_{oi}^2 - 1} \tag{D10}$$

where

F_{oi} ratio of tubing OD to ID

Δp_t change in tubing pressure at packer (psi)

Δp_{an} change in annulus pressure at packer (psi).

D5 PACKER CALCULATIONS [2]

The differential pressure across a packer is the sum of general contributions:

1. Tubing weight in thousands of pounds, value found in Table D3a (U.S. units) or tubing weight in thousands of decanewtons, value found in Table D3b (metric units).

2. Add to (1) above, the pressure due to annular fluid (pressure per foot (or meter) height multiplied by depth of packer) (Tables D4a, b).

3. Subtract the pressure due to the tubing fluid from 1 and 2 above (Table D5).

Tables D3a-b Equalizing pressures across a packer [2].

Table D3a				Table D3b		
OD casing (in.)	Pressure/ 1 000# 2-in. EUE (psi)	Tubing weight 21/2-in. EUE (psi)		OD casing (mm)	Pressure/ 1 000 daN 50.8 mm (kPa)	Tubing weight 63.5 mm (kPa)
4 1/2	106	127		114.3	1 643	1 969
5	82	94		127	1 271	1 457
5 1/2	65	72		139.7	1 008	1 116
6	52	57		152.4	806	884
6 5/8	42	45		168.3	651	698
7 (*)	34	36		177.8 (*)	527	558
7 (**)	39	42		177.8 (**)	605	651
7 5/8	30	31		193.7	465	481
8 5/8	23	23		219.1	356	257
9 5/8	17	18		244.5	289	279
10 3/4	13	14		273.1	200	217
11 3/4	11	11		298.5	170	171
13 5/8	9	9		339.7	140	140

(*) 17 – 26 lb/ft
(**) 26 – 38 lb/ft

(*) 26.3 kg/m – 38.69 kg/m
(**) 38.69 kg/m – 56.5 kg/m

Tables D4a-b Conversion table for various weight muds [2].

Table D4a		Table D4b	
Pounds/gallon	Pressure/ft of depth (psi)	Specific gravity	Pressure/m of depth (kPa)
10.0	0.519	1.20	11.7
10.6	0.551	1.27	12.5
11.2	0.581	1.34	13.1
11.8	0.602	1.41	13.9
12.4	0.644	1.49	14.6
13.0	0.675	1.56	15.3

Table D5 Hydrochloric acid table [2].

% HCl	psi/ft of depth	kPa/m of depth
5	0.4437	10.04
10	0.4547	10.29
15	0.4654	10.53
20	0.4764	10.78
25	0.4876	11.03
30	0.4991	11.29

▼ Example

- *In metric units*

Consider a packer set on 73 mm EUE tubing, and in 177.8 mm OD, 25.3 kg/m casing at 1 524 meters. The annulus is filled with 1 342 kg/m mud/tubing filled with 15% acid, 4 448 daN tubing force on packer.

The differential pressure would be:

1. (from Table D3b)
   4.448 × 558   =   2.5 MPa
2. (from Table D4b)
   13.17 × 1 524   = 20.1 MPa
   Total downward pressure   = 22.6 MPa
3. (from Table D5)
   10.53 × 1 524   = − 16 MPa
   Total differential pressure =   6.6 MPa

6.6 MPa is needed on the tubing to balance the pressure.

D

- *In U.S. units*

Consider a packer set on 21/2-in. EUE tubing, and in 7-in. OD, 17 lb/ft casing at 5 000 feet. The annulus is filled with 11.2 lb/ft mud/tubing filled with 15% acid, 10 000 lbs tubing weight on packer.

The differential pressure would be:

1. (from Table D3a)

   10 × 36 = 360 psi

2. (from Table D4a)

   0.582 × 5 000 = 2 910 psi

   Total downward pressure = 3 270 psi

3. (from Table D5)

   0.4654 × 5 000 = – 2327 psi

   Total differential pressure = 943 psi

   943 psi is needed on the tubing to balance the pressure. ▲

D6 ISO 14310 [3]

This product standard has been developed by users/purchasers and suppliers/manufacturers of packers and bridge plugs intended for use in the petroleum and natural gas industry worldwide. This product standard is intended to give requirements and information to both parties in the selection, manufacture, testing and use of packers and bridge plugs. Further, this standard addresses supplier/manufacturer requirements which set the minimum parameters with which suppliers/manufacturers must comply, in order to be able to claim conformity with this standard.

This product standard has been structured to allow for grades of increased requirements in quality control and design validation. These standards allow the user/purchaser to select the grades required for a specific application.

6.1 Quality control

There are three quality control grades which provide the user/purchaser with the choice of requirements to meet their preference or application. Quality control grade 3 is the minimum grade of quality offered by this product standard. Quality control grade 2 provides additional inspection and verification steps, and quality control grade 1 is the highest grade provided. Additional quality upgrades can be specified by the user/purchaser as supplemental requirements.

6.2 Design validation

There are six design validation grades which provide the user/purchaser with the choice of requirements to meet their preference or application. Design validation grade 6 is the minimum grade and represents equipment where the validation method has been defined by the suppliers/manufacturers. The complexity and severity of the validation testing increases as the grade number decreases.

Design validation grade 6 and quality control grade 3 are intended to represent equipment designed and manufactured consistent with minimum industry practice. These grades are sufficient for a number of applications. However, some applications could require and justify the higher grades of quality control and design validation defined by this specification.

This design standard is not intended to inhibit a supplier/manufacturer from offering, or the user/purchaser from accepting, alternative equipment or engineering solutions. Where an alternative is offered, the supplier/manufacturer should identify any variations from this product standard and provide details.

REFERENCES

1   Bradley HB (1992) *Petroleum Engineering Handbook*. Chapter 4, Patton LD, Production Packers, 4-1 to 4-11, Third Printing. Society of Petroleum Engineers, Richardson, TX
2   Dowell Schlumberger (1982) *Field Data Handbook*
3   ISO 14310 (1998) *Petroleum and Natural Gas Industries Completion Equipment. Packers and Bridge Plugs*. Geneva, Switzerland.

E

Pressure Losses

Pressures Losses

E1 TYPES OF FLOW [1]

Production fluids are in general Newtonian fluids where the shear stress is directly proportional to the shear rate.

The flow is continuous, and may be of the following types.

1.1 Laminar flow

Any laminar layer of the fluid is displaced, with respect to other laminar layers, in parallel to the direction of flow, and is moving at its specific speed.

1.2 Turbulent flow

Small eddies are formed throughout the volume of the fluid.

1.3 Determination of the type of flow

The type of flow of a fluid is characterized by the Reynolds number Re, which is then compared to a critical value Re_c:

$$Re_c = 2\ 100$$

Equations E1 and E2 show the expressions for the Reynolds number and it is generally accepted critical values for fluids of various rheological models circulating in pipes or in annulars.

- **Inside casing and tubing**

$$Re = \frac{VD\rho}{\mu} \tag{E1}$$

where, in SI units

  V fluid circulation velocity (m/s)
  D inner diameter of casing or tubing (m)
  ρ density of fluid (kg/m^3)
  μ dynamic viscosity (Pa·s).

- **Annular**

$$\text{Re} = \frac{0.8165(D_o - D_i)V\rho}{\mu} \qquad (E2)$$

where, in SI units
  - D_o outer diameter of annular (m)
  - D_i inner diameter of annular (m).

- **Critical velocity**

  The critical velocity V_c, is the velocity at the critical Reynolds number.
  If $V \le V_c$, the flow is of the laminar type.

- **Density of fluid**

$$\rho = \frac{141.5}{131.5 + °\text{API}} \qquad (E3)$$

- **Viscosity**

  Figure M1 (Chapter M) represents viscosity of crude oil versus temperature.

E2 PRESSURE LOSSES IN CASING AND TUBING [1]

2.1 In SI units

- **Critical velocity:** V_c (m/s)

$$V_c = \frac{2\,100\mu}{D\rho} \qquad (E4)$$

- **Critical flow rate:** Q_c (m³/s)

$$Q_c = \frac{525\pi D\mu}{\rho} \qquad (E5)$$

- **Pressure loss inside casing or tubing:** ΔP_i (laminar flow) (Pa)

$$\Delta P_i = \frac{32LV\mu}{D^2} = \frac{128LQ\mu}{\pi D^4} \qquad (E6)$$

- **Pressure loss inside casing or tubing:** ΔP_i (turbulent flow) (Pa)

$$\Delta P_i = \frac{0.1L\rho^{0.8}V^{1.8}\mu^{0.2}}{D^{1.2}} = \frac{1.2126L\rho^{0.8}Q^{1.8}\mu^{0.2}}{\pi^{1.8}D^{4.8}} \qquad (E7)$$

where
  - L length (m)
  - V circulation velocity (m/s)
  - Q flow rate (m³/s)
  - D inner diameter (m)
  - μ dynamic viscosity (Pa·s)
  - ρ density of fluid (kg/m³).

2.2 In U.S. units

- **Critical velocity:** V_c (ft/min)

$$V_c = \frac{135.82\mu}{D\rho} \qquad (E8)$$

- **Critical flow rate:** Q_c (gal/min)

$$Q_c = \frac{5.54D\mu}{\rho} \qquad (E9)$$

- **Pressure loss inside casing or tubing:** ΔP_i (laminar flow) (psi)

$$\Delta P_i = \frac{LV\mu}{89\,775D^2} = \frac{LQ\mu}{3\,663D^4} \qquad (E10)$$

- **Pressure loss inside casing or tubing:** ΔP_i (turbulent flow) (psi)

$$\Delta P_i = \frac{L\rho^{0.8}V^{1.8}\mu^{0.2}}{3\,212\,923D^{1.2}} = \frac{L\rho^{0.8}Q^{1.8}\mu^{0.2}}{10\,141D^{4.8}} \qquad (E11)$$

where
 - L length (ft)
 - V circulation velocity (ft/min)
 - Q flow rate (gal/min)
 - D inner diameter (in.)
 - μ dynamic viscosity (cp)
 - ρ density of fluid (lbm/gal).

2.3 In practical metric units

- **Critical velocity:** V_c (m/min)

$$V_c = \frac{4.96\mu}{D\rho} \qquad (E12)$$

- **Critical flow rate:** Q_c (l/min)

$$Q_c = \frac{2.514D\mu}{\rho} \qquad (E13)$$

- **Pressure loss inside casing or tubing:** ΔP_i (laminar flow) (bar)

$$\Delta P_i = \frac{LV\mu}{120\,967D^2} = \frac{LQ\mu}{61\,295D^4} \qquad (E14)$$

- **Pressure loss inside casing or tubing:** ΔP_i (turbulent flow) (bar)

$$\Delta P_i = \frac{L\rho^{0.8}V^{1.8}\mu^{0.2}}{306\,529D^{1.2}} = \frac{L\rho^{0.8}Q^{1.8}\mu^{0.2}}{90\,163D^{4.8}} \tag{E15}$$

E

where
- L length (m)
- V circulation velocity (m/min)
- Q flow rate (l/min)
- D inner diameter (m)
- μ dynamic viscosity (mPa.s)
- ρ density of fluid (kg/dm³).

E3 PRESSURE LOSSES IN ANNULAR [1]

3.1 In SI units

- **Critical velocity:** V_c (m/s)

$$V_c = \frac{2572\mu}{(D_o - D_i)\rho} \tag{E16}$$

- **Critical flow rate:** Q_c (m³/s)

$$Q_c = \frac{643\pi(D_o + D_i)\mu}{\rho} \tag{E17}$$

- **Pressure loss in annular:** ΔP_a (laminar flow) (Pa)

$$P_a = \frac{48LV\mu}{(D_o - D_i)^2} = \frac{192LQ\mu}{\pi(D_o + D_i)(D_o - D_i)^3} \tag{E18}$$

- **Pressure loss in annular:** ΔP_a (turbulent flow) (Pa)

$$\Delta P_a = \frac{0.1275L\rho^{0.8}V^{1.8}\mu^{0.2}}{(D_o - D_i)^{1.2}} = \frac{1.5465L\rho^{0.8}Q^{1.8}\mu^{0.2}}{\pi^{1.8}(D_o + D_i)^{1.8}(D_o - D_i)^3} \tag{E19}$$

where
- L length (m)
- V circulation velocity (m/s)
- Q flow rate (m³/s)
- D_o & D_i outer and inner diameter (m)
- μ dynamic viscosity (Pa·s)
- ρ density of fluid (kg/m³).

3.2 In U.S. units

- **Critical velocity:** V_c (ft/min)

$$V_c = \frac{165.35\mu}{(D_o - D_i)\rho} \tag{E20}$$

- **Critical flow rate:** Q_c (gal/min)

$$Q_c = \frac{6.79(D_o + D_i)\mu}{\rho} \tag{E21}$$

- **Pressure loss in annular:** ΔP_a (laminar flow) (psi)

$$\Delta P_a = \frac{LV\mu}{59\,851(D_o - D_i)^2} = \frac{LQ\mu}{2\,442(D_o + D_i)(D_o - D_i)^3} \tag{E22}$$

- **Pressure loss in annular:** ΔP_a (turbulent flow) (psi)

$$\Delta P_a = \frac{L\rho^{0.8}V^{1.8}\mu^{0.2}}{2\,519\,939(D_o - D_i)^{1.2}} = \frac{L\rho^{0.8}Q^{1.8}\mu^{0.2}}{7\,952(D_o + D_i)^{1.8}(D_o - D_i)^3} \tag{E23}$$

where

L	length (ft)
V	circulation velocity (ft/min)
Q	flow rate (gal/min)
D_o & D_i	outer and inner diameter (in.)
μ	dynamic viscosity (cp)
ρ	density of fluid (lbm/gal).

3.3 In practical metric units

- **Critical velocity:** V_c (m/min)

$$V_c = \frac{6.08\mu}{(D_o - D_i)\rho} \tag{E24}$$

- **Critical flow rate:** Q_c (l/min)

$$Q_c = \frac{3.08(D_o + D_i)\mu}{\rho} \tag{E25}$$

- **Pressure loss in annular:** ΔP_a (laminar flow) (bar)

$$\Delta P_a = \frac{LV\mu}{80\,645(D_o - D_i)^2} = \frac{LQ\mu}{40\,863(D_o + D_i)(D_o - D_i)^3} \tag{E26}$$

- **Pressure loss in annular:** ΔP_a (turbulent flow) (bar)

$$\Delta P_a = \frac{L\rho^{0.8}V^{1.8}\mu^{0.2}}{240\,415(D_o - D_i)^{1.2}} = \frac{L\rho^{0.8}Q^{1.8}\mu^{0.2}}{70\,696(D_o - D_i)^{1.8}(D_o - D_i)^3} \qquad (E27)$$

where
L	length (m)
V	circulation velocity (m/min)
Q	flow rate (l/min)
D_o & D_i	outer and inner diameter (m)
μ	dynamic viscosity (mPa.s)
ρ	density of fluid (kg/dm^3).

REFERENCE

1 Publication de la Chambre Syndicale de la Recherche et de la Production du Pétrole et du Gaz Naturel (1982) *Drilling Mud and Cement Slurry Rheology Manual*, 24–28. Éditions Technip, Paris.

F

Fundamentals of Petroleum Reservoirs

F

Fundamentals of Petroleum Reservoirs

F1 CHARACTERIZATION OF RESERVOIR ROCKS [1, 2]

The methods used to characterize reservoir rocks are essentially core analysis and well logging.

1.1 Porosity

1.1.1 Definition

The porosity Φ is:

$$\Phi = \frac{V_{\text{pores}}}{V_{\text{total}}} \text{ (expressed in \%)}$$

where

V_{pores} pore volume of rock sample
V_{total} total volume of rock sample.

The effective porosity of rocks varies between less than 1% and over 40%. It is often stated that the porosity is as shown in Table F1.

Table F1

Low	$\Phi < 5\%$
Mediocre	$5\% < \Phi < 10\%$
Medium	$10\% < \Phi < 20\%$
Good	$20\% < \Phi < 30\%$
Excellent	$\Phi > 30\%$

Porosity decreases with increasing depth.

1.1.2 Effect of pressure [1, 2]

The overall compressibility (c_t) of a pore volume unit is due to the sum of all its compressible components:

$$c_t = c_o S_o + c_w S_w + c_p \qquad \text{(F1)}$$

where

c_e equivalent compressibility
c_o oil compressibility
c_w water compressibility
S_o oil saturation
c_p pore compressibility
S_w water saturation.

The reservoir is modeled by:
- An incompressible porous rock, and
- A fluid of equivalent compressibility c_e:

$$c_e = \frac{c_o S_o + c_w S_w + c_p}{S_o} \qquad \text{(F2)}$$

1.1.3 Compressibility coefficient

A compressibility coefficient is defined by:

$$C = -\frac{1}{V} \times \frac{dV}{dP} \qquad \text{(F3)}$$

The order of magnitude for compressibility is as follows:
- **for oil:** $C_o = 1 \text{ to } 3 \times 10^{-4} \text{ bar}^{-1}$ or $7 \text{ to } 20 \times 10^{-6} \text{ psi}^{-1}$ (F4)
- **for water:** $C_w = 0.4 \text{ to } 0.6 \times 10^{-4} \text{ bar}^{-1}$ or $3 \text{ to } 5 \times 10^{-6} \text{ psi}^{-1}$ (F5)
- **for pore spaces:** $C_p = 0.3 \text{ to } 1.5 \times 10^{-4} \text{ bar}^{-1}$ or $2 \text{ to } 10 \times 10^{-6} \text{ psi}^{-1}$ (F6)

1.2 Permeability [1]

Permeability is determined by Darcy's law. Let us consider a sample of:
- Length dx
- Cross-section A
- A fluid of dynamic viscosity μ
- A flow rate Q (measured in the conditions of interval dx)
- The upstream pressure P, and the downstream pressure $(P - dP)$
- A permeability coefficient k.

The oil, water and gas saturations are:

$$S_o = \frac{V_o}{V_p} \qquad S_w = \frac{V_w}{V_p} \qquad S_g = \frac{V_g}{V_p} \tag{F11}$$

expressed in percent, with $S_o + S_w + S_g = 100\%$.

1.4.2 Interfacial tension σ

Interfacial tension σ can be defined as the force per unit length necessary to maintain contact between the two lips of an incision assumed in the interface.

Orders of magnitude are given in Table F3.

Table F3

Oil/gas	0 to 15 mN/m	
Water/oil	15 to 35 mN/m	Reservoir
Water/gas	35 to 55 mN/m	
Air/water	72 mN/m	
Air/mercury	480 mN/m	Laboratory

1.4.3 Capillary pressure

Capillary pressure P_c is calculated as follows:

In a tube (Fig. F3):

$$P_c = \frac{2\sigma \cos\theta}{r} \tag{F12}$$

where
- σ interfacial tension
- θ connection angle of the interface with the solid
- r tube radius.

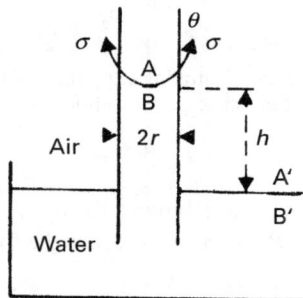

Figure F3 Capillary pressure in a cylindrical tube [1].

F

In a crack:

$$P_c = \frac{2\sigma\cos\theta}{w} \tag{F13}$$

where w is the thickness of the crack.

1.4.4 Average capillary properties of a reservoir

A curve representing the average capillary pressures of a geological formation can be obtained from the P_c curves available, by two methods:

- Determination of an average capillary pressure curve from the permeability distribution results on the field
- Plot of the capillary pressure function:

$$J\left(S_{wf}\right) = \frac{P_c}{\sigma\cos\theta}\sqrt{\frac{k}{\Phi}} \tag{F14}$$

where S_{wf} is the saturation by wetting fluid (Fig. F4).

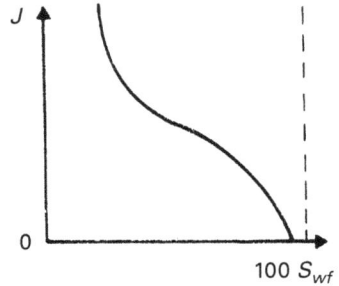

Figure F4 J curve [1].

1.5 Effective and relative permeabilities [4]

The permeability concept was defined in Paragraph 1.2 for a single-phase homogeneous fluid moving through a porous rock. The great majority of oil reservoirs in practice contain at least two fluids, namely, connate water and oil; if free gas is also present, there will be three fluids in the reservoir. Fig. F5 illustrates an oil–water mixture passing through a core.

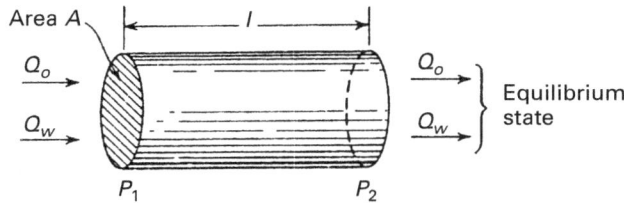

Figure F5 Diagram illustrating experimental determination of effective permeabilities [4].

After experimental conditions have stabilized, the upstream (injection) pressure P_1 and the downstream (outlet) pressure P_2 are measured, and the following two quantities are calculated:

$$k_o = \frac{\mu_o Q_o l}{A\left(P_1 - P_2\right)} \tag{F15}$$

$$k_w = \frac{\mu_w Q_w l}{A(P_1 - P_2)}$$ (F16)

where

k_o and k_w	effective permeabilities to oil and water (darcy)
μ_o and μ_w	viscosities of oil and water (cP or mPa·s)
Q_o and Q_w	oil rate and water rate (cm^3/s)
P_1 and P_2	injection pressure and outlet pressure (bar)
l	length of core (cm)
A	area of core (cm^2).

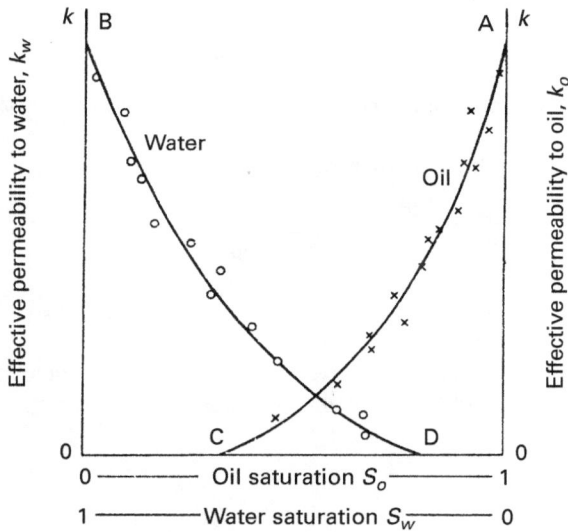

Figure F6 Typical effective permeability curves (oil–water system) [4].

Three important points should be noted about the effective permeability curves of an oil–water system:

1. k_o drops very rapidly as S_w increases from zero. Similarly k_w drops sharply as S_w decreases from unity.

2. k_o drops to zero while there is still considerable oil saturation in the core (point C of Fig. F6).

3. The values of both k_o and k_w are always less than k (except at points A and B).

1.6 Well logging [1]

Purposes of electric well logs are:

- Identification of the reservoirs: lithology, porosity, saturations
- The dip of the beds
- Survey of the well: diameter, inclination, casing cementing, formation/hole connection (perforations)
- Comparison among several wells, by "electric" correlations which highlight variations in depth, thickness, facies, etc.

1.6.1 Electric logs

Resistivity log

The characteristics obtained are a function of the porosity and saturation (water/hydrocarbons).

They are derived from the empirical equations:

$$F = \frac{R_o}{R_w} \approx \frac{a}{\Phi^m} \tag{F17}$$

with

$a = 1$ and $m \approx 2$, in general

F formation factor (constant for a given sample), sometimes denoted F_R

R_o resistivity of rocks 100% saturated with water of resistivity R_w

Φ porosity.

Archie's equation

$$S_w = \frac{1}{\Phi} \sqrt[n]{\frac{R_w}{R_t}} \tag{F18}$$

with

$n = 2$ for formations without fractures or vugs

R_t calculated resistivity of the rock whose water saturation is S_w.

This equation is satisfied for clean reservoirs (with very little shale).

Note that Archie's equation can confirm Eq. F17:

$$100\% \approx \frac{1}{\Phi} \sqrt{\frac{R_w}{R_o}} \tag{F19}$$

hence:

$$\frac{R_o}{R_w} = \frac{1}{\Phi^2} \tag{F20}$$

1.6.2 Radioactivity logs

Neutron log N

N depends on the quantity of hydrogen and, accordingly, on Φ:

$$N = a - b \log \Phi \qquad \text{(F21)}$$

Density log (gamma/gamma) D

The formations are irradiated by gamma rays that are received as a function of the density of the formation:

$$D = \Phi D_f + (1 - \Phi)D_m \qquad \text{(F22)}$$

where
D total density read on the log (also noted ρ)
D_f fluid density (filtrate)
D_m density of rock matrix.

D_m values are:
sand/sandstones   2.65
shales   2.65 to 2.70
limestones   2.71
dolomites   2.85

which leads to the calculation of the porosity Φ.

1.6.3 Sonic (or acoustic) logs

These logs consist of the transmission and reception of sound waves:

$$\frac{1}{V} = \frac{\Phi}{V_f} + \frac{1 - \Phi}{V_m} \qquad \text{(F23)}$$

where
V measured velocity
V_f transit velocity of the saturating fluid
V_m transit velocity of the matrix material.

Table F4 Transit velocity [1].

Saturating fluid	V_f		Matrix material	V_m	
	(m/s)	(ft/s)		(m/s)	(ft/s)
Air	335	1 100	Shale	1 600 to 4 800	5 200 to 15 700
Oil	1 300	4 250	Sandstone	5 500	18 000
Water	1 500 to 1 800	5 000 to 6 000	Limestone	< 7 000	< 23 000

F2 OIL PROPERTIES [6, 7]

This section includes:
- Oil specific gravity
- Bubble-point pressure P_b
- Oil viscosity μ_o
- Oil formation volume factor B_o.

2.1 Oil specific gravity [6]

The oil specific gravity γ_o can be calculated from the API gravity from the relationship:

$$\gamma_o = \frac{141.5}{131.5 + °\text{API}} \tag{F24}$$

or

$$°\text{API} = \frac{141.5}{\gamma_o} - 131.5 \tag{F25}$$

where
γ_o oil specific gravity
$°\text{API}$ API gravity.

2.2 Bubble-point pressure [6]

Bubble-point pressure P_b is the fluid pressure in a system at its bubble point. Bubble-point pressure is used synonymously with saturation pressure:

$$P_b = 18.2 \left[\left(\frac{R_s}{\gamma_g} \right)^{0.83} \times 10^{(0.00091 T_R - 0.0125 °\text{API})} - 1.4 \right] \tag{F26}$$

where
R_s produced gas/oil ratio, (scf/bbl)
γ_g gas specific gravity
T_R reservoir temperature (°F).

If the separator temperature and pressure are known, the gas gravity γ_g should be corrected for separator conditions, using the following equation:

$$\gamma_{gcr} = \gamma_g \left(1 + \frac{5.912}{10^5} \times °\text{API} \times T_{sp} \times \log \frac{P_{sp}}{114.7} \right) \tag{F27}$$

where
γ_{gcr} corrected gas gravity
T_{sp} separator temperature (°F)
P_{sp} separator pressure (psi).

2.3 Oil viscosity [6]

Two correlations are provided to help in estimating oil viscosity μ_o.

Estimation of dead oil (gas free) viscosity μ_{do}

$$\mu_{do} = \left[0.32 + \frac{1.8 \times 10^7}{°API^{4.53}}\right]\left(\frac{360}{T_R + 200}\right)^D \tag{F28}$$

with
$$D = 10^{[0.43 + (8.33/°API)]} \tag{F29}$$

where T_R is the reservoir temperature (°F).

Above the bubble-point pressure

$$\mu_o = \mu_{ob} + Y\frac{P_R - P_b}{1\ 000} \tag{F30}$$

and
$$Y = e^A \tag{F31}$$

with
$$A = -2.68 + 0.98 \ln\mu_{ob} + 0.091(\ln\mu_{ob})^2 \tag{F32}$$

where
μ_o	undersatured oil viscosity (cp)
μ_{ob}	oil viscosity at bubble-point pressure (cp)
P_R	reservoir pressure (psi)
P_b	bubble-point pressure (psi).

Below the bubble-point

The viscosity of oil in a hydrocarbon reservoir increases with decreasing pressure, as shown in Fig. F7, at saturated conditions (below the bubble-point), due to the release of the solution gas.

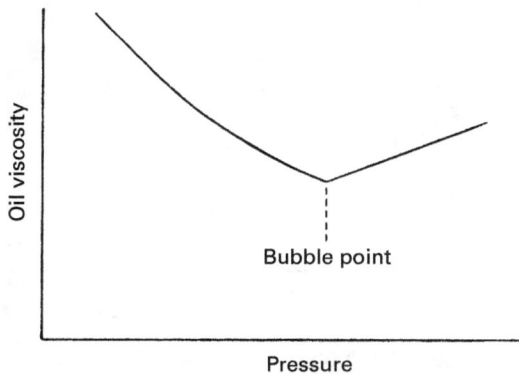

Figure F7 Oil viscosity as function of pressure [7].

2.4 Oil formation volume factor [6, 7]

Oil formation volume factor B_o is defined as the ratio of the liquid volume at stock-tank (standard) conditions.

This factor is used to convert reservoir barrels to stock-tank barrels. Fig. F8 or Eq. F33 developed by Standing is used to estimate B_o, at or below the bubble-point pressure:

$$B_{ob} = 0.972 + \frac{1.47}{10^4}\left[R_s\sqrt{\frac{\gamma_g}{\gamma_o}} + 1.25T_R\right]^{1.175} \tag{F33}$$

Above the bubble-point pressure, the oil formation volume factor is calculated from B_o at the bubble point:

$$B_o = B_{ob}\, e^{-c_o(P_b - P_R)} \tag{F34}$$

where

B_{ob} oil formation volume factor at bubble point (reservoir bbl/standard bbl)
B_o oil formation volume factor above bubble point (reservoir bbl/standard bbl)
c_o undersaturated oil compressibility (psi^{-1})
R_s solution gas (scf/bbl)
γ_g gas specific gravity
γ_o oil specific gravity
T_R reservoir temperature (°F)
P_R reservoir pressure (psi)
P_b bubble-point pressure (psi).

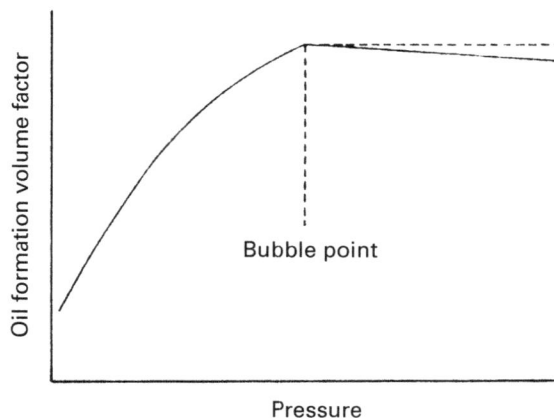

Figure F8 B_o as function of pressure [7].

F3 WATER PROPERTIES [1, 7]

This section includes:
- Water viscosity μ_w
- Water formation volume factor B_w
- Water compressibility c_w.

3.1 Water viscosity [1, 7]

Water viscosity μ_w is primarily a function of temperature; salinity also has a slight influence on μ_w. A correlation for estimating water viscosity μ_w at reservoir temperature is:

$$\mu_w = 4.33 - 0.07T_R + 4.73\times10^{-4}T_R^2 - 1.415\times10^{-6}T_R^3 + 1.56\times10^{-9}T_R^4 \qquad (F35)$$

where
  μ_w water viscosity (cp)
  T_R reservoir temperature (°F).

- *In metric units*
$$\mu_w = 0.3 \text{ to } 0.7 \text{ cP } (3.10^{-4} \text{ to } 7 \cdot 10^{-4} \text{ Pa·s}).$$

3.2 Water formation volume factor [7]

Water formation volume factor B_w is primarily a function of temperature, and to a lesser degree, pressure. A simple correlation is:

$$B_w = -1.485\times10^{-6}P_R + 0.952 + 10^A \qquad (F36)$$

where
  B_w water formation volume factor, (reservoir bbl/standard bbl)
  P_R reservoir pressure, (psi)
  $A =$ 0.001996 $(T_R - 100) - 1.2676$.

3.3 Water compressibility [1, 7]

Water compressibility c_w may be estimated from a correlation by Meehan:

$$c_w = \left(A + BT_R + CT_R^2\right)\times10^{-6} \qquad (F37)$$

where
  $A =$ $3.8546 - 1.34\times10^{-4}P_R$
  $B =$ $-1.052\times10^{-2} + 4.77\times10^{-7}P_R$
  $C =$ $3.9267\times10^{-5} - 8.8\times10^{-10}P_R$
  T_R reservoir temperature (°F)
  P_R reservoir pressure (psi).

- *In metric units*

Water compressibility is about:

$$c_w = 0.5 \times 10^{-4} \text{ bar}^{-1}$$

F4 GAS PROPERTIES [1, 5, 7]

4.1 Practical equation of state [1]

The equation employed is the ideal gas equation, with a factor Z indicating the difference in behavior between the real gas and an ideal gas:

$$PV = Z_n RT = Z \frac{m}{M} RT \qquad \text{(F39)}$$

where
- P absolute pressure (measured from vacum)
- T absolute temperature (measured from absolute zero):
  - – in metric units: $T(°K) = T(°C) + 273$
  - – in U.S. units: $T(°R) = T(°F) + 460$
- V volume occupied at P and T by n moles of gas
  (volume occupied at 15°C and 1 bar by 1 mol.g of gas: 23 957 cm^3)
- m mass of gas considered
- M molecular weight of gas
- R universal ideal gas constant:
  - – in metric units: $R = 8.315 \times 10^{-5}$ (m^3, bar) / mol·g
  - – in U.S. units: $R = 10.7$ mol·lb (P in psi, T in °R, V in cu ft)
- Z compressibility factor (see Paragr. 4.3).

4.2 Pseudocritical and pseudoreduced temperature and pressure [7]

To determine gas properties pseudocritical and pseudoreduced temperature and pressure must be calculated. These values may be computed either from a gas analysis (if available) or from the gas specific gravity.

When a gas analysis is available, the pseudocritical properties are calculated from the relationships:

$$T_{pc} = \Sigma Y_i T_{ci} \qquad \text{(F40)}$$

and

$$P_{pc} = \Sigma Y_i P_{ci} \qquad \text{(F41)}$$

where
- T_{pc} pseudocritical temperature (°R)
- Y_i mole fraction of the ith component

T_{ci} critical temperature of the ith component (°R)
P_{pc} pseudocritical pressure (psi)
P_{ci} critical pressure of the ith component (psi).

Table F5 shows molecular weights, gas specific gravity, critical temperatures and critical pressures.

Table F5 Critical properties [7].

Component	Molecular weight MW_i	Gas specific gravity γ_g	Critical temperature T_{ci}		Critical pressure P_{ci}	
			(°K)	(°R)	(bar)	(psi)
Methane	16.04	0.554	191.2	344	46.41	673
Ethane	30.07	1.038	305.7	550	48.94	712
Propane	44.09	1.522	370.1	666	42.57	617
n-Butane	58.12	2.006	408.3	766	37.97	551
n-Pentane	72.15	2.491	469.7	847	33.75	485
n-Hexane	86.17	2.975	507.8	914	30.34	435
n-Heptane	100.20	3.459	540.3	972	27.36	397
n-Octane	114.22	3.943	568.7	1025	24.97	362
Water	18.02	0.622	647.5	1365	221.18	3 206
Carbon dioxide	44.01	1.519	304.3	548	73.84	1 073
Nitrogen	28.02	0.967	126.3	227	33.92	492
Hydrogen sulfide	34.08	1.176	373.7	673	90.05	1 306
Air	28.96	1.000	132.6	239	37.72	547

Additionally, the gas specific gravity γ_g (Table F5, third column) may be calculated from:

$$\gamma_g = \frac{\Sigma(Y_i MW_i)}{28.9} \tag{F42}$$

where MW_i is the molecular weight of the ith component.

4.3 Gas formation volume factor [1, 7]

All hydrocarbon gases under typical conditions will deviate from the ideal gas law, thus requiring a correction factor. This factor is known as the gas compressibility factor Z. It is determined either from laboratory measurements or from the chart of Standing and Katz, as shown in Fig. F9.

The gas formation volume factor is denoted by the symbol B_g, and may be calculated from:

$$B_g = 5.035Z\frac{T_R + 460}{P_R} \tag{F43}$$

where

B_g	gas formation volume factor (reservoir bbl/mil cu ft)
Z	gas compressibility factor
T_R	reservoir temperature (°F)
P_R	reservoir pressure (psi)
P_{pr}	pseudoreduced pressure (psi) = P_R/P_{pc}
P_{pc}	pseudocritical pressure (psi).

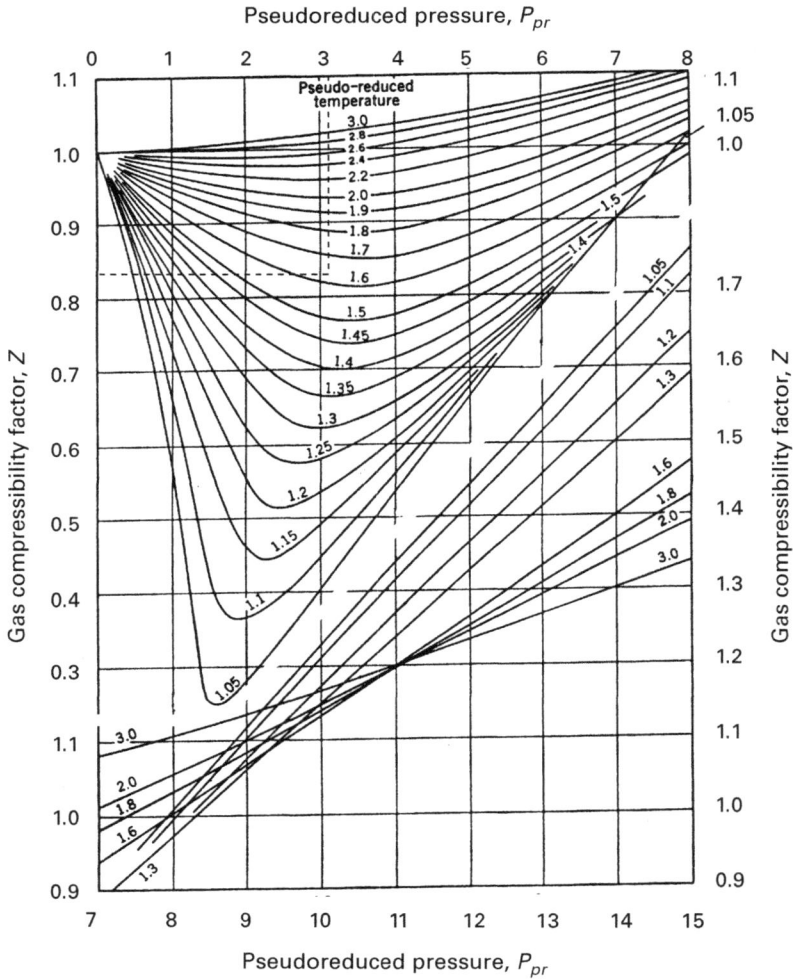

Figure F9 Natural gas deviation factor (after Standing and Katz) [1, 4, 7].

▼ **Example**

Find Z for:
$P_R = 2000$ psi
$P_{pc} = 650$ psi
$T_R = 200°F$ (660°R)
$T_{pc} = 410°R$

Solution

1. $P_{pr} = \dfrac{P_R}{P_{pc}} = \dfrac{2\,000}{650} = 3.07$

2. $T_{pr} = \dfrac{T_R}{T_{pc}} = \dfrac{660}{410} = 1.61$

3. Enter abscissa (top) at 3.07 (P_{pr}). Go down to T_{pr} of 1.61, between 1.6 and 1.7 lines.
4. $Z = \mathbf{0.828}$ ▲

4.4 Gas viscosity [7]

The viscosity of natural gas is primarily a function of pressure.
As pressure decreases, gas viscosity decreases, as illustrated in Fig. F10.
The usual range of gas viscosity is from 0.01 to 0.04 cp.

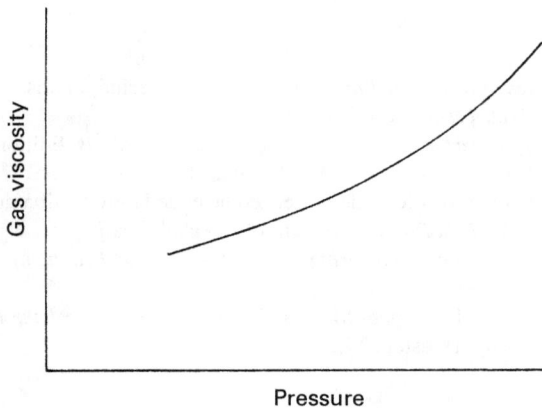

Figure F10 Gas viscosity as function of pressure [7].

When unavailable from laboratory data, gas viscosities may be estimated from the following equations developed from work by Lee et al. [7]:

$$MW_g = 28.9\gamma_g \qquad (F44)$$

$$Y1 = \frac{\left(9.4 + 0.02 MW_g\right)\left(T + 460\right)^{1.5}}{209 + 19 MW_g + T + 460} \tag{F45}$$

$$Y2 = 3.5 + 0.01 MW_g + \frac{986}{T + 460} \tag{F46}$$

$$Y3 = 2.4 - 0.2 Y2 \tag{F47}$$

$$Y4 = 0.007532 \frac{MW_g}{B_g} \tag{F48}$$

$$Y5 = Y2(Y4)^{Y3} \tag{F49}$$

$$\mu_g = \frac{Y1 \times e^{Y5}}{10^4} \tag{F50}$$

where
M_{wg} gas molecular weight
γ_g gas specific gravity
T temperature (°F)
B_g gas formation volume factor (reservoir bbl/mil cu ft)
μ_g gas viscosity (cp).

REFERENCES

1 Cossé R (1993) *Basics of Reservoir Engineering*. Editions Technip, Paris
2 Bourdarot G (1998) *Well Testing*. Editions Technip, Paris
3 Monicard RP (1980) *Properties of Reservoirs Rocks: Core Analysis*. Editions Technip, Paris
4 Nind TEW (1964) *Principles of Oil Well Production*. McGraw-Hill Book Co, New-York
5 Publication de la Chambre Syndicale de la Recherche et de la Production du Pétrole et du Gaz Naturel (1970) *Formulaire du Producteur*. Editions Technip, Paris
6 Mian MA (1992) *Petroleum Engineering Handbook for the Practicing Engineer*. Vol. 1. PennWell Books, Tulsa, OK
7 Koederitz LF, Harvey AH, Honarpour M (1989) *Introduction to Petroleum Reservoir Analysis*. Gulf Publishing Company, Houston, TX.

G

Well Productivity

G

Well Productivity

G1 OIL FLOW AROUND WELLS

1.1 Types of flow [2]

- **Radial circular, steady-state flow**
  r_e is the distance to the constant pressure boundary
  $A = \pi r_e^2$ is the drainage area
  Top and bottom of formation are tight. h is the height.
- **Lateral rectangular, steady-state flow**
  The drainage area is a rectangle; $A = a \times b$ where a is the width, b is the length.
  Flow occurs through the $(b \times h)$ walls.
  The other boundaries are tight.
- **Bottom-water drive (or gas-cap drive), steady-state flow**
  Various possible drainage area for the vertical well; rectangular drainage area for the horizontal well.
  Constant pressure boundary at water-oil contact (WOC) or gas-oil contact (GOC). Displacement of contact is negligible.
- **No flux boundaries, pseudosteady-state flow**
  Various possible drainage area (reservoir) for the vertical well; rectangular reservoir for the horizontal well.
  Confined reservoir (geological limits, faults...) or zero flux limits (production patterns boundaries).

1.2 Skin effects [1]

The global skin S is the result of:
- Damage skin factor S_d
- Perforation skin factor S_p
- Partial penetration factor S_{pp}
- Inclination skin factor S_i
- Injection skin factor S_{inj}.

1.2.1 Formation damage skin factor [9]

The introduction of the damage skin factor S_d allows to integrate the effect of wellbore damage. It is simply defined by the Hawkins' formula:

$$S_d = \left(\frac{k}{k_s} - 1\right)\ln\frac{r_s}{r_w} \tag{G1}$$

where

- k permeability of reservoir
- k_s permeability of damage zone
- r_s penetration of damage zone
- r_w well radius.

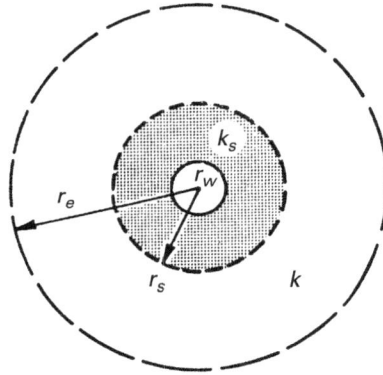

Figure G1 Scheme of a well with a damage zone [5].

▼ **Example. Permeability impairment versus damage penetration** [9]

Assume that a well has a radius r_w equal to 10 cm (0.328 ft) and a penetration of damage 3 ft beyond the well (i.e., $r_s = 3.328$ ft). What would be the skin effect if the permeability impairment results in k/k_s equal to 5 and 10, respectively. What would be the required penetration of damage to provide the same skin effect as the latter case but with $k/k_s = 5$?

Solution

From Eq. G1, $k/k_s = 5$, and the given r_s and r_w

$$S_d = (5-1)\ln\frac{3.328}{0.328} = 9.3 \tag{G2}$$

For $k/k_s = 10$ and $r_s = 3.328$ then, similarly, $S_d = 20.9$.
However, if $S_d = 20.9$ and $k/k_s = 5$, then:

$$r_s = r_w e^{20.9/4} = 61 \text{ ft} \tag{G3}$$

This example suggests that permeability has a much larger effect on the value of the skin effect than the penetration of damage. Except for a phase change-dependent skin effect, a penetration of damage such as the one calculated in Eq. G3 is impossible. Thus, skin effects derived from wells tests (frequently ranging between 5 and 20) are likely to be caused by substantial permeability impairment very near the well. **This is a particular important point in the design of matrix stimulation treatments.**   ▲

1.2.2 Perforation skin factor S_p [1]

- **Parameters influencing S_p:**
    - Quantity of perforations per foot
    - Penetration depth of perforation in the formation
    - Repartition of perforations around the well
    - Diameter of perforations.
- **Determination of the type of configuration :**
    - Simple or staggered
    - Figure G2 determines the configuration type depending on:
        - height of repeating pattern h, in.
        - phasing angle $\theta°$.

From charts of Fig. G4 and G5, representing the type of configuration, determine S_p, according to the following example in plotting:

1. Height of repeating pattern h
2. Well diameter d_w
3. Permeability anisotropy ratio k_v/k_h
4. Phasing angle $\theta°$
5. Total core penetration a_p
6. Perforation skin factor S_p.

Eventually, use corrected chart on Fig. G5.

1.2.3 Partial penetration skin factor S_{pp}

This skin is also called choke skin S_c.

- Parameters influencing S_{pp}:
    - Ratio between the heights of perforated zone and reservoir bed
    - Positioning and number of perforated zones.
- Evolution of the flowing type (Fig. G3).

G

Pattern number	Simple pattern		h (in.)	θ (degrees)	Equivalent staggered pattern
	Top view	Front view			
1-12			12	0	
2-12			12	180	
3-12			12	120	
4-12			12	90	
1-6			6	0	
2-6			6	180	
3-6			6	120	
4-6			3	90	
1-3			3	0	
2-3			3	180	
3-3			3	120	
4-3			3	90	

Figure G2 Height h and phasing angle θ° for various perforation patterns [1].

Well Productivity

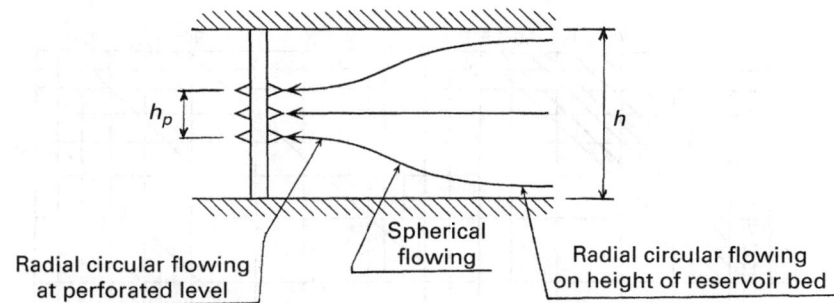

Figure G3 Flowing types [1].

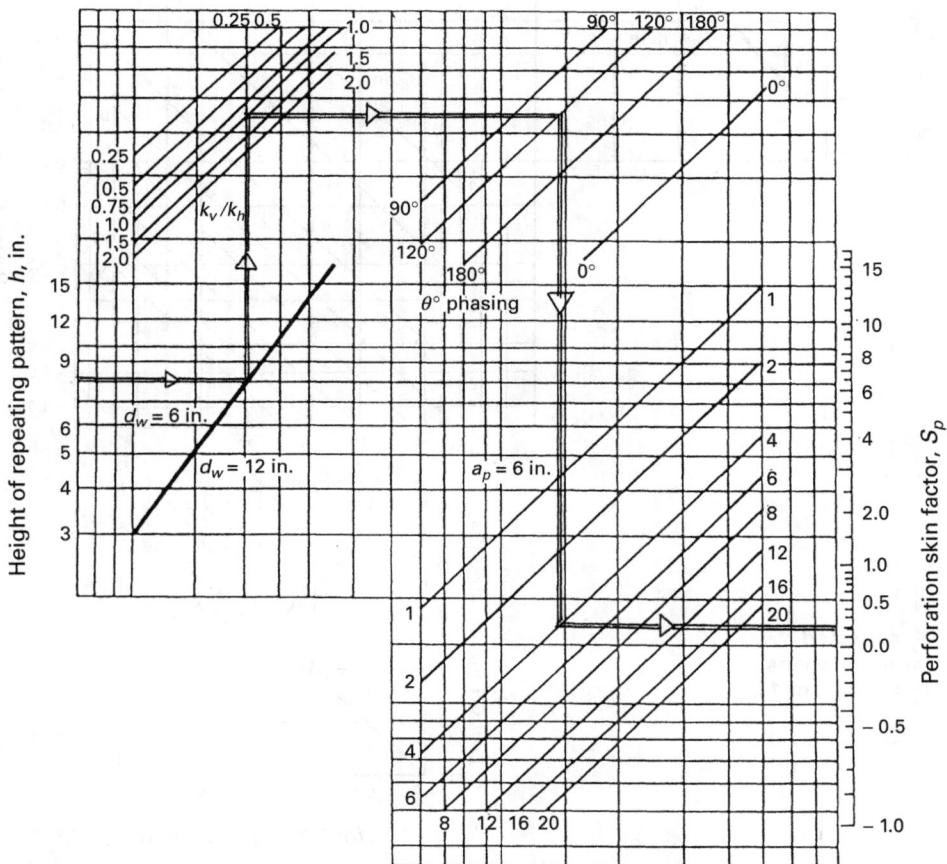

Figure G4 S_p chart for a simple configuration (perforations of 1/2 in.) [1].

G

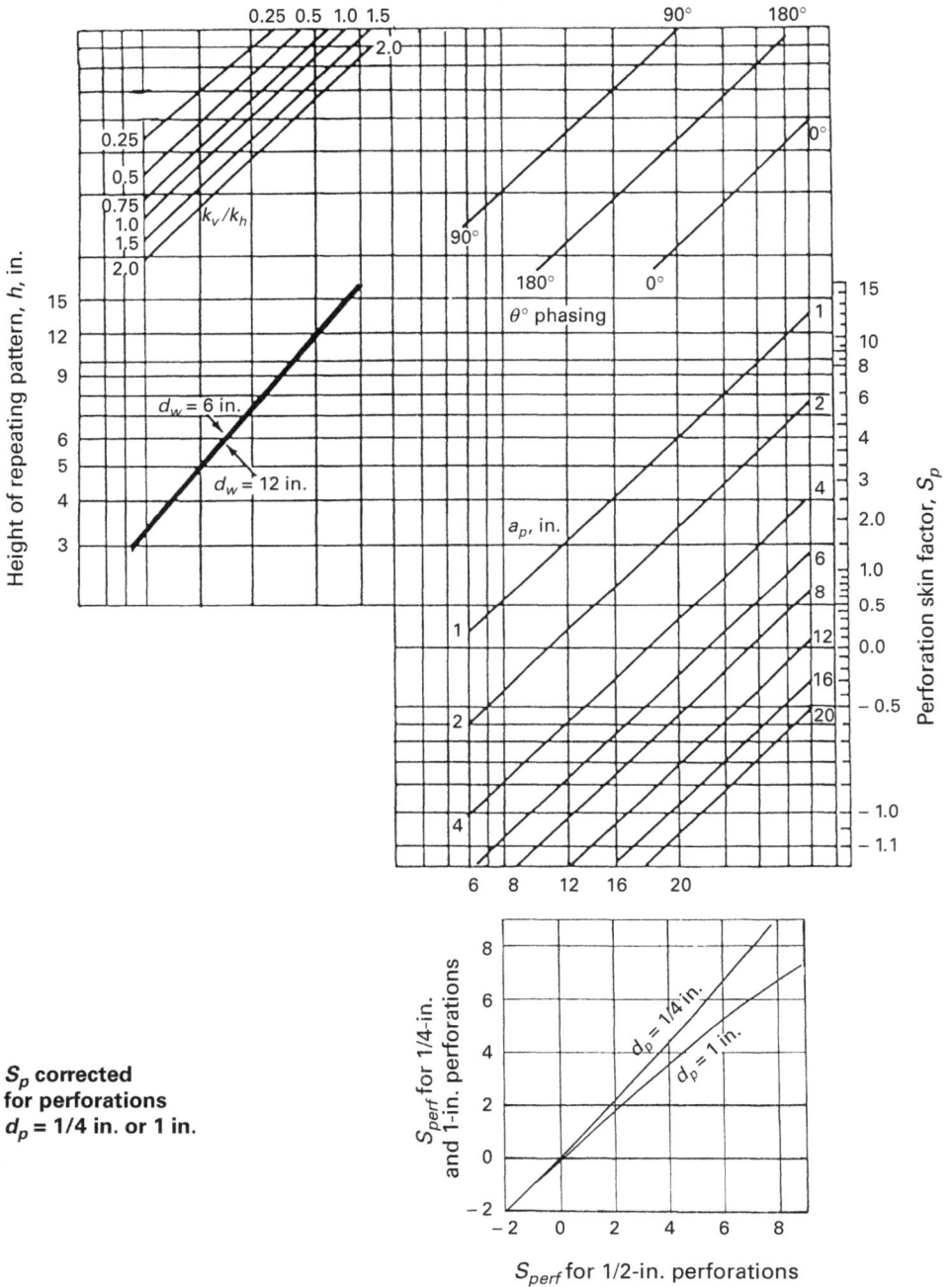

Figure G5 S_p chart for staggered configuration (perforations of 1/2 in.) [1].

- **Determination of** S_{pp}
  From chart of Fig. G6, S_{pp} is depending on:

$$b = \frac{h_p}{h}$$

$$\frac{h'}{r_w} \sqrt{\frac{k_v}{k_h}}$$

where

h height of reservoir bed
h_p perforated height
n number of "interface" between perforated and non-perforated sections
h' h/n
r_w well radius.

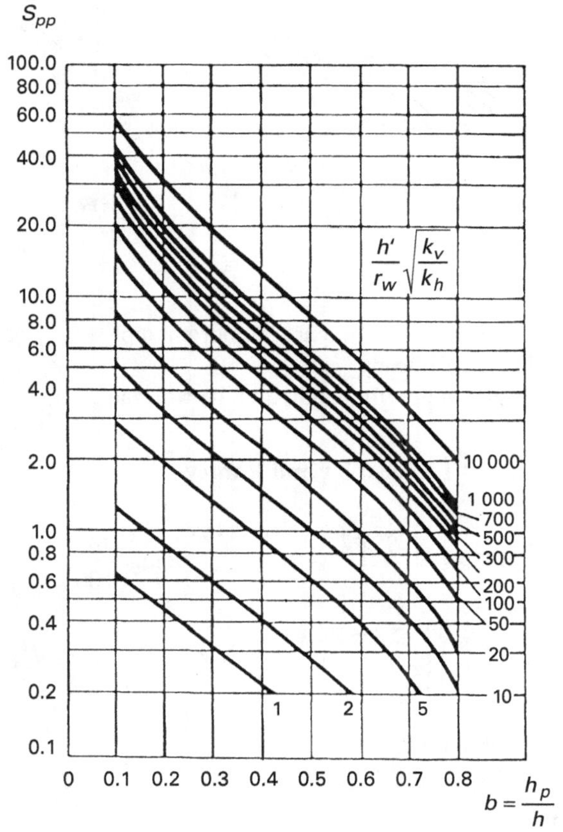

Figure G6 S_{pp} chart [1].

1.2.4 Inclination skin factor S_i

- **Parameters influencing S_j**
  They are represented on Fig. G7.

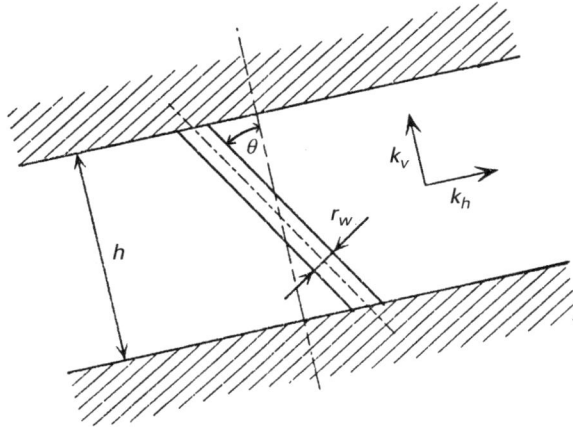

Figure G7 [1].

where

θ well inclination angle compared to normal dip

k_v/k_h permeability anisotropy

h/r_w ratio between the useful height and the well radius.

- **Evaluation of S_i after Cinco, Miller and Ramey**
  S_i is a function of:

$$S_i = -\left(\frac{\theta'}{41}\right)^{2.06} - \left(\frac{\theta'}{56}\right)^{1.865} \times \log\left(\frac{h_D}{100}\right)$$

with

$$\theta' = \text{arctg}\left(\text{tg } \theta \sqrt{\frac{k_v}{k_h}}\right) \qquad \text{and} \qquad h_D = \frac{h}{r_w}\sqrt{\frac{k_h}{k_v}} \qquad \text{(G4)}$$

This relation is available for: $0 < \theta' < 75°$.

Figure G8 représents S_i versus h_D and θ'.

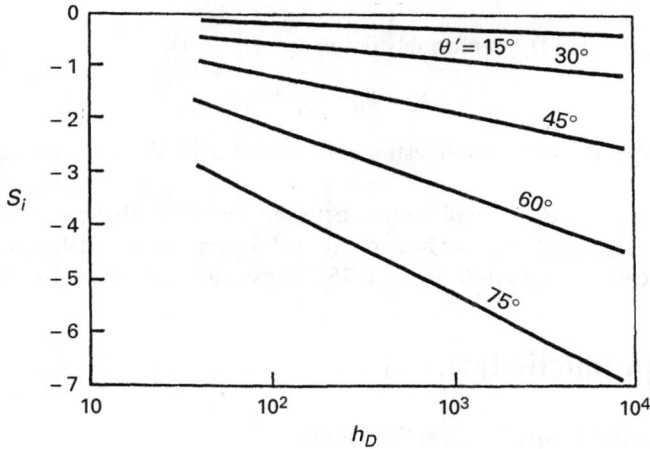

Figure G8 S_i versus h_D and θ' [1].

1.2.5 Injection skin factor S_{inj}

- **Generation of two zones by injection**

  1. Zone flooded by the injected fluid V_i; its radius r_i is:

$$r_i = \sqrt{\frac{V_i}{\pi h \Phi (S_w - S_{wi})}} \tag{G5}$$

where

    h reservoir height
    Φ porosity of the reservoir
    S_w saturation after injection
    S_{wi} initial water saturation.

  2. Zone containing the fluid of origin.

- **Determination of injection skin factor S_{inj}**

$$S_{inj} = \left(\frac{k_o}{\mu_o} \times \frac{\mu_w}{k_w} \right) \ln \frac{r_i}{r_w} \tag{G6}$$

where

    k_o permeability of reservoir
    k_w permeability near the well
    μ_o oil viscosity of reservoir
    μ_w injected fluid viscosity
    r_w well radius.

G

1.2.6 Global skin factor S

It is the sum of the skin factors formely defined:

$$S = S_d + S_p + S_{pp} + S_i + S_{inj} \tag{G7}$$

S is obtained directly from well testing. The other partial skins are deduced from charts or formulas.

Damage skin factor S_d is the difference between S and the other partial skin factors.

Mathematically, the skin effect has no physical dimension. It can therefore be added to $\ln(r_e/r_w)$ for steady state, to $[\ln(r_e/r_w) - 0.75]$ for pseudosteady state (Paragr. 1.3).

1.3 Well productivity [1, 6–8]

1.3.1 Steady-state flow in a vertical well

As fluid flows in a vertical well, the streamlines converge and the area for flow decreases.

Productivity Index PI

The Productivity Index PI is defined as follows:

- *In all unit systems (general case)*

$$PI = \frac{q}{P_e - P_w} \tag{G8}$$

- *In SI units*

$$PI = 2\pi \times \frac{hk}{B\mu\left(\ln\dfrac{r_e}{r_w} + S\right)} \tag{G9}$$

- *In field metric units*

$$PI = 0.05365 \times \frac{hk}{B\mu\left(\ln\dfrac{r_e}{r_w} + S\right)} \tag{G10}$$

- *In U.S. units*

$$PI = 0.00708 \times \frac{hk}{B\mu\left(\ln\dfrac{r_e}{r_w} + S\right)} \tag{G11}$$

1.3.2 Pseudosteady-state flow to a vertical well

In this case, the reservoir produces by decompression of its entire mass. The pressure drop is the same at every point, and the pressure difference between two given points remains constant (example: sandstone lenses in shale).

Using the same units as in Paragraph 1.3.1:

- *In SI units*

$$\text{PI} = 2\pi \times \frac{hk}{B\mu\left(\ln\dfrac{r_e}{r_w} + S - 0.75\right)} \tag{G12}$$

- *In field metric units*

$$\text{PI} = 0.05365 \times \frac{hk}{B\mu\left(\ln\dfrac{r_e}{r_w} + S - 0.75\right)} \tag{G13}$$

- *In U.S. units*

$$\text{PI} = 0.00708 \times \frac{hk}{B\mu\left(\ln\dfrac{r_e}{r_w} + S - 0.75\right)} \tag{G14}$$

1.3.3 Flow efficiency

Flow Efficiency, noted FE, is defined as the ratio of the actual productivity index to the ideal productivity index:

$$\text{FE} = \frac{PI_{\text{actual}}}{PI_{\text{ideal}}} \tag{G15}$$

- Steady-state flow:

$$\text{FE} = \frac{\ln r_e/r_w}{\ln r_e/r_w + S} \tag{G16}$$

- Pseudosteady-state flow:

$$\text{FE} = \frac{\ln r_e/r_w - 0.75}{\ln r_e/r_w - 0.75 + S} \tag{G17}$$

G2 WELL TEST ANALYSIS METHOD [1, 4, 5, 8]

2.1 Nomenclature

Symbols and units are defined in Table G1.

Table G1 Nomenclature.

Symbols	Parameters	SI units	Field metric units	Field U.S. units
B	Formation volume factor, res bbl/std bbl	Dimensionless	Dimensionless	Dimensionless
c_t	System total compressibility	Pa^{-1}	bar^{-1}	psi^{-1}
h	Formation thickness	m	m	ft
k	Permeability	μm^2	$10^{-3}\,\mu m^2$ ($\approx$ millidarcy)	md (millidarcy)
K	Diffusivity	$\mu m^2 \cdot Pa/Pa \cdot s$	$10^{-3}\,\mu m^2$ .bar/ mPa.s	md·psi/cp
m	± Slope of linear portion of semilog plot of pressure transient data	Pa/cycle	bar/cycle	psi/cycle
PI	Productivity Index	$m^3/s \cdot Pa$	$m^3/(d \cdot bar)$	bbl/(D-psi)
P_{1hr}	Pressure on straight-line portion of semilog plot 1 hour after beginning a transient test	Pa	bar	psi
P_i	Initial pressure	Pa	bar	psi
P_{wf}	Flowing bottomhole pressure	Pa	bar	psi
P_e	Pressure at the boundaries	Pa	bar	psi
P_w	Wellbore pressure	Pa	bar	psi
q	Volumetric flow rate	m^3/d	m^3/d	bbl/D
S	Skin factor	Dimensionless	Dimensionless	Dimensionless
r_e	Effective radius	m	m	ft
r_i	Initial radius	m	m	ft
r_w	Wellbore radius	m	m	ft
t	Time	s	h	h
T	Temperature	°K	°K	°R
μ	Viscosity	Pa·s	mPa.s	cp
Φ	Porosity	Fraction	Fraction	Fraction

2.2   Test type: drawdown or injectivity

2.2.1 Pressure drawdown or injectivity

The pressure drawdown is expressed:

- *In SI units*

$$P_i - P_{wf} = \frac{qB\mu}{4\pi kh}\left(\ln \frac{Kt}{r_w^2} + 0.81 + 2S \right) \quad \text{and} \quad r_i = 2\sqrt{Kt} \qquad (G18)$$

where the diffusivity is:

$$K = \frac{k}{\Phi\mu c_t} \quad \text{and} \quad kh = \frac{\mu qB}{4\pi m}$$

- *In field metric units*

$$P_i - P_{wf} = 21.5 \times \frac{qB\mu}{kh}\left(\log t + \log \frac{k}{\Phi \mu c_t r_w^2} - 3.10 + 0.87S\right) \qquad \text{(G19)}$$

and:

$$r_i = 0.038 \sqrt{\frac{kt}{\Phi \mu c_t}}$$

- *In field U.S. units*

$$P_i - P_{wf} = 162.6 \times \frac{qB\mu}{kh}\left(\log t + \log \frac{k}{\Phi \mu c_t r_w^2} - 3.23 + 0.87S\right) \qquad \text{(G20)}$$

and:

$$r_i = 0.032 \sqrt{\frac{kt}{\Phi \mu c_t}}$$

2.2.2 Permeability equation

Equations of Paragraph 2.2.1 show that bottomhole pressure varies logarithmically versus time. If the pressure measured at the bottomhole is plotted on a graph (Fig. G9) versus the logarithm of time, a straight line with a slope m can be observed once the wellbore storage effect has ended.

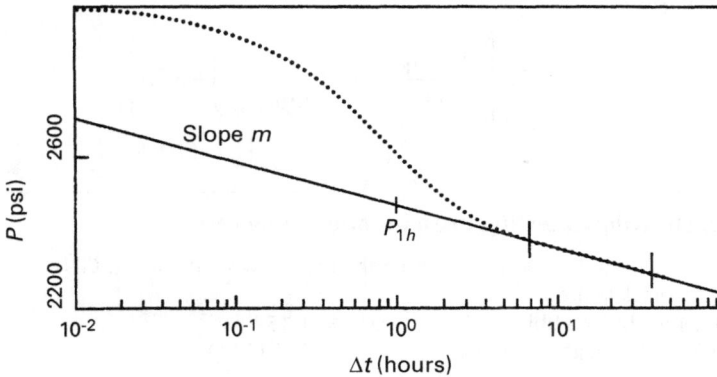

Figure G9 [8].

The slope m is used to determine the permeability of the reservoir:

- *In SI units*

$$k = \frac{2.303}{4\pi} \times \frac{qB\mu}{mh} \qquad \text{(G21)}$$

139

- *In field metric units*

$$k = 21.5 \times \frac{qB\mu}{mh} \tag{G22}$$

- *In field U.S. units*

$$k = 162.6 \times \frac{qB\mu}{mh} \tag{G23}$$

G

2.2.3 Skin-factor equation

The skin-factor value is usually computed using the pressure measurement at 1 hour on the semi-log straight line; at this point, $\log t = 0$.

- *In SI units*

$$S = 0.5 \times \left(\frac{P_i - P_{1hr}}{m} - \ln \frac{Kt}{r_w^2} - 0.81 \right) \tag{G24}$$

- *In field metric units*

$$S = 1.15 \times \left[\frac{P_i - P_{1hr}}{m} - \log \left(\frac{k}{\Phi \mu c_t r_w^2} \right) + 3.10 \right] \tag{G25}$$

- *In field U.S. units*

$$S = 1.15 \times \left[\frac{P_i - P_{1hr}}{m} - \log \left(\frac{k}{\Phi \mu c_t r_w^2} \right) + 3.23 \right] \tag{G26}$$

▼ Example. Drawdown testing in an infinite-acting reservoir

Estimate oil permeability and skin factor from the drawdown data of Fig. G10.
Known reservoir data are:

B	= 1.14 res bbl/std bbl	P_i	= 1 154 psi
c_t	= 8.74×10^{-6} psi^{-1}	q	= 348 bbl/D
h	= 130 ft	r_w	= 0.25 ft
m	= -22 psi/cycle (Fig. G10)	μ	= 3.93 cp
P_{1hr}	= 954 psi (Fig. G10)	Φ	= 20 percent

Using Eq. G23 and Eq. G26:

$$k = \frac{(162.6)(348)(1.14)(3.93)}{(22)(130)} = 89 \text{ md}$$

$$S = 1.15 \left\{ \left(\frac{1154 - 954}{22} \right) - \log \left[\frac{89}{(0.2)(3.93)(8.74 \times 10^{-6})(0.25)^2} \right] + 3.23 \right\} = 4.6$$

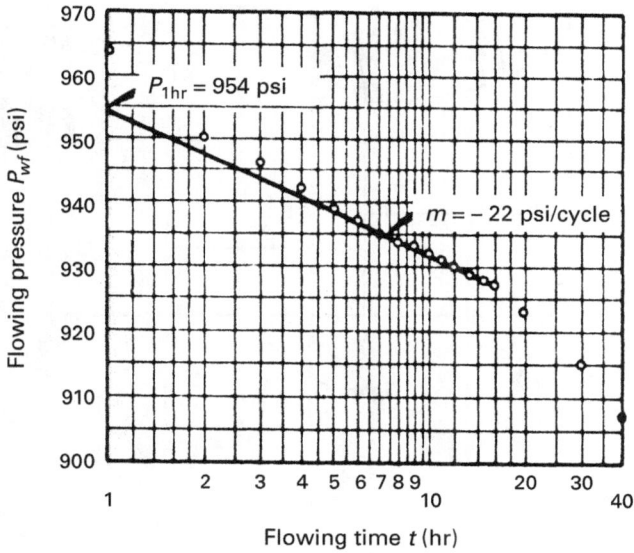

Figure G10 Semilog data plot for the drawdown test
described in the example [4].

▲

2.3 Test type: buildup or falloff

2.3.1 Pressure buildup or falloff

Pressure buildup and falloff are shown in Fig. G11, where:

$P_{wf}(t)$ flowing pressure: time is counted from when the well is open

$P_{ws}(\Delta t)$ pressure during the buildup phase: time is counted from when the well is shut
        out, t_p.

The pressure buildup is expressed:

- *In SI units*

$$P_i - P_{ws}(\Delta t) = \frac{qB\mu}{4\pi kh} \times \ln\frac{t_p + \Delta t}{\Delta t} \qquad \text{and} \qquad r_i = 2\sqrt{K\Delta t \frac{t_p}{t_p + \Delta t}} \qquad \text{(G27)}$$

where the diffusity is:

$$K = \frac{k}{\Phi\mu c_t} \qquad \text{and} \qquad kh = \frac{\mu qB}{4\pi m}$$

G

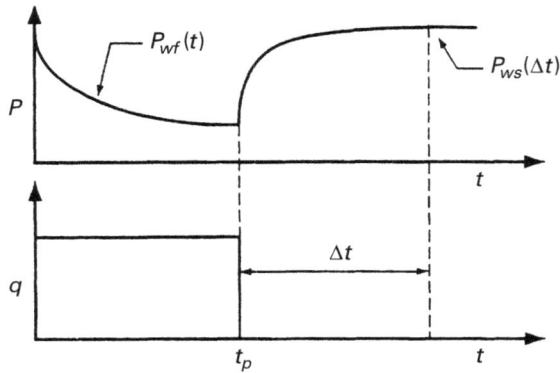

Figure G11 [8].

- *In field metric units*

$$P_i - P_{ws}(\Delta t) = 21.5 \times \frac{qB\mu}{kh} \times \ln\frac{t_p + \Delta t}{\Delta t} \quad \text{and} \quad r_i = 0.038\sqrt{\frac{k\Delta t}{\Phi\mu c_t} \times \frac{t_p}{t_p + \Delta t}} \quad (G28)$$

- *In field U.S. units*

$$P_i - P_{ws}(\Delta t) = 162.6 \times \frac{qB\mu}{kh} \times \ln\frac{t_p + \Delta t}{\Delta t} \quad \text{and} \quad r_i = 0.032\sqrt{\frac{k\Delta t}{\Phi\mu c_t} \times \frac{t_p}{t_p + \Delta t}} \quad (G29)$$

2.3.2 Permeability equation

Equations of Paragraph 2.3.1 show that bottomhole pressure varies logarithmically versus time. If the pressure measured at the bottomhole is plotted on a graph (Fig. G12) versus the logarithm of $(t_p + \Delta t)/\Delta t$, a straight line with a slope of m can be observed once the wellbore storage effect has ended.

The slope m is used to determine the permeability of the reservoir:

- *In SI units*

$$k = \frac{2.303}{4\pi} \times \frac{qB\mu}{mh} \quad (G30)$$

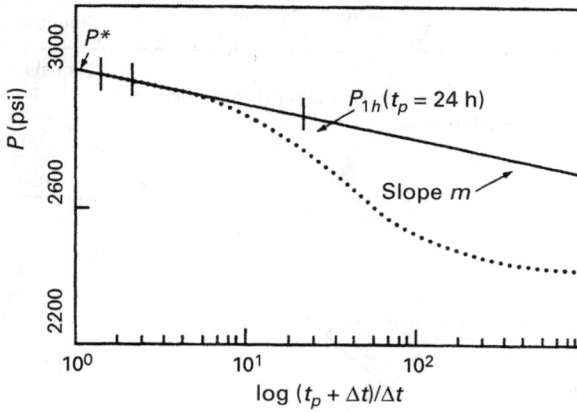

Figure G12 [8].

- *In field metric units*

$$k = 21.5 \times \frac{qB\mu}{mh} \tag{G31}$$

- *In field U.S. units*

$$k = 162.6 \times \frac{qB\mu}{mh} \tag{G32}$$

2.4 Pressure-buildup curve shapes

Conceptually, graphs of pressure buildup, drawdown, injectivity, or falloff behavior in individual wells can be divided into three areas:

1. Front-end effects (wellbore storage, fractures, damage)
2. Semilog straight line for which most analysis techniques apply
3. Boundary effects.

Those three parts are illustrated for a pressure buildup case in Fig. G13.

143

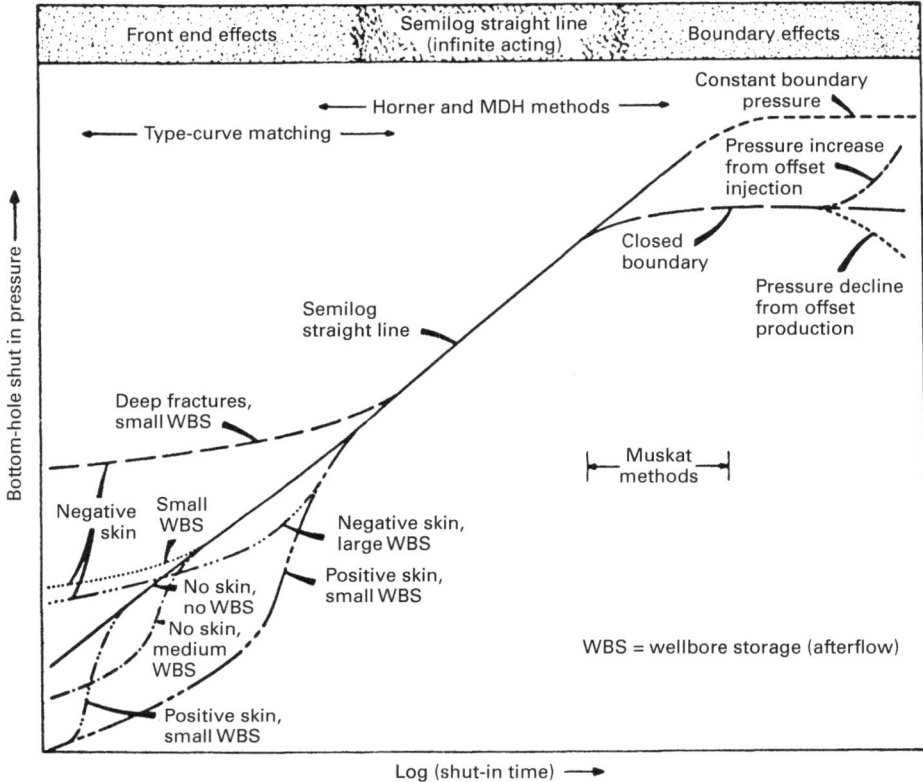

Figure G13 Typical bottomhole pressure-buildup curve shapes.
For production at pseudosteady state before shut-in [4].

G3 INFLOW PERFORMANCE RELATIONSHIP (IPR). PRODUCTIVITY INDEX OF AN OIL WELL PRODUCING UNDER THE BUBBLE POINT; VOGEL'S METHOD [8, 11, 12, 13]

The Inflow Performance Relationship (IPR) for a well is the relationship between flow rate into the wellbore and wellbore flowing pressure P_{wf}. The IPR is illustrated graphically by plotting P_{wf} versus flow rate.

In a well producing a reservoir whose pressure is lower than the bubble point, the relationship between the well flow rate and the difference between the average pressure and the bottomhole pressure is not linear.

3.1 Principal hypotheses [11]

The IPR is established on the following hypotheses:
- Drainage mechanism by expansion of dissolved gas
- The reservoir pressure is the bubble pressure of oil
- Reservoir radial circular, uniform, isotrope, of finite length
- No skin
- Oil and gas at the same pressure and constant characteristics (viscosity)
- No water flowing.

3.2 IPR curve [11, 12]

Vogel reported the results of a study in which he used a mathematical reservoir model to calculate the IPR for oil wells producing from saturated reservoirs. The final equation for Vogel's method was based on calculations made for 21 reservoir conditions.

- **Empirical equation**
  After plotting dimensionless IPR curves for all the cases considered, Vogel arrived at the following relationship between dimensionless flow rate and dimensionless pressure:

$$\frac{q_o}{(q_o)_{max}} = 1 - 0.2\frac{P_{wf}}{P_{r=b}} - 0.8\left(\frac{P_{wf}}{P_{r=b}}\right)^2 \tag{G33}$$

  where

q_o	oil flow rate
P_{wf}	well flowing pressure
$P_{r=b}$	reservoir pressure = oil bubble pressure.

- **Limits and remarks**
  - A difference can be significant between the previous and actual flow rates if:
    - oil is very viscous
    - reservoir pressure is superior to bubble point
    - there is no skin.
  - Otherwise, the difference is less than 20%, and 10% in general.
  - BSW < 10%.
  - As the maximal flow rate decreases with the depletion of the reservoir, it is necessary to periodically readjust the parameters.

- **Graphical representation of $P_{wf}/P_{r=b}$ versus $q_o/(q_o)_{max}$**
  A plot of the dimensionless IPR represented by the Vogel's equation (G33) is shown in Fig. G14, which can be used in lieu of Eq. G33.

- **Utilization**
  To use this method, it is necessary to know experimentally the flow rate corresponding to the bottomhole pressure (the fitting will be better if this pressure is weak versus the bubble point).

G

Using the graph (or equation), first calculate $(q_o)_{max}$, and then calculate the flow rate for a given bottomhole pressure.

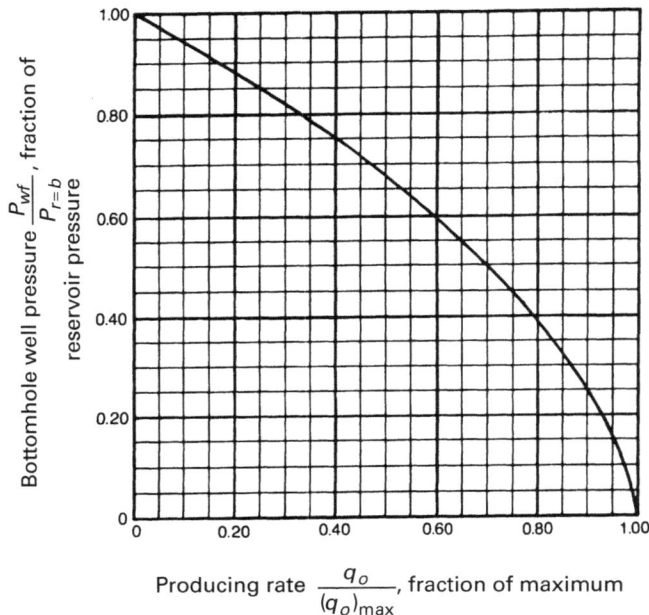

Figure G14 Inflow performance relationship for solution-gas drive reservoirs. Graphical representation of $P_{wf}/P_{r=b}$ versus $q_o/(q_o)_{max}$ [11, 13, 14].

▼ Example using Vogel's method [8]

Assuming a well produces under the following conditions:

q_o = 65 bbl/D
P_{wf} = 1 500 psi
P_r = 2 000 psi

Determine the absolute open-flow potential of the well and the flow rate that would be attained if a gas lift installation (mechanical pumping) could bring the the bottomhole pressure down to 500 psi.

Based on Eq. G33, the data can be used to calculate:

$$q_{max} = 162 \text{ bbl/D}$$

Using Eq. G33 with a bottomhole pressure of 500 psi gives:

$$\frac{q_o}{(q_o)_{max}} = 0.9 \quad \text{then} \quad q_o = 146 \text{ bbl/D}$$

▲

These same results could have been obtained by using Fig. G14 rather than Eq. G33.

REFERENCES

1 Perrin D (1998) *Évaluation des skins de colmatage, de perforation, de pénétration partielle, d'inclinaison et d'injection*. ENSPM Formation Industrie, Pau, France

2 Dupuy JM (1998) *Well productivity*. IFP internal report, Rueil-Malmaison, France

3 Cossé R (1993) *Basics of Reservoir Engineering*. Editions Technip, Paris

4 Earlougher RC Jr (1977) *Advances in well test analysis*. SPE monograph Vol. 5, Chapter 3 and Appendix E. Society of Petroleum Engineers of AIME, Dallas, TX

5 Joshi SD (1991) *Horizontal Well Technology*, Chapter 2. PennWell Books, Tulsa, OK

6 Butler RM (1994) *Horizontal Wells for the Recovery of Oil, Gas and Bitumen*. Petroleum Society of CIM, Monograph 2, Calgary, Canada

7 Chevron paper, *Formation Damage and Pseudo Skin Factors. Their Efect on the Productive Capacity of Wells*. Unpublished paper

8 Bourdarot G (1998) *Well Testing: Interpretation Methods*. Editions Technip, Paris

9 Economides MJ, Hill AD, Ehlig-Economides C (1994) *Petroleum Production Systems*. PTR Prentice-Hall, New-Jersey

10 Brooks JE, (1997) A simple method for estimating well productivity. SPE Europeen Formation Damage Conference, *Paper SPE 38148*, The Hague

11 Perrin D (1998) *Détermination de la courbe IPR d'un puits*. ENSPM Formation Industrie, Pau, France

12 Dale Beggs H (1991) *Production Optimization Using Nodal Analysis*. OGCI Publications

13 Vogel JV (1968) Inflow performance relationships for solution-gas drive wells. *Journal of Petroleum Technology*, January 1968

14 HC "Slip" Slider. *Worldwide Practical Petroleum Reservoir Engineering Methods*. Pennwell Publishing Company, Tulsa, Oklahoma.

G

H

Formation Damage Control

Formation Damage Control

H1 FORMATION DAMAGE PREVENTION [1]

H

Almost every field operation is a potential source of damage to well productivity.

1.1 Recognized forms of formation damage

Some recognized forms of formation damage are:
- Drilling mud solids invasion into the formation.
- Drilling mud filtrate invasion into the formation.
- Cement filtrate invasion into the formation.
- Inaquate perforations: size, number, or penetration.
- Perforation crushing and compaction of formation matrix.
- Solids in completion or workover fluids invading into the formation or plugging perforations.
- Invasion of completion or workover fluids into the formation.
- Plugging of formation with native clays.
- Asphaltene or paraffin precipitation in the formation or perforations.
- Scale precipitation in the formation or perforations.
- Creation of, or injecting, an emulsion in the formation.
- Injection of acids or solvents into a formation which contain solids or precipitate solids.
- Sand fill in the wellbore.
- Injection of oil wetting surfactant into the formation.
- Excessive drawdown which may cause fines movement, compaction of weak formation, or influx of water production.

1.2 Formation damage during drilling operations [1]

- **Formation damage mechanisms are:**
  (a) Invasion of solids from drilling fluid
  (b) Invasion of fluids from the drilling mud
  (c) Effect of fluids invasion.
  Table H1 indicates some preventions for formation damage mechanisms.

151

Table H1 Formation damage during drilling operations [1].

Formation damage mechanisms	Formation damage prevention
Invasion of solids from drilling fluid:	**Minimize solids invasion:**
Clays	Wide particle size in muds
Cuttings	Low spurt loss
Weighting materials	Mud conditioning
Loss circulation materials	High bit weight and low RPM
Fluid loss additives	Acid, water, or oil soluble additives
	Barite minimizing
Invasion of fluids from the drilling mud:	**Minimize fluids invasion:**
Water or brine	Low invasion fluids:
Surfactants	Drilling time minimizing
Oil	Low overbalance
Emulsion	Air, foam or gas drilling
Viscosified fluid	
Effect of fluids invasion:	**Minimize effect of fluids invasion:**
Clay swelling	Inverted mud
Clay mobilization	Oil based mud
Water blocking	Salinity of formation matching
Creation of emulsion	
Oil wetting	
Precipitation of scales	

- **Effect of drilling practices:**
  (a) Scrapping mud cake with bit trips
  (b) Erosion of mud cake by high circulation rate
  (c) Increased invasion by high overbalance and long drilling time.

1.3 Formation damage by perforating

1.3.1 Perforation damage [2]

Productivity and injectivity profiles commonly show that only a small fraction of perforations are flowing. Fig. H1 shows the damaged zone around perforation. The extent and degree of damage are dependent on formation type, permeability and porosity, type of shaped charge, and direction and level of differential pressure when shooting.

1.3.2 Causes and remedies [2]

- **Empirical evidence of perforation damage.** Laboratory studies indicate that the zone permeability can be as low as 10% to 20% of that of the undamaged formation, even when shot underbalanced in clean, compatible fluids.
- **Perforation cleanup.** Laboratory tests demonstrate that the effects of the perforation-damaged can be mitigated by flow through the perforations. The cleanup process and final Core Flow Efficiency CFE are affected by flowing differential pressure (Fig. H2).

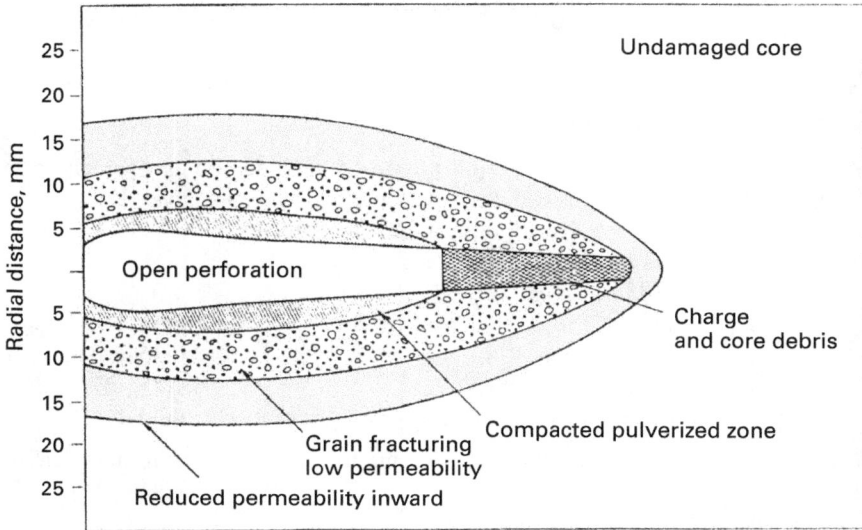

Figure H1 Effect of perforation depth and shot density on well-productivity ratio in a formation with drilling damage [2, 6].

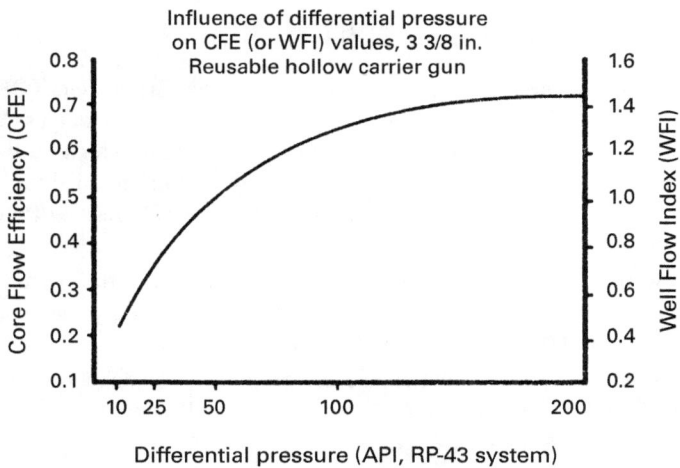

Figure H2 Effect of differential pressure on final CFE [2].

Because perforation damage is inevitable, optimum cleanup is essential. Primary methods involve underbalanced perforating or overbalanced perforating followed by surge flowing or washing of perforations.

1.3.3 Predicting well performance

Simulation studies and analytical and semianalytical techniques are used widely for completion design, evaluation of perforated completions, and system analysis [2].

The following example is given for drilling-damage effects.

The various input parameters (Fig. H3) are as follows:

L_p	length of perforation	1 ft
L_{dd}	length of drilling-damaged zone	0.5 ft
r_w	wellbore radius	0.3646 ft
k	formation permeability	100 md
k_{dd}	formation drilling-damaged-zone permeability	40 md

Drilling damage around the wellbore requires modification of the perforation length, L_p, and the wellbore radius, r_w.

The modified perforation length L'_p is:

$$L'_p = L_p\left(1 - \frac{k_{dd}}{k}\right)L_{dd} = 1 - \left(1 - \frac{40}{100}\right)(0.5) = 0.7 \text{ ft} \tag{H1}$$

and the modified wellbore radius r'_w is:

$$r'_w = r_w + \left(1 - \frac{k_{dd}}{k}\right)L_{dd} = 0.3646 + \left(1 - \frac{40}{100}\right)(0.5) = 0.6646 \text{ ft} \tag{H2}$$

1.3.4 Effectiveness of gun perforating [8]

1. Damage factors indicated in linear cores from Core Flow Efficiencies (**CFE**), as determined according to API RP 43, can be related to Well Flow Efficiencies (**WFE**) for the model described. WFE is defined as the ratio of flow rate into a cased well through a real perforation in a zone that has been damaged by perforating, drilling, or workover to the flow rate into the same well through a clean, ideal perforation of the same depth in an undamaged zone.

2. Permeability in a 1/2-in.-thick damaged zone resulting from the perforating process ranges from about 0.3 of the undamaged-formation permeability for good perforating conditions to about 0.01 for adverse perforating conditions.

3. In a radial system with formation damage and no perforation damage, WFE is substantially reduced until the perforation penetrates substantially beyond the damage zone.

4. In a radial system with both formation damage and perforation damage, WFE remains considerably below that for an undamaged system even when the perforation penetrates substantially through the zone of formation damage.

5. The major effect of permeability damage around a perforation occurs from the damage within the first 1/2 in. of the perforation.

(a)

(b)

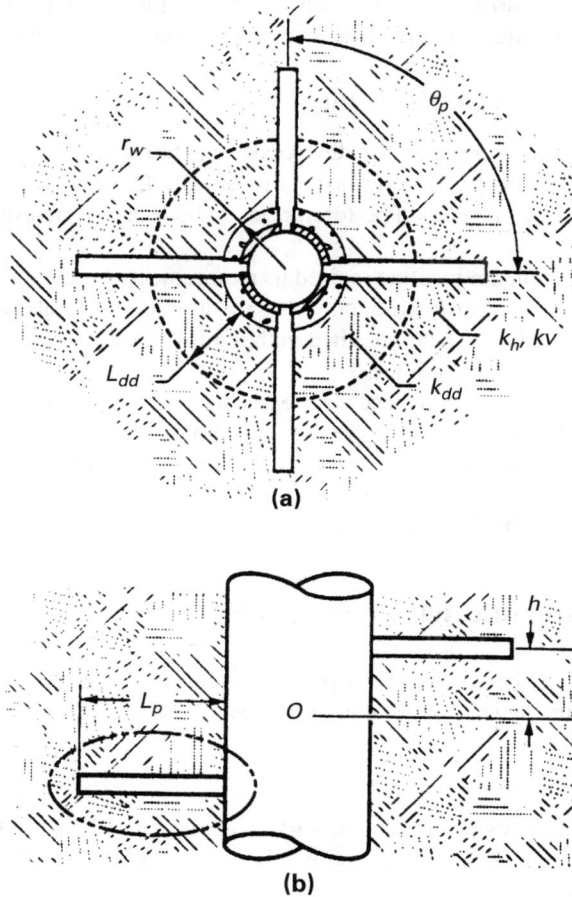

Figure H3 Well parameters [2].

6. Productivity may range from as low as 5 to 90 percent of undamaged, open-hole productivity, depending upon the nature of drilling and perforating operations. Therefore, every precaution should be taken to avoid permeability damage to the formation during drilling, workover, and perforating.

7. If formation damage is avoided during the drilling process, a perforation depth of 12 in. or more is required to overcome the loss in productivity from damaged perforations that is indicated for many commercial guns by standard API RP 43 tests (CFE ≈ 0.7-0.8). Increasing shot density from 4 per foot to 8 per foot has about the same effect as doubling penetration from 6 in. to 12 in.

8. Perforation quality is more important than either shot density or penetration. The effect of severe damage in the perforations cannot be overcome by increasing either shot density or depth of penetration within the limits of present-day and economics.

9. In completions with drilling (or workover) and perforating damage, a few deeply pene-trating perforations are more effective than many shallow perforations.

1.3.5 API RP-43, Fifth Edition [9, 10]

The American Petroleum Institute has developed recommended industry test procedures for oilwell perforators. These are described in the API RP-43, which recommends standard test procedures for the evaluation of perforators under both surface and simulated well conditions.

The American Petroleum Institute revised its requirements for charge certification with the Fifth Edition published in 1991. Section 1 is required for perforator certification.

The Section 1 gun/charge data presented in the following sections were obtained under API RP-43, Fifth Edition, procedures.

Section 2 provides a standard test for measuring perforator performance in stressed Berea sandstone with simulated wellbore pressure applied.

Section 3 evaluates perforating systems at elevated temperature and atmospheric pressure and to test the hardware at elevated pressure.

Section 4 measures the flow performance of a perforation.

1.4 Completion and workover fluids [1]

Formation damage can be minimized by selecting fluids that do not damage the formation or do not create damage which cannot be corrected.

1.4.1 Surfactants

The two primary functions of surfactants (surface active agents) are to change the surface tension of fluids and to affect the wettability of rock. The main rules that should be memorized are as follows:

- Anionic surfactants tend to water wet sand.
- Cationic surfactants tend to oil wet sand.
- Anionic surfactants tend to oil wet carbonates.
- Cationic surfactants tend to water wet carbonates.
- Anionic surfactants tend to emulsify oil-in-water and break water-in-oil emulsions.
- Cationic surfactants tend to emulsify water-in-oil and break oil-in-water emulsions.
- Anionic surfactants tend to disperse clays in water.
- Cationic surfactants tend to flocculate clays in water.
- Anionic and cationic surfactants are not compatible with each other.

1.4.2 Salts

- Salts are used for both weight control and prevention of clay problems.
- Fluid densities obtainable with these salts are indicated in Table H2.

Table H2 Salt densities.

Salts	SI units	U.S. units (lb/gal)
Sodium chloride (NaCl)	1 to 1.17	8.34 to 9.8
Potassium chloride (KCl)	1 to 1.16	8.34 to 9.7
Calcium chloride (CaCl$_2$)	1 to 1.39	8.34 to 11.6
Calcium chloride and sodium chloride combination	1.20 to 1.40	10.0 to 11.7
Calcium bromide (CaBr$_2$) – Calcium chloride	1.40 to 1.81	11.7 to 15.1
Calcium bromide	1.70 to 1.87	14.2 to 15.6
Calcium chloride – Calcium bromide – Zinc bromide (ZnBr$_2$)	1.82 to 2.30	15.2 to 19.2

H2 REMEDIAL CLEAN UP [3–5]

In any case of production index decline, it is most important to have all available facts and to make the best possible analysis from these facts as to the actual factors contributing to this decline. Scale deposits usually form as a result of crystallization and precipitation of minerals from water.

Many treatments are specifically designed for the removal of a blocking agent such as gyp, drilling mud, paraffin, wax, formation silicate particles, or other materials on the wellbore face and in the formation immediately adjacent.

2.1 Reperforation

It has been found quite easy in some cases to reperforate a well in the same zone in which it was originally perforated. The detonation of the gun has a loosening effect on any blocking materials in the formation adjacent to the well and in the previous perforations, while at the same time creating more drainage holes into the wellbore.

2.2 Abrasive jet cleaning

One or more streams of sand-laden fluid are forced through a hardened, specially designed nozzle at pressures of 7 MPa (1 000 psi) and up, to impinge against the wall of the hole. These jets loosen and break up gyp deposits and even penetrate the formation in critical zone. By moving the tool up and down during the job, the hole can be completely cleaned up.

2.3 Mud removal

The material most used is a mud-dissolving acid consisting of inhibited hydrochloric acid with a fluoride added.

157

H

Acid clean-up solutions, containing special surfactants to increase penetration and provide special mud-dispersing properties, are also used when an infiltration of mud into the formation is suspected.

Other solutions containing phosphates or other chemicals are used to loosen and disperse mud particles so that they can be more easily flushed from their position in or adjacent to the wellbore.

2.4 Chemical removal

2.4.1 Water-soluble scale

The most common water-soluble scale is sodium chloride which can be readily dissolved with relatively fresh water. If gyp scale is newly formed and porous, it may be dissolved by circulating water containing 55 g/l NaCl past the scale.

2.4.2 Acid-soluble scale

- **Calcium carbonate ($CaCO_3$)**
  It is the most prevalent of all scale compounds and is soluble in acids such as:
  – Hydrochloric acid
  – Acetic acid
  – Formic acid
  – Sulfamic acid.
- **Application downhole in pumping wells**
  – Acetic acid, when it is desired to leave chrome-plated or alloy pump.
  – Chrome surface: no damage with acetic acid at temperatures below 100°C, but HCl may severely damage chrome.
- **Iron scale**
  HCl plus a sequestring agent is normally used to remove iron scale. Sequestring agents most commonly used include:
  – Citric acid
  – Acetic acid
  – Mixtures of citric and acetic acids
  – EDTA (ethylenediaminetetraacetic acid)
  – NTA (nitrilotriacetic acid).
- **Acid required for $CaCO_3$ and iron scales**
  The amount of acid required for $CaCO_3$ and iron scales is given below in Table H3.

Table H3

Type of acid-soluble scale	Gallons of 15% HCl/cu ft of scale	Liters of 15% HCl/dm^3 of scale
$CaCO_3$	95	13
Fe_2O_3	318	43
FeS	180	24

2.4.3 Acid-insoluble scales

The only acid-insoluble scale which is chemically reactive is calcium sulfate or gypsum. Table H4 shows the relative solubility of gyp in some of the chemicals normally used for gyp conversion.

Table H4 Gyp solubility tests.

Type of solution	Percent gyp converted	
	24 h	**72 h**
NH_4HCO_3	87.8	91.0
Na_2CO_3	83.8	85.5
Na_2CO_3-NaOH	71.2	85.5
KOH	67.6	71.5

If waxes, iron carbonate and gyp are present, the following scale removal procedure is suggested:

- Degrease with a solvent such as kerosene, or xylene, plus a surfactant.
- Remove iron scales with a sequestered acid.
- Convert gyp scales to $CaCO_3$ or $Ca(OH)_2$.
- Dissolve converted $CaCO_3$ scale with HCl or acetic acid. Dissolve $Ca(OH)_2$ with water or weak acid.

2.5 Paraffin removal

Solvents can be circulated past the affected parts of the wellbore or dumped in the hole and allowed to soak for a period of time opposite the trouble area.

In the past, many paraffin solvents have contained chlorinated materials having an organic chloride ion.

Hot-oil treatments are also commonly used to remove paraffin. In such a treatment, heated oil is pumped down the tubing and into the formation. The oil dissolves the paraffin deposits and carries them out of the wellbore when it is produced.

A recent development has been the use of paraffin inhibitors, designed to provide a hydrophilic surface on metal well equipment.

2.6 Large-volume injection treatments

A simple technique often used to free or open up blockages of any kind in a formation consists simply of pumping large volumes of crude oil, kerosene, or distillate into the formation. These treatments are especially effective when the formation is blocked by fine silicates or other solids. Pumping the oil into the formation rearrange these fine particles so that flow channels to the wellbore are reopened. Sometimes it is advisable to add surface-active agents or emulsion-breaking agents to the oil.

2.7 Steam injection

In some areas where low-gravity crude is produced, steam is used to heat and reduce the viscosity of the oil and thereby allow the oil to move more easily to the wellbore. Two types of steam injection are used. In some areas, steam is injected into a central injection well and the oil produced from adjacent or surrounding wells.

The other type of injection is often referred to as "huff-and-puff". This consists of alternate steam-injection and oil-production cycles from the same well.

2.8 General comments

In any case of production index decline, it is important to have all available facts and to make the best possible analysis from these facts as to the factors contributing to the decline. Also, whenever a fluid is to be pumped into a specific part of a zone, some chemical or mechanical method should be used to ensure that the fluid enters the proper zone.

H3 RECOMMENDED PRACTICE FOR FORMATION DAMAGE TESTING [7]

A recommended practice for damage testing has been developed which covers all aspects of the test methodology from core selection and preparation to the writing of the final report and interpretation of the results. It is hoped that this recommended practice will now gain acceptance throughout the industry and that the process of evaluating fluids for formation damage potential will become simpler and more efficient through the greater use and validity of comparisons between data from different sources.

The following procedure has been designed to provide a methodology for assessing formation damage in a variety of testing situations:
• Core preparation and characterization
• Fluid preparation
• Test procedures
• Reporting: A written report should be produced using Table H5 as a guide to composition.

This report should contain all the section headings listed in the left-hand column of this table. The content of each section cannot be specified to cover all situations but suggested elements, which should be included to facilitate interpretation and use of the results obtained, are included in the right-hand column of the table.

Table H5 List of elements to be included in the report [7].

Section headings	Suggested elements to be included
Date of issue	
Reporting authors	
Type of formation material used in tests	e.g. core/outcrop/synthetic
Objectives of the test and background	Requesting party Well/formation to be investigated Origin of formation material Origin of brines/crudes/fluids Project history
Mineralogy of material used in test	Pore throat size and method of determination used Results of X-Ray diffraction (XRD), scanning electron microscopy (SEM), and thin section analyses before and after treatment(s)
Characteristics of core plug	Preserved or restored. If restored method used for core restoration Cutting method and lubricants used Water saturation, S_{wi} Method of preparation to base saturation Ageing time (restoration) Cleaning methods (detail to include equipment, solvents and procedures used)
Fluids used in test	Muds: full compositional information and preparation methods, field or laboratory sourced, details of additional solids/contaminants. Brines: chemical analysis Oil: crude (live / dead), synthetic Gas: type, humidified or not
Test conditions	Plug dimensions and orientation relative to bedding Surface area of plug face exposed to fluids Core holder dimensions and rig schematic (including diagram) Core and fluid temperatures (e.g. if lower than core temperature during dynamic phase) Pore pressure Overburden pressure Overbalance pressure Drawdown pressure / flow rate Volume of fluid produced during drawdown Exposure period (including dynamic/static periods) Flow rate during dynamic period (if applicable) Shear rates at core face during dynamic period (if applicable)
Results	Core plug reference Depth taken (if known/applicable) Fluids used in test Test sequence (e.g. flood/drawdown/breaker/drawdown) Permeability, initial and final Flow rate with time Cumulative volume flowed with time Viscosity of fluids Fluids loss vs. time Temperature and pressure at which fluid collected Break-through pressure (if applicable) Comments on test by investigator
Deviations from recommended practice	
Interpretation	

REFERENCES

1   Sparlin DD, Hagen RW Jr. (1983) *Formation Damage Prevention Manual*. SPE Seminar

2   Bell WT, Sukup RA, Tariq SA (1995) *Perforating*. Monograph, Volume 16 SPE. Henry L. Doherty Series

3   Fitzgerald PE, Martinez SJ (1962) Remedial clean-up and other stimulation treatments. *Petroleum Production Handbook*, Thomas C. Frich, B. William Taylor, SPE of AIME

4   Coulter AW Jr., Martinez SJ, Fischer KF (1988) Remedial clean-up and other stimulation treatments. *Petroleum Production Handbook*, H.B. Bradley, SPE of AIME

5   Allen TO, Robert AP (1997) *Production Operations*. OGCI

6   Krueger RF (1986) An overview of formation damage and well productivity in oilfield operations. *Journal of Petroleum Technology*, February 1986, 131–152

7   Marshall DS, Gray R, Byrne M (1997) Development of a recommended practice for formation damage testing. *Paper SPE 38154* presented at the 1997 SPE European Formation Damage Conference held in The Hague, The Netherlands

8   Klotz JA, Krueger RF, Pye DS (1974) *Effect of Perforating Damage on Well Productivity*. JPT, Nov. 1974, 171–182

9   API Recommended Practice RP 43 (1991) *Evaluation of Well Perforators*. Fifth Edition, January 1991

10 Schlumberger, Perforating Services (1995) Document SMP-7043-1.

I

Sand Control

Sand Control

I1 SAND FORMATION PROPERTIES AND GEOLOGY [1]

1.1 Introduction

Most oil and gas wells produce through sandstone formations that were deposited in a marine or detrital environment. Marine-deposited sands, where most of the hydrocarbons are found, are often cemented with calcareous or siliceous minerals and may be strongly consolidated. In contrast, Miocene and younger sands are often unconsolidated or only partially consolidated with soft clay or silt. These structurally weak formations may not restrain grain movement. When produced at high flow rates, they may produce sand along with the fluids.

1.2 Methods of sand control

Four general types of sand-control methods have been developed to reduce or to prevent the movement of formation sands with produced fluids.

1. In some cases sand production can be prevented merely by restricting the production rate and thus reducing the drag forces on the sand grains. This simple approach is usually uneconomical. Increasing perforation size and density along with the use of clean, nondamaging completion fluids will help to decrease fluid velocity and drawdown pressure at higher production rates.
2. Gravel packing is the oldest, simplest, and most consistently reliable method of sand control. It has wide applications on both land and offshore.
3. Sand consolidation plastic treatments inject resins into the producing interval, binding the formation sand grains together while leaving the pore spaces open. With use of special preflush systems and diverting agents, intervals up to 10 m (30 ft) thick have been successfully consolidated and provided with the strength necessary to allow high production rates.
4. Resin-coated gravel packing places gravel coated with a resin both inside and outside the perforations and in the casing. As the resin cures, the sand grains are bound together. A strong, highly permeable, synthetic sandstone filter results. After curing, the excess resin-coated gravel is drilled from the casing, resulting in a full-open wellbore. The gravel pack can be used with or without a screen, in primary or remedial work, and through coiled or concentric tubing.

165

I2 GRAVEL PACKING

2.1 General

The ideal gravel pack (Fig. I1) would consist of:
- Formation sand replaced by gravel to a distance of 1 ft (30 cm) or more radial from well-bore.
- A sharp gravel to sand interface and prevent invasion of formation sand into the gravel.
- The gravel being tightly packed in the annulus, perforations, and cavities behind the casing.
- The screen completely stopping the gravel with all slots or wire spacings open to flow.
- The gravel must not be polluted. Fluid and pumping system must be clean.

Figure I1 represents the idealized concept of the result of gravel packing.

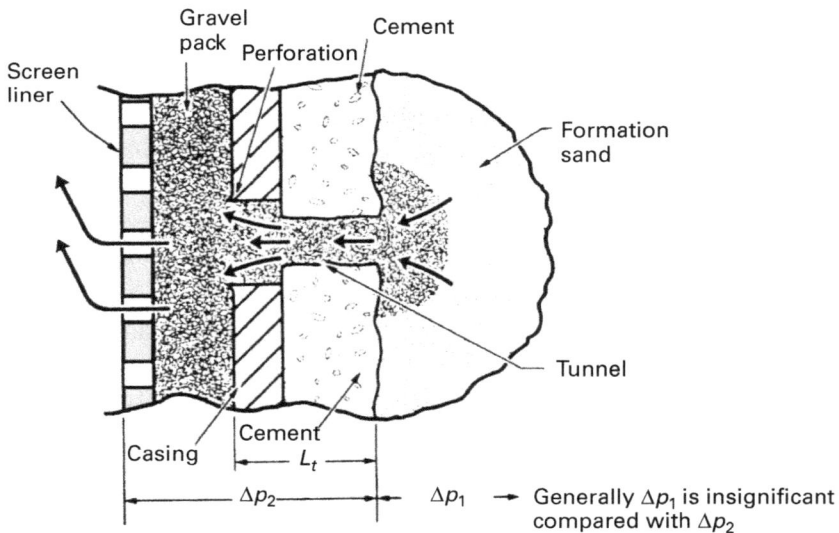

Figure I1 Idealized gravel-pack configuration [2, 5].
Generally, Δp_1 is insignificant compared with Δp_2.

2.2 The perforations

Every effort should be made to design and perform a gravel pack so that the fluid flowing through the perforations is in laminar flow.

2.2.1 **Perforation geometry** [5]

Figure I2 shows a typical geometry for a perforated natural completion.

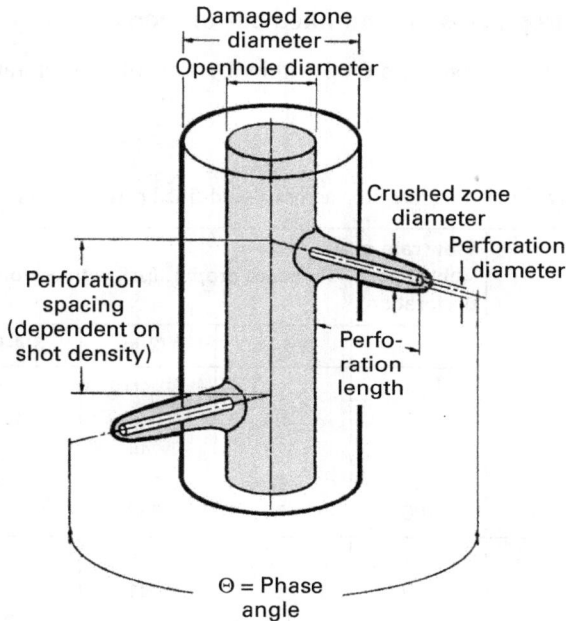

Figure I2 Typical well geometry [5].

Major parameters influencing completion efficiency

These parameters are:
- Effective shot density (actual number of producing perforations per foot)
- Perforation penetration into formation
- Angular phasing
- Perforation diameter.

Influence of geometric factors

Generally, geometric factors influence productivity ratio for natural completion in an ideal, isotropic reservoir:
- Productivity increases as shot density increases.
- Productivity increases with increases in perforation penetration.
- Penetration increase effect is more significant for shallow penetrations than for deep ones.

- Angular phasings other than 0° increase productivity by reducing the interference with flow resulting from the presence of the wellbore.
- Perforation diameter plays a relatively minor role in determining productivity.

2.2.2 Pressure drop across a sand-filled perforation [1]

Table I1 indicates that the pressure drop within a gravel-filled perforation tunnel can be quite significant.

Table I1 Pressure drop across a sand-filled perforation [1].

Gravel	Flow rate q (bbl/D per perforation)	Pressure drop Δ_p (psi) with perforation diameter of:			
		3/8-in.	1/2-in.	3/4-in.	1-in.
U.S. 10/20 mesh	1	0.6	0.2	0.1	0.05
500 darcies	10	24	8	2.3	1
	25	132	44	10	4
	50	495	175	37	13
	100	2 079	666	137	48
U.S. 20/40 mesh	1	2	1	0.4	0.2
119 darcies	10	55	21	6	3
	25	272	99	25	11
	50	983	357	81	31
	100	4 037	1298	282	104
U.S. 40/60 mesh	1	6	3	1.3	0.7
40 darcies	10	177	67	20	9
	25	893	324	80	33
	50	3 250	1178	260	98
	100	13 400	4 360	927	323
Formation sand	1	450	190	64	32
1 darcy	10	27 760	9 280	2091	808

2.3 The screen [2]

The gravel is retained in an annular region within the wellbore by a mechanical device such as:

   (a) Slotted liners
   (b) Wire screens
   (c) Prepacked screens.

2.3.1 Slotted liners

The slotted liners usually have vertical slots spaced uniformly about the casing. The width of the slots can vary from 0.020 in. (0.5 mm) to as large as desired, depending on the gravel size and the particular situation. Fig. I3 illustrates various slot combinations for slotted liners. Their use is usually confined to typically long completion intervals or low-productivity wells.

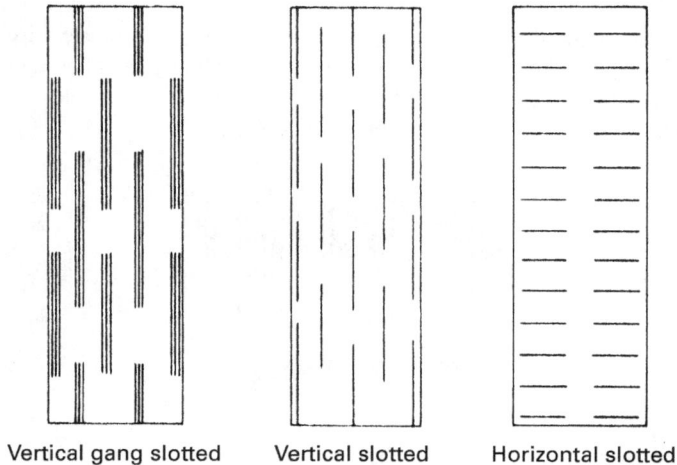

Vertical gang slotted   Vertical slotted   Horizontal slotted

Figure I3 Slotted-liner slot geometry [2].

2.3.2 Wire screens

Wire screens are available in three basic constructions:
- (a) Welded 1 or 2
- (b) Wire wrapped
- (c) Rod base.

- **Welded 1.** The wire is resistance welded to longitudinal rods on a lathe-type machine. The screen (jacket) is manufactured separately, then placed over a drilled pipe base and welded into position.
- **Welded 2.** It is a variation of Welded 1 construction. In this case, the screen is wrapped and the resistance welded directly on the pipe base, with the longitudinal rods as the only separation.
- **Wire wrapped.** In these screens, the wire is wrapped directly on the pipe base, which may be drilled, slotted and grooved.
- **Rod base.** These screens are similar to the jackets on welded construction screens but contain no pipe base.

Figure I4 shows the design of a wire-wrapped screen [12].

Figure I4 Wire-wrapped screen [12].

2.3.3 **Prepacked screens** [2, 12]

These screens are made by filling the annulus between two concentric wire screens with properly sized resin-coated gravel. The gravel pack is actually installed and consolidated at the surface and then is lowered into the well.

Figure I5 shows the design of a prepacked screen [12].

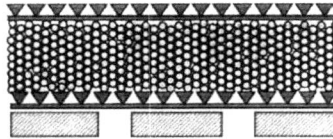

Figure I5 Prepacked screen [12].

2.3.4 **Characteristics and considerations** [2]

Tensile strength

1. The primary tensile failure of pipe-based screens and slotted liners occurs in the 10-round API coupling. The failure stress is normally about 520 MPa (75 000 psi) for Grade J-55 tubing.
2. Rod-based screens usually fail at the point where the screen is welded to the pin connection and have about one-half the strength of pipe-based screens.
3. Pipe-based screens are about twice as strong as rod-based screens under loading conditions.

Collapse strength

1. Collapse failures of Welded 1 screens relate primarily to the excessive standoff of the screen from the pipe base. If the standoff is low, the collapse resistance is high; if the standoff is high, the collapse resistance is low.
2. Welded 2 and wire-wrapped screens usually exhibit higher collapse resistance than Welded 1 screens do.
3. The collapse failures of wire-wrapped screens relate to the shape of the openings in the pipe base. Round holes support loads more effectively than slots do, creating a screen that has about twice as much collapse resistance.

Flow capacity

1. The flow capacities of screens, slotted liners, and gravel packs do not pose significant restrictions to flow unless they become plugged.
2. Rod-based screens offer no significant advantage over pipe-based screens in terms of flow capacity.

Prepack screens

1. Prepacked screens provide excellent sand control but are easily plugged with formation materials. As a consequence, their completion lives may be limited owing to low well rates.
2. Gravel packing around the prepacked screens is an alternative to achieving sand control and sustained productivity, provided that the gravel-packing operation can be conducted without impairing productivity.
3. Prepacked screens have been used as the top joint opposite the upper perforations to help prevent screen erosion if the gravel pack settles and exposes the screen.

2.4 Flow streams [4]

Figure I6 represents flow into a perforated, gravel-packed, cased hole completion.

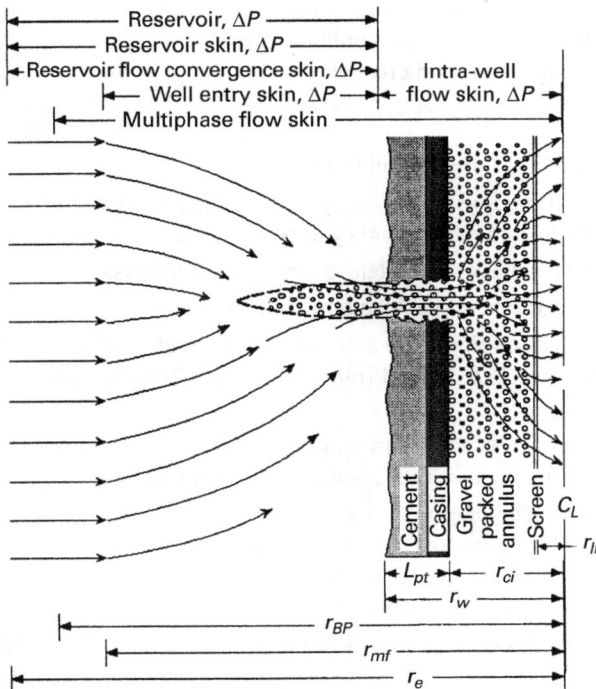

Figure I6 Scheme of the flow system [4] (after R.C. Burton).

In the scheme, the parameters are noted as below:

L_{pt} perforation tunnel length from r_w to r_{ci}
r_{li} internal radius of screen
r_{ci} internal radius of casing
r_w wellbore radius
r_{BP} bubble point radius
r_{mf} mud filtrate invasion radius
r_e drainage radius.

2.5 Guide to designing a gravel pack [3]

1. Obtain a sample of formation material. In preferential order:
   A. Pressurized core.
   B. Rubber sleeve core.
   C. Conventional core.
   D. Sidewall core.
   E. Bailed sample.
   F. Produced sample.
2. Run a sieve analysis and plot.
3. If core material is available, then perform:
   A. X-Ray diffraction on fines and bulk material.
   B. SEM (scanning electron microscope).
   C. HCl solubility.
   D. HCl-HF solubility.
   E. Flow tests with completion fluids and acids.
4. Use this information, logs and geology to determine the optimum perforating method. Usually underbalanced tubing conveyed perforating.
5. Check CBL for zone isolation, plan squeeze jobs if necessary.
6. Design pre and post gravel pack acid treatments.
7. Determine completion fluid type and weight, pills and filter equipment. Design flowrate = 0.25 to 0.5 gal/(min-ft^2) or 10 to 20 l(min.m^2).
8. Select gravel size:
   A. 5–6 times the 50 percentile formation sand size.
   B. Round off to the next smaller commercially available size.
9. Pick screen size based on gravel size: [*]
   A. For 20–40 U.S. mesh, use 0.012-in. screen.
   B. For 40–60 U.S. mesh, use 0.008-in. screen.
   (*) Meets or exceeds API specifications.
10. Run a well flow analysis to determine optimum perforation size and density, as well as tubing size.
11. Using the logs, pick the completion interval and PBTD (Plugged Back Total Depth) or sump packer setting depth, as determined by the lower tell-tale length:

    A. Tell-tale length = perforated interval, ft × 0.1 (3 cm);
       maximum 30 ft (or 9 m), minimum 5 ft (1.5 m).

12. Pick the screen length and diameter:
    A. Overlap the entire perforated interval 5 ft (1.5 m) minimum on each side.
    B. Run screen all the way through the interval and to PBTD; minimize blank areas below screen.
    C. Use 3/4 to 1-in. radial clearance between screen and casing.

13. Determine the centralizer spacing; well deviation:
    0–45°   = 15 ft (4.5 m)
    45°–60° = 10 ft (3 m)
    +60°   = 5 ft   (1.5 m).

14. Determine the blank pipe size and length:
    A. Circulating pack, minimum (90 ft or 27 m of blank).
    B. Squeeze pack, no circulation:
       B1. 10 ppg slurry, blank length = 5 × screen length.
       B2. 15 ppg slurry, blank length = 4 × screen length.

15. Size wash pipe body OD = 0.7 × screen pipe base ID.

16. Select the gravel-pack packer setting depth.

17. Calculate the annular volume in cu ft (or m^3) from the bottom of the hole to 60 ft (18 m) above the top of the screen.

18. Add 1/4–1/2 cu ft (7 dm^3–14 dm^3) for each net feet (30 cm) of perfs.

19. Determine the gravel concentration (ppg) and add the annular volume plus the excess to determine the total slurry volume.

20. Round up the slurry volume to the next higher convenient batch size.
    (Ex: 7 bbl calculated, mix 10 bbl).

21. Determine the pad volumes:
    A. Straight squeeze pack, 3–5 bbl (0.5–0.8 m^3) lead pad.
    B. Circulate pack, 10 bbl (1.6 m^3) minimum lead pad.
    C. Openhole, 25 bbl (4 m^3) minimum lead pad.

22. Calculate the slurry pump rate:
    Q bbl/min = 0.485 d^2, d = workstring diameter (in.), or
    Q l/min = 11.95 d^2, d = workstring diameter (cm).

23. Calculate the displacements:
    A. Pad at tool.
    B. Pad at formation (top perf).
    C. Slurry at tool.
    D. Slurry at formation.
    E. Slurry at lower telltale.
    F. Slurry tail at tool.
    G. Slurry tail at formation (top perf).

24. Calculate the reverse out pressure (slurry remaining in workstring).

Table 12 Commercial gravels [2].

U.S. mesh	Average diameter		Permeability
	(in.)	(mm)	*k* (darcy)
3–4	0.226	5.74	8 100
4–6	0.160	4.06	3 700
6–8	0.113	2.87	2 900
6–10	0.106	2.68	2 703
8–10	0.0865	2.197	
8–12	0.080	2.030	1 969
10–14	0.0675	1.71	800
10–16	0.063	1.60	
10–20	0.056	1.42	652
10–30	0.051	1.295	
12–18	0.053	1.410	
12–20	0.050	1.26	510
16–20	0.040	1.014	333
16–30	0.035	0.88	273
20–40	0.025	0.889	170
30–40	0.01975	0.502	110
40–50	0.014	0.358	
40–60	0.013	0.334	69
50–60	0.01075	0.273	43
40–70	0.013	0.330	
40–100	0.012	0.300	
60–70	0.009	0.230	

13 SCREEN INSTALLATION AND GRAVEL-PACK PROCEDURE GUIDELINES [2]

3.1 Introduction

The techniques for installing screens and performing gravel packs fall into four categories:
- (a) Perforation cleaning
- (b) Prepacking the perforation
- (c) Conventional gravel packing
- (d) Crossover gravel packing.

Individual circumstances and well conditions may require departures from the procedures outlined.

3.2 Perforation cleaning

3.2.1 Wash tool. Reverse circulation type

1. Make up the tool going in the hole, with the distance between the upper swab cups equal to the length of the perforated interval plus 3 m (10 ft).
2. With the lower swab cup in the bypass position, run the work-string or assembly into the hole slowly to 3 m (10 ft) above the top perforation.
3. Close the bypass around the lower swab cup and test the tool to 3.5 MPa (500 psi) by applying annulus pressure with the annular preventer closed.
4. With the annular preventer closed and 1.4 MPa (200 psi) pressure applied down the annulus, lower the tool to the first perforation. Mark the workstring at the top perforation, indicated by a pressure loss or the start of circulation.
5. Begin reverse circulation with filtered fluid at the minimum rate required to transport sand up the workstring to the surface.
6. Wash the perforation with a 5 m³/m (10 bbl/ft) interval or until pressures stabilize. Brine is the preferred wash fluid.
7. Continue washing until the bottom perforation is reached, which is signaled by an increase in pressure as the lower swab cup passes below the bottom perforation.
8. Release the bypass around the lower swab cup and reverse circulation until the wellbore is clean.
9. Repeat steps 5 through 8 from the top of the perforations downward.
10. With the bypass open around the lower swab cup, perform a muleshoe.

Note: Perforation washing rates should be high enough to lift the wash debris from the well. A lift of 30 m/min (100 ft/min) is usually sufficient.

3.2.2 Wash tool. Conventional circulation type

First make up the wash tool and check the operation valve. The up position of tubing indicates that the valve is closed; the down position, that it is open. (Rotational-type valves are not recommended because left-hand torque is usually required to open the tool.)
   Proceed with the perforation washing operation in the following sequence.

1. Run the workstring in the hole to below the perforations. Raise it slightly and test the tool to 3.5 MPa (500 psi) down the workstring.
2. With 1.4 MPa (200 psi) on the tubing, slowly raise the workstring until a loss of pressure indicates that the bottom perforation has been reached. Mark the workstring.
3. Initiate circulation down the workstring at 160 to 240 l/min (1 to 1.5 bbl/min) using filtered fluid. Increase the pump rate as much as possible without causing fracturing.
4. Wash perforations with 5 m³ fluid/m (10 bbl fluid/ft). Brine is the preferred wash fluid.
5. Slack off and open the circulating valve. Reverse circulation to clean the well.
6. Repeat steps 4 and 5 until the entire perforated interval is cleaned.
7. Slowly pull the workstring from the hole (300 to 450 m/h or 1 000 to 1 500 ft/h), taking extreme care to maintain the hole full of fluid.

Note: Specific situations may dictate higher pump rates and larger wash-fluid volumes. Annular velocities should be about 30 m/min (100 ft/min) for efficient removal of wash debris.

3.2.3 Surge tool

When the surge-tool procedure is implemented, one should design the tool string with the following considerations in mind.
- Atmospheric-chamber collapse strength must be capable of withstanding full hydro-static load plus 7 MPa (1 000 psi) packer test pressure, including a safety factor of 1.25.
- Atmospheric-chamber volume should create a 7 MPa (1 000 psi) differential pressure into the wellbore.
- Packer-setting depth should be such that the casing storage volume below the packer is greater than the atmospheric chamber volume.

Then make up the tool string in the following order:
1. Check valve (flow allowed, injection prevented).
2. Pup joint.
3. Mechanically set packer (must hold pressure from either direction and set with right-hand rotation).
4. Pup joint.
5. Surge valve (straight pull or left-hand rotation to open).
6. Atmospheric chamber with length designed according to considerations listed above.
7. Landing nipple with plug (preferably wireline-retrievable) installed.
8. One joint of tubing.
9. Ported collar or sliding sleeve.
10. Tubing, as required, to space out the packer to calculated setting depth. (Test above and below the atmospheric chamber to 35 MPa. Torque turn or externally test the atmospheric-chamber tubing.)

Then take the following steps:
1. Enter the hole to the required packer-setting depth. Fill the tubing every two or three stands.
2. Set the packer and test the annulus to 7 MPa (1 000 psi).
3. Connect the kill manifold to the tubing annulus and test the lines to working pressure.
4. Open the surge valve as dictated by the type of valve used (rotation or pull).
5. Rig up the wireline unit, equalize, and pull the plug from the landing nipple above the chamber.
6. If pressure exits on the tubing, try to bleed it off. Do not bleed more than 320 or 480 liters (2 or 3 bbl). If the pressure does not bleed down, open the ported collar and reverse the tubing clean of produced fluids. Use the fluid weight, as required, to control surface pressure.

7. Release the packer and reverse circulate through the kill manifold and choke.

8. When clean fluids return and the well is completely under control, lower the pipe and reverse circulate to clean the well to below perforations.

9. Slowly pull out of the hole.

3.3   Prepacking the perforations

This practice is essential for long life, high-productivity, cased-hole gravel packs. The Muleshoe method is proposed:

1. Enter the hole with open-ended tubing to 1.50 m (5 ft) below perforations. Close annular preventers and reverse circulate with filtered fluids.

2. Test injectivity at 160 to 320 l/min (1 to 2 bbl/min) and record pressure on the annulus. If the pressure level is high (and the rate is low), perform a mud-acid treatment.

3. Initiate gravel circulation at a concentration of 0.9 to 1.2 kg/l (3/4 to 1 lbm/gal) in filtered fluid.

4. As the sand slurry nears the end of the tubing, close the annulus return line and begin injection.

5. Alternately move the pipe up 1.5 m (5 ft) and down 1.2 m (4 ft) for about 1 minute. Continue reciprocating until the entire perforated interval is packed and the wellbore is full of sand to above the perforations, which is indicated by a high injection pressure.

6. Slowly reverse the wellbore clean to below the perforations and estimate the volume of sand placed outside them.

7. Repeat steps 3 through 6 until no additional sand is placed outside the perforations.

8. Reverse circulate the well around at least two tubing volumes.

9. Pick the workstring up above the perforations and shut the well in for 30 minutes.

10. Lower the pipe and tag the bottom. If more than 30 dm^3 (1 cu ft) of gravel has fallen back into the wellbore, repeat prepack steps 3 through 9. If the wellbore is clean, pull out of the hole. If gravel continues to fall into the wellbore, spot a high-viscosity pill across the perforated interval.

11. Slowly pull out of the hole to prevent swabbing prepack gravel into the wellbore.

3.4   Conventional gravel-pack methods

3.4.1 Washdown gravel-pack method

This procedure should not be used if the screen setting is longer than the working height of the rig.

1. Clean the perforations with one of the washing-tool or surge-tool procedures described earlier.

2. Prepack the perforations, following steps 1 through 7 of the previous section.

3. After completing the prepack until gravel is no longer being placed outside the perforations, leave gravel inside the wellbore to a height calculated by:

$$h_g = \frac{C_{an} L_s}{C_c}$$

where

h_g	height to gravel left in casing
C_{an}	capacity of annulus between screen OD and casing ID
L_s	length of screen and blank liner setting
C_c	capacity of casing ID.

4. Make up the screen setting in the following sequence:
   - Shoe (standard or turbojet)
   - Screen with centralizers
   - Blank liner with centralizers
   - Backoff subs.
5. Hang the screen off in slips and make up the wash pipe.
6. Space out the wash pipe using a slip-joint releasing tool so that the pipe seals into the seat shoe.
7. Make up the tubing to go in the hole with the screen setting to the top of the gravel fill.
8. Initiate circulation down the tubing at 160 to 320 liters per minute (1 to 2 bbl/min).
9. Slowly lower the screen to the required completion depth while circulating the cleaning fluid.
10. Stop the circulation and wait 15 minutes.
11. Apply annulus pressure and check for complete gravel coverage of the screen. (If required, reverse gravel down the annulus to fill in around the screen, according to the procedure described in the reverse circulation section).
12. With 22 kN (5 000 lbf) pull above the pipe weight, rotate the tubing to the right and release it from the backoff sub.
13. Lift the screen 0.9 to 1.5 m (3 to 5 ft) and clean the well with reverse circulation.
14. Pull out of the hole with the tubing.
15. Run a tieback overshot with a mechanically set packer to the top of the screen.
16. Locate and space out so that the overshot will seal onto the top of the screen setting. Test the annulus at the final packing pressure and watch for circulation, indicating that the overshot is not sealed.
17. Set the packer and test the annulus to 10 MPa (1 500 psi).
18. Release the packer or nipple up the tree, as production requirements dictate.

3.4.2 Reverse-circulation method

First, clean the perforations according to one of the wash-tool or surge-tool methods, followed by the muleshoe prepack method. The gravel-pack procedure is normally performed as stated below.

1. Make up the screen setting as follows:
   – Bladed bull plug
   – Screen with centralizers
   – Blank liner with centralizers
   – Telltale screen joint, with a 0.6 m (2 ft) screen located 3 to 4.5 m (10 to 15 ft) below the top of the setting
   – Backoff subs.
2. Set the screen setting in slips and make up the wash pipe and releasing tool. Space the wash pipe to within 0.3 m (1 ft) of the bottom of the screen and make up the releasing tool to the backoff sub.
3. Enter the hole to the required setting depth.
4. Initiate reverse circulation with a gravel slurry at 1.2 kg/l (1 lbm/gal) in filtered fluid.
5. After filling the annulus to above to above the screen, calculate the volume required to fill the annulus to the telltale.
6. Reverse circulate 75 to 80% of the calculated gravel volume and follow that with clean fluid.
7. If the telltale is not covered, reverse circulate a 160 or 320 liters (1 or 2 bbl) gravel slurry followed by clean fluid.
8. Repeat step 7 until the telltale is covered, as noted by a large pressure drop. Do not exceed the blank-liner collapse rating.
9. Release the workstring from the screen and reverse the well clean. Pull out of the hole.
10. Run a tieback overshot with a mechanical packer and space out so that the overshot is over the backoff sub.
11. Test the tieback seal at the final pressure and watch for circulation.
12. Set the packer and test to 10 MPa (1 500 psi).
13. Release the packer or nipple up the tree, as dictated by production requirements.

14 GUIDELINES FOR CHOOSING A SAND-CONTROL METHOD [2]

In deciding upon the most appropriate sand-control, the engineer must first be aware of the treatment options available (see Fig. I7) and then select one according to the geometric factors and completion requirements of the particular well.

4.1 Sand-control options

4.1.1 Gravel packing

Gravel packs covering 6 m to 60 m (20 to 200 ft) intervals are common, and many treatments covering intervals of 150 m (500 ft) or more have been performed.

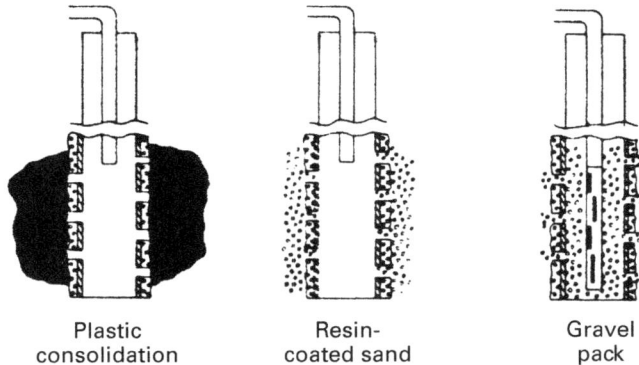

Plastic consolidation   Resin-coated sand   Gravel pack

Figure 17 Sand-control techniques [2].

Gravel packing offers an economical method of controlling sand, but it has several disadvantages:

- While initial installation is economical, a remedial treatment to replace a failed screen may involve an expensive fishing job.
- In overpressured reservoirs that cannot be controlled with a calcium chloride brine, use of a special gravel-packing fluid is required.
- Productivity impairment caused by perforation tunnels being filled with sand and/or invasion of the gravel pack by formation fines is an important consideration. This problem is especially severe in small casings where very small screens must be used.

4.1.2 Bare screens, slotted liners, or prepacked screens

They may have utility in certain wells where a short-term solution to a sand-control problem is needed and for economical or other reasons where a gravel pack is not a satisfactory solution.

4.1.3 Consolidation

The objective of chemical consolidation is to treat the formation in the immediate vicinity of the wellbore with a material that will bond the sand grains together at their points of contact. For these treatments to be effective, two requirements must be met:

   (a) The formation must be treated (consolidated) outside all perforations.
   (b) The consolidated sand mass must remain permeable to well fluids.

Advantages

- Future workovers are simplified because no mechanical equipment is left in the wellbore.

- Because all the formation material is cemented together for several feet surrounding the wellbore, a productivity decline associated with the migration of fine particles toward the wellbore does not occur because the flow velocities are reduced at a treatment radius of 0.9 to 1.2 m (3 to 4 ft) from the well.
- Consolidation can be performed without a rig on wells that have not sanded up.

Disadvantages
- Cost: a minimum of U.S. $ 5 000 per foot of perforated interval.
- Coverage: the problems arise from injecting insufficient resin into low-permeability zones of highly stratified reservoirs. Consolidation success is reduced in perforated intervals longer than 5 m (15 ft).

4.1.4 Resin-coated gravel

This technique consolidates gravel placed inside and immediately surrounding the perforation. The gravel (usually 40/60 or 20/40 U.S. mesh) is coated with a resin at the surface and then pumped into the well as a slurry. This slurry is then squeezed through the perforations to fill a region behind the casing. After hardening, the consolidated gravel prevents formation sand from entering the wellbore. Excess resin-coated gravel is removed from inside the casing by either drilling or washing.

Advantage
- Reduced cost: 25 to 50% of the cost of a conventional consolidation treatment or gravel pack, making it economically attractive in longer intervals.

Disadvantage
- Short lifetime.

4.2 Wellbore configuration

Wellbore geometric factors (casing sizes and hole deviation) can play an important role in the effectiveness of the various sand-control methods.

4.2.1 Conventional completions

Figure I8 shows the success, in terms of cumulative fluid production, of openhole (OHGP) and cased-hole gravel packing (CHGP), and plastic consolidation.

Openhole gravel packing is the most successful sand-control technique. Unfortunately, only a few wells usually meet the long-life, productivity, and minimum recompletion requirements necessary for successful openhole production.

Because the formation fines are consolided for several feet around the wellbore, the productivity decline with time is less with a plastic consolidation treatment than with a gravel pack.

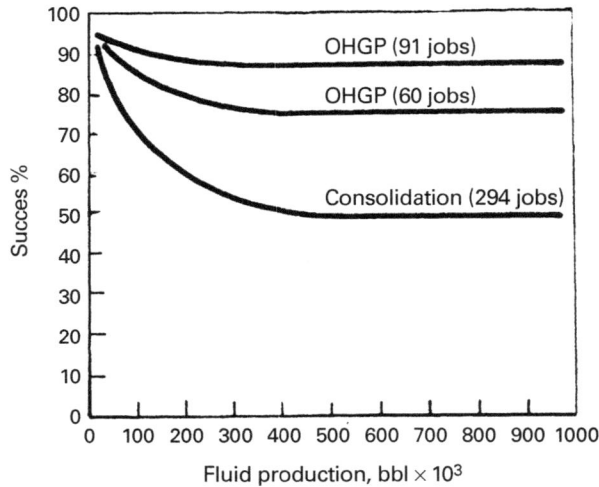

Figure I8 Sand-control success in conventional completions [2].

4.2.2 Well deviation

In gravel packs, the basic problem is completely filling the annulus between the casing and screen with gravel. For wells inclined more than 45° from vertical, totally filling the annulus with gravel may not be possible unless proper placement designs are implemented. However, special tools and placement techniques designed to overcome this problem have been developed.

4.3 Reservoir properties

The properties of reservoirs where sand must be controlled will influence the type of sand-control method use. Interval length and sand quality are often the deciding factors, but reservoir temperature, pressure, and should also be considered.

4.3.1 Interval length

Figure I9 compares gravel-packing and consolidation success rates in intervals having less than 12 ft (3.6 m) perforations.

4.3.2 Sand quality

A good-quality sand has a narrow grain-size distribution and a low nonsilica content (i.e., 5 to 15%). The permeability of such sand is generally high (1 darcy or greater).

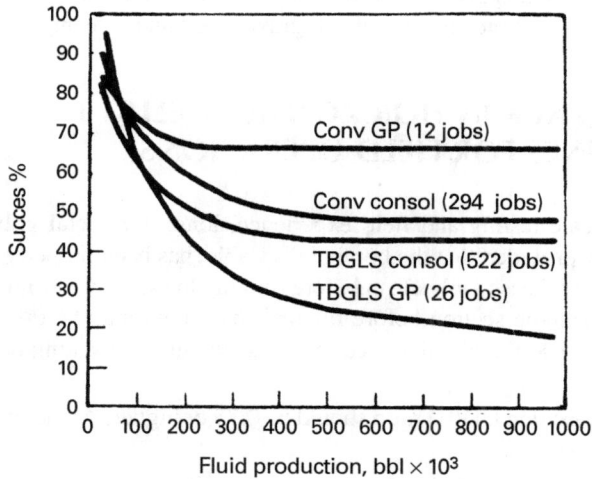

Figure 19 Sand-control success in intervals less than 12 ft (3.6 m) [2].
GP: gravel packing. TBGLS: tubingless.

A poor-quality sand, on the other hand, generally has a wide particle-size distribution, an appreciable fraction of which will pass through a 400 U.S. mesh screen. The nonsilica content of these sands can be as high as 50%, and permeability may be reduced to 100 md or less.

4.3.3 Reservoir temperature

Reservoir temperature does not play an important role in the selection of a sand-control technique. With few exceptions, all techniques can be used in reservoirs, with a static bottomhole temperature ranging from 50 to 100°C (120 to 200°F), a range that probably encompasses 95% of the all sand-producing wells.

4.3.4 Reservoir pressure

Workover fluids are commonly available to control formation gradients from 7 to 14 kPa/m (0.30 to 0.60 psi/ft).
    Special completion fluids may be required in abnormally pressured reservoirs.

4.3.5 Reservoir fluid

The type of fluid (oil, gas, or water) produced from the reservoir should be considered during technique selection. Fine particles entrained in a high-velocity gas stream may severely erode mechanical equipment.

The erosion problem is even worse when gas is in turbulent flow, which it may well be (depending on rate) as it moves through the gravel pack and screen.

I5 GRAVEL PACK IN HORIZONTAL WELLS [7]
GUIDELINES FOR FIELD OPERATIONS

Based on field-scale testing and field experience, long, horizontal gravel packs can be placed effectively provided that the borehole is stable, has been properly cleaned, and that returns to surface are reasonably clean before running the screen. Drilling, displacing, and stabilizing the openhole section before running the screen may be the key to successful gravel placement. Additional guidelines for the actual gravel packing operation are listed below.

- The wash-pipe-screen OD/ID ratio should be 0.75 or higher. Avoid large diameters that cannot be easily fished.
- Run the wash pipe to the end of the prepacked screen section.
- Use completion brine for the gravel-pack transport fluid.
- Perform circulation tests before pumping gravel to ensure that a superficial velocity of at least 30 cm/s (1 ft/s) can be maintained based on returns through the wash pipe.
- Do not break down (fracture) the formation with high pump pressure.
- Do not begin pumping gravel until the above conditions have been met.
- Pump gravel at mix ratios that are about 1.2 kg/l (1 lbm/gal):
  - For return rates that are about to exceed 30 cm/s (1 ft/s), gravel-mix ratios can be increased slightly.
  - For return rates that are less than 30 cm/s (1 ft/s), reduce the gravel-mix ratio.
- Avoid gravel-mix ratios that exceed 1.8 kg/l (1.5 lbm/gal).

I6 UNIFORM INFLOW COMPLETION SYSTEMS
IN HORIZONTAL WELLS

Horizontal wells are now largely used to improve recovery efficiency, increase reservoir drainage area, delaying water and gas coning, and increasing production rate. But in long highly deviated and horizontal wells the non-uniform production inflow profile along the lateral length can result in premature water and/or gas production. Special sand screen with controlled inflow production device are proposed in the market to solve such problem

Norsk Hydro has patented US5435393 "Procedure and production pipe for production of oil or gas from an oil or gas reservoir" and Institut Français du Pétrole has a concept patented US7100686B2 "Controlled pressure drop liner". Baker Oil tools markets such screens named "Equalizer" Reservoir-Optimized Completion System.

From the basic concept the aim is to control the fluid pressure drop using specific sand screens in order to achieve a uniform production per unit length of long horizontal well sections. For each sand screen section, the calculated pressure drop is obtained with Baker Equalizer by helical flow channel to provide fluid resistance and by IFP concept by tubes adapted in size and length located between the wire screen and the base central liner. Both concepts, fluids arriving from the reservoir are passing through the wired screens, through helical flow channel or tubes generating the pressure drop and arriving inside of the base pipe collecting the production to the production liner

The system can be used for both production and injection wells. The expected gains are an uniform repartition of the production in the producing zone or a better repartition of the injection particularly for long horizontal sections. The expected results are to minimize the contribution of productive zones close to the casing shoe and the activation process and to allow production of reservoir zones at the end of the liner

The final result will be to produce with minimal water and gas break-through and to optimise the entire length of the productive zone and the length of the well drilled in the reservoir for a better lifetime management of the well and reservoir.

Cutaway drawing of Baker Oil Tools Equalizer ICD Screen
paper Baker presented OMC 04/04/2006 [13]

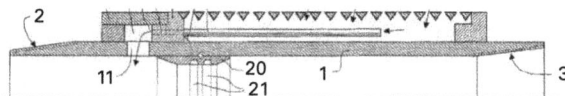

IFP Concept patented US7100686B2 "Controlled pressure drop liner" [14]

Figure I10 Special sand screen with controlled inflow production device [13, 14].

REFERENCES

1 Coulter A W Jr., Martinez SJ, Fischer KF (1988) Remedial clean-up and other stimulation treatments. *Petroleum Production Handbook*, Chap. 56, H.B. Bradley, SPE of AIME
2 Penberthy WL Jr., Shaughnessy CM (1992) Sand Control. *SPE Series on Special Topics*, Volume 1. Henry L. Doherty Series
3 Ledlow LB, Sauer CW, Till MV (1985) Sand control: recent design, placement, and evaluation techniques lead to improved gravel pack performance. SPE Reprint series No. 43, *Paper SPE 14162* presented at the 60th ATCE of SPE held in Las Vegas, NV
4 Burton RC (1998) Use of perforation-tunnel permeability as a means of assessing cased-hole gravel-pack performance. *Paper SPE 39455* presented at the 1998 SPE International Symposium on Formation Damage Control in Lafayette, Louisiana
5 Bell WT, Sukup RA, Tariq SM (1995) Perforating. *Monograph volume 16 SPE*, Henry L. Doherty Series
6 Restarick H (1995) Horizontal completion options in reservoirs with sands problems. *Paper SPE 29831* presented at the SPE Middle East Oil Show held in Bahrain

7   Penberyhy WL Jr., Bickham KL, Nguyen HT, Paulley TA (1996) Gravel placement in horizontal wells. *Paper SPE 31147* presented at the SPE International Symposium on Formation Damage Control held in Lafayette, Louisiana

8   Malbrel C (1999) Screen sizing rules and running guidelines to maximise horizontal well productivity. *Paper SPE 54743* presented at the SPE European Formation Damage Conference held in The Hague, The Netherlands

9   *Oil Well Screen Selection Consideration*. Presented by Well Completion Technology (1999), Houston, Texas

10 Procyk A, Whitlock M, Ali S (1998) Plugging-induced screen erosion difficult to prevent. *Oil & Gas Journal*, July 20, 1998

11 Ballard T, Kageeson-Loe N, Mathisen AM (1999) The development and application of a method for the evaluation of sand screens. *Paper SPE 54745* presented at the SPE European Formation Damage Conference held in The Hague, The Netherlands

12 Mabrel C (1997) Crépines et contrôle des venues de sables. Développements récents. *Pétrole et Techniques*, n° 411, 69, Nov-Déc. 1997

13 Cutaway drawing of Baker Oil Tools Equalizer ICD Screen. *Paper Baker* presented OMC 2006, April 4th.

14 IFP Concept patented US 7100686B2 "Controlled pressure drop liner".

J

Stimulation

Stimulation

J1 INTRODUCTION

The primary methods for well stimulation are hydraulic fracturing and matrix acidizing. The coverage of these two subjects ranges from fundamental principles to design techniques and new concepts, especially in acid fracturing and matrix acidizing. The examples worked out in this chapter are relevant to industrial practice as well as for promoting a fundamental understanding of well stimulation. The new technique "Frac-packs" is introduced.

J2 HYDRAULIC FRACTURING

2.1 Basic equations [3]

The chosen hypotheses are:
  (a) The borehole is vertical and parallel to one of the principal components of the geostatic stress field.
  (b) The rock is assumed to be poroelastic, linear and isotropic.

Then:
* The fracture is a vertical plane parallel to the major geostatic horizontal stress σ_H.
* Fracturing is performed in open hole.
* Considering the short injection times, poroelastic effects are neglected.

The rupture–reopening sequence is shown on Fig. J1. During the first cycle, the fracture is initiated for a pressure p_b (breakdown pressure) then propagated at a lower value p_p (propagation pressure). Once a certain volume of fluid has been injected, pumping is stopped without bleeding-off of the lines. The fracture which is no longer supplied begins to close. Two different phases are observed during closure. Just after shut-in there is a sharp pressure drop.

Then, after a well marked inflexion point called ISIP (Instantaneous Shut-In Pressure), pressure decreases at a lower rate depending on the rock permeability. When pressure reaches σ_h, the crack completely closes and the fluid can theoretically only flow in the reservoir through the borehole wall, which has a much smaller surface than the fracture.

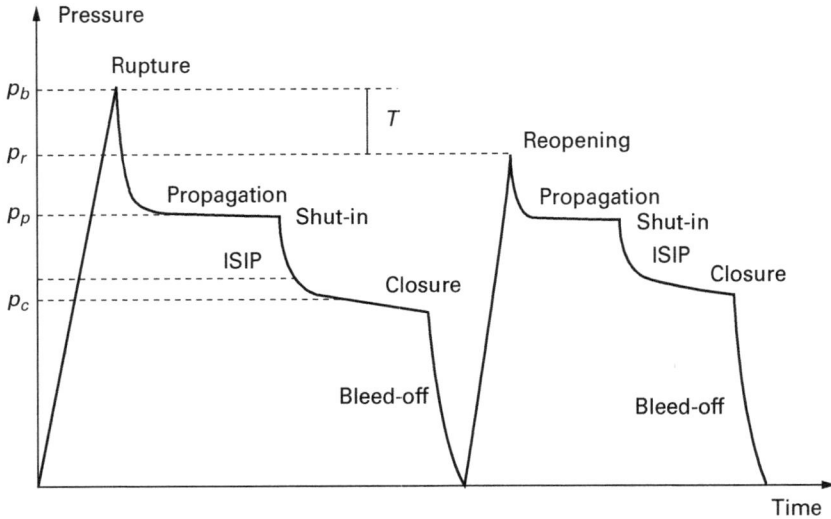

Figure J1 Rupture–reopening sequences. Definition of characteristic pressures [3].

A new inflexion point called closure pressure p_c and theoretically equal to σ_h is associated with this second transition.

During a rupture–reopening sequence, reference pressures and stresses are related through the following expressions (ignoring poroelastic effects):

- Breakdown pressure:

$$p_b = -\sigma_H + 3\sigma_h - T \qquad \text{(J1)}$$

- Reopening pressure:

$$p_r = -\sigma_H + 3\sigma_h \qquad \text{(J2)}$$

- Propagation pressure:

$$p_p = \sigma_h + \Delta p_k + \Delta p_p \qquad \text{(J3)}$$

- Instantaneous shut-in pressure:

$$ISIP = \sigma_h + \delta_p \qquad \text{(J4)}$$

- Closure pressure:

$$p_c = \sigma_h \qquad \text{(J5)}$$

where

σ_H major geostatic horizontal stress
σ_h geostatic horizontal stress
Δp_k cohesive resistance of the material
Δp_p pressure drop in the fracture
δp extra pressure needed to keep the fracture open after shut-in.

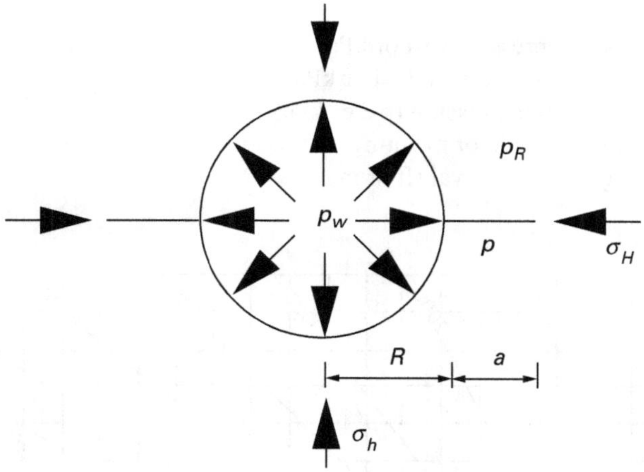

Figure J2 Initiation of a hydraulic fracture in a well [3].

2.2 General fracturing treatment formulas [4]

- Bottom-hole fracturing pressure gradient:

$$G_F = \frac{P_S + P_h - P_{tf} - P_{pf}}{D} \tag{J6}$$

The fracture gradient G_F expected for horizontal fractures (above 2 000 ft or 650 m) is of the order of 1.0 psi/ft (23 kPa/m) or higher, and for vertical fractures (below 4 000 ft or 1 300 m), it is 0.7 psi/ft (16 kPa/m) or lower.

- Bottom-hole fracturing pressure:

$$P_F = G_F \times D \tag{J7}$$

- Instantaneous shutdown pressure:

$$P_i = P_F - P_h \tag{J8}$$

- Total surface pressure:

$$P_S = P_F + P_{tf} + P_{pf} - P_h \tag{J9}$$

- Hydraulic horsepower:

$$HHP = 0.0245 \times P_S \times Q \tag{J10}$$

with P_S (psi) and Q (bbl/min)

- Input power WP (kW):

$$WP = 0.017 \times P_S \times Q \tag{J11}$$

with P_S (kPa) and Q (m^3/min)

where

P_h total hydrostatic pressure (psi or kPa)
P_{pf} perforation friction pressure (psi or kPa)
P_{tf} total tubular friction pressure (psi or kPa)
Q injection rate (bbl/min or m³/min)
D depth of producing interval (ft or m).

J

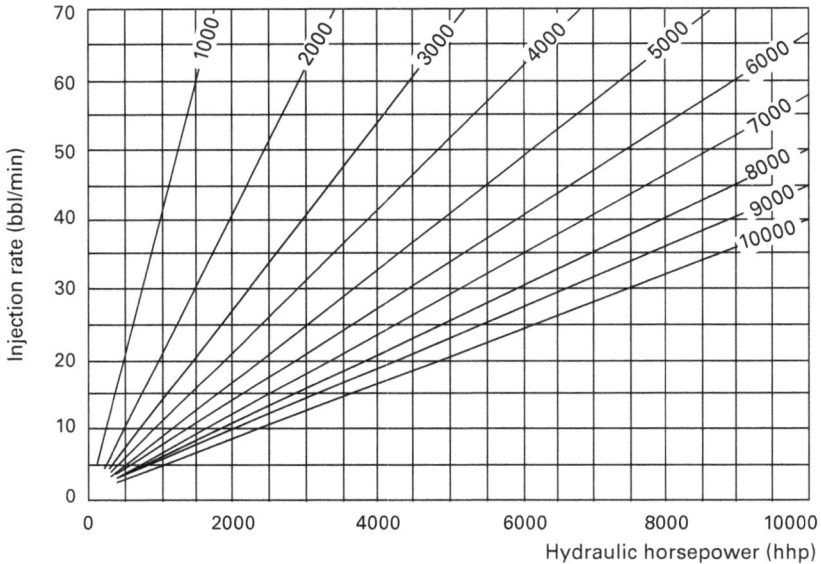

Figure J3 Hydraulic horsepower chart [4].

2.3 Simple calculation of fracture dimensions [8]

Approximate calculations can be done on a hand calculator or simple computer to arrive at fracture dimensions during pumping. The following equations should be programmed to obtain approximate fracture dimensions for a Perkins-Kern-Nordgren (PKN) fracture. The PKN model should be used in all situations, except shallow wells where fracture height is greater than fracture length. In this case, the Geertsma-de Klerk (GdK) model should be used.

Additional simple calculations are presented for the estimation of final propped fracture dimensions. All expressions for fracture length and volume refer to that quantity for one wing of the total fracture; i.e., fracture length is the distance from the wellbore to one of the fracture tips. All equations contain dimensionalizing constants so that if the dependent variables are put in with the units given in the nomenclature, the independant variables will also have the units indicated in the nomenclature.

2.3.1 Nomenclature

Symbol	Designation	Metric unit	U.S. unit
a	Nordgren length constant	m	ft
B	Nordgren time constant	min	min
C	Fluid-loss coefficient	m/min$^{1/2}$	ft/min$^{1/2}$
e	Nordgren width constant	cm	in.
G	Shear modulus of elasticity	kPa	psi
h_g	Gross fracture height	m	ft
h_n	Net permeable sand thickness	m	ft
K	Power-law constant	Pa·s^n	lbf-s^n/ft^2
L_D	Dimensionless fracture length		
q_i	Flow rate into one wing of a vertical fracture	m^3/min	bbl/min
t	Job pumping time	min	min
t_D	Dimensionless job time		
$\bar{w}$	Volumetric average fracture width	cm	in.
w_D	Dimensionless fracture width		
w_{wb}	Fracture width at wellbore	cm	in.
μ_e	Effective non-Newtonian fracture-fluid viscosity	mPa·s	cp
ν	Poisson's ratio, dimensionless		

2.3.2 Equations

The following series of equations needs to be solved iteratively on a hand calculator or simple computer. The solution is an approximation, so that the effects of non-Newtonian fluids and net sand less than fracture height can be included in the calculations.

Equations J12 to Eq. J21 are solved iteratively with an initial guess for the wellbore maximum fracture width. Calculated values of maximum fracture width are used in subsequent iterations until the calculations converge.

$$L = aL_D \qquad (J12)$$

$$w_{wb} = ew_D \qquad (J13)$$

$$\mu_e = 47.880K\left(\frac{80.842q_i}{h_g\bar{w}^2}\right)^{n-1} \qquad (J14)$$

$$t_D = \frac{t}{B} \qquad (J15)$$

$$L_D = 0.5809t_D^{0.6295} \qquad (J16)$$

and

$$w_D = 0.78t_D^{0.1645} \qquad (J17)$$

where

$$\bar{w} = \left(\frac{\pi}{4}\right)^2 w_{wb} \qquad (J18)$$

193

$$B = 1.7737 \times 10^{-4} \left[\frac{(1-v)\mu_e q_i^2}{32C^5 h_g G} \left(\frac{h_g}{h_n} \right)^5 \right]^{2/3} \tag{J19}$$

$$e = 5.0872 \times 10^{-2} \left[\frac{16(1-v)\mu_e q_i^2}{C^2 h_g G} \left(\frac{h_g}{h_n} \right)^2 \right]^{1/3} \tag{J20}$$

and
$$a = 7.4768 \times 10^{-2} \left[\frac{(1-v)\mu_e q_i^5}{256C^8 h_g^4 G} \left(\frac{h_g}{h_n} \right)^8 \right]^{1/3} \tag{J21}$$

2.3.3 Calculation steps

The algorithm for using Eq. J1 through Eq. J21 at a particular pumping time consists of the following steps:

1. Guess an initial maximum wellbore width. Usually 0.10 in. (0.25 cm) will suffice.
2. Calculate overall average fracture width from Eq. J18.
3. Calculate the effective viscosity from Eq. J14.
4. Calculate B from Eq. J19 and t_D from Eq. J15.
5. Use the t_D value to calculate w_D from Eq. J17.

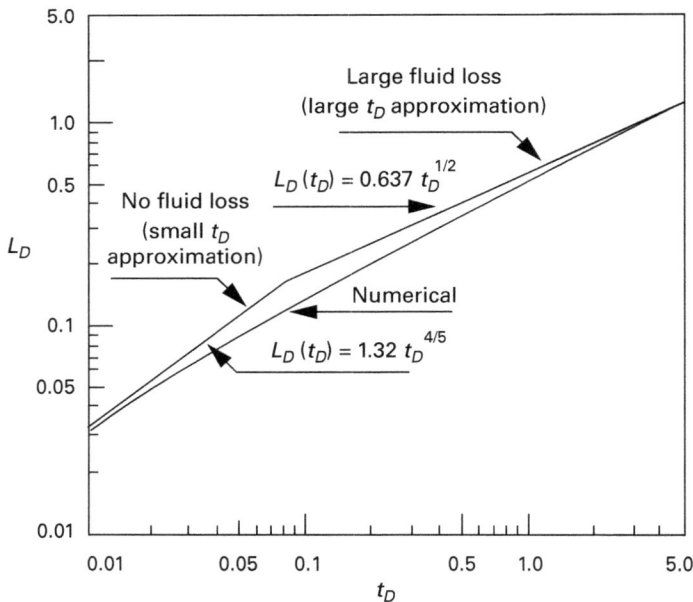

Figure J4 Dimensionless fracture length vs. dimensionless time [12].

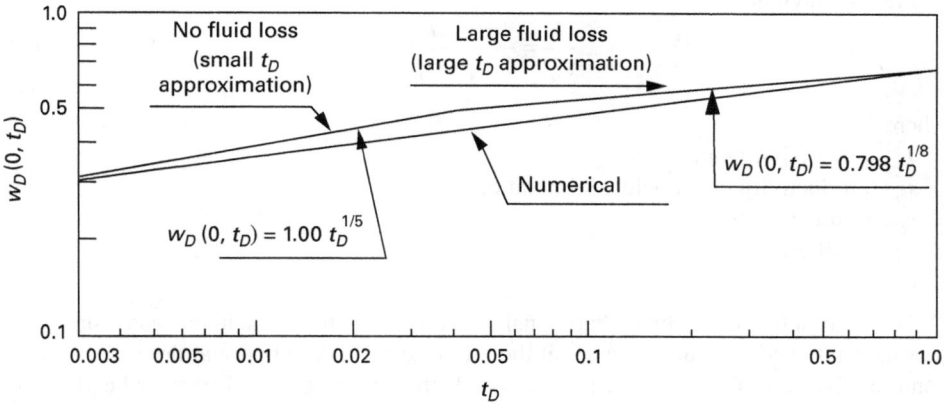

Figure J5 Dimensionless maximum fracture width at well
vs. dimensionless time [12].

6. Use Eq. J20 to calculate e and Eq. J13 to calculate maximum wellbore width, and compare with initial guess. Repeat steps 1 through 6 as necessary until the calculated width value does not significantly change. Use an average of the n and $n - 1$ values of maximum wellbore width for the $n + 1$ guess.

7. Once the wellbore width value is determined, use Eq. J21 to calculate a and Eq. J12 to calculate fracture length.

Equation J14 evaluates the viscosity of a non-Newtonian fluid at the wall shear rate determined by the average fracture width with the full, single-fracture-wing injection rate. It is at best an estimate of the effective viscosity value. For very high fluid-loss rates, a lesser value of flow rate would probably be more representative.

Equations J16 and J17 are log-log linear approximations to the numerical curves presented in Figs. J4 and J5.

Fracture volume is approximately given by Eq. J22, which is based on the assumption that the fracture is strictly elliptical in both the vertical and horizontal planes:

$$V = \overline{w}h_g L \qquad\qquad (J22)$$

2.4 Improvement of productivity index [8]

For both oil and gas wells, the effect of fracturing can be represented conveniently as the ratio of productivity index after and before fracturing, J/J_o.

2.4.1 Prats' method

The most easily applicable technique for determining productivity-index ratio is Prats' method. It is the simplest, but its weakness is the highly idealized conditions of applicability.

Prats found that:

$$\frac{J}{J_o} = \frac{\ln(r_e / r_w)}{\ln(r_e / 0.5L_f)} \tag{J23}$$

where
J productivity index
J_o productivity index before stimulation
r_e drainage radius
r_w wellbore radius
L_f fracture half-length.

The assumptions on which Prats' analytic solution is based include steady-state flow (constant rate and constant pressure at the drainage radius), cylindrical drainage area, incompressible fluid flow, infinite fracture conductivity, and propped fracture height equal to formation height.

▼ Example

A gas well was fractured and then produced at constant bottomhole pressure (BHP) for almost three years. Fracture and formation properties include the following:
r_e = 2 106 ft (642 m)
r_w = 0.354 ft (10.8 cm)
L_f = 500 ft (152 m)

$$\frac{J}{J_o} = \frac{\ln(2\ 106 / 0.354)}{\ln[2\ 106 / (0.5)(500)]} = 4.08$$

The well stabilized at about 490 days; the stabilized PI ratio is about 5.3. The estimate from the Prats' method gives a result in moderate agreement.   ▲

2.4.2 McGuire-Sikora chart

This chart (Fig. J6) is based on the assumptions of pseudosteady-state flow (constant-rate production with no flow across the outer boundary), square drainage area, compressible fluid flow, and a fracture propped throughout the entire productive interval.

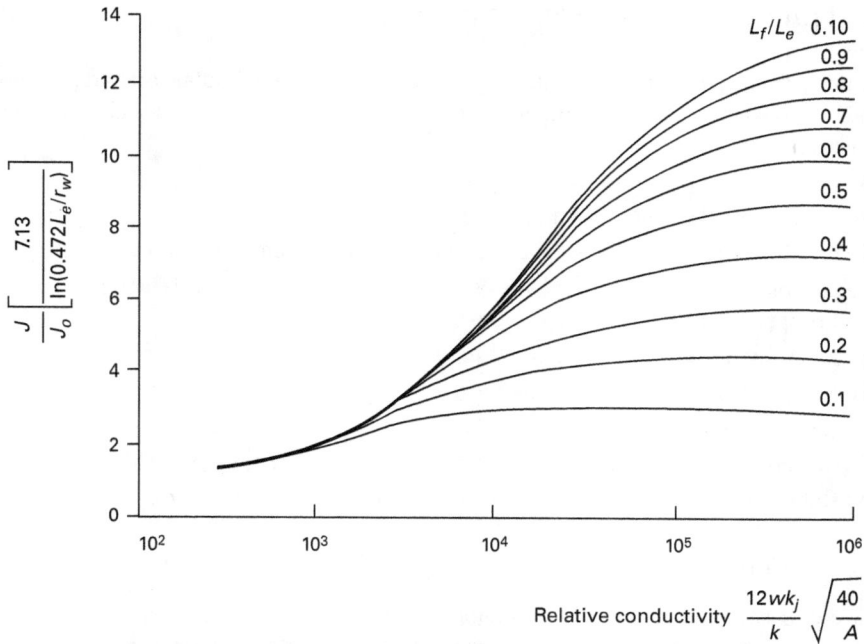

Figure J6 Graph showing increase in productivity from fracturing Holditch's modification of McGuire and Sikora chart [8].

▼ Example

Fracture and formation properties include the additional following characteristics:

A = 320 acres (square)
L_e = 1 867 ft (distance to side of square)
wk_f = 2 200 md-ft
k = 0.1 md

therefore: $L_f/L_e = 500/1\ 867 = 0.268$

and:

$$\frac{12wk_f}{k}\sqrt{\frac{40}{A}} = \frac{(12)(2\ 200)}{(0.1)}\sqrt{\frac{40}{320}} = 9.33 \times 10^4$$

from the Holditch's modification of the the McGuire-Sikora chart:

$$\frac{J}{J_o}\left[\frac{7.13}{\ln(0.472L_e/r_w)}\right] \cong 4.8$$

$$\frac{J}{J_o} = \frac{4.8\ln[(0.472)(1867)/(0.354)]}{7.13} = 5.3$$

The estimate from the modified McGuire-Sikora chart agrees closely with the result from the stimulator. ▲

2.5 Fracturing fluids and additives [8]

Fracturing fluids are pumped into underground formations to stimulate oil and gas production. To achieve successful stimulation, the fracturing fluid must have certain physical and chemical properties. It should:

* Be compatible with the formation material
* Be compatible with the formation fluids
* Be capable of suspending proppants and transporting them deep into the fracture
* Be capable, through its inherent viscosity, to develop the necessary fracture width to accept proppants or to allow deep acid penetration
* Be an efficient fluid (i.e., have low fluid loss)
* Be easy to remove from the formation
* Have low friction pressure
* Have a preparation of the fluid: simple and easy to perform in the field
* Be stable so that it will retain its viscosity through out the treatment.

2.5.1 Water-based fluids

The water-based fluids are used in the majority of hydraulic fracturing treatments today. Typical products available from service companies and comparative costs are shown in Table J1.

Table J1 Comparative costs of polymers [8].

Water-based polymers	Comparative costs
Guar	1.0
Hydroxypropyl guar (HPG)	1.29
Carboxymethylhydroxypropyl guar (CMHPG)	1.40
Carboxymethylcellulose (CMC)	1.62
Hydroxyethylcellulose (HEC)	1.62
Carboxymethylhydroxyethylcellulose (CMHEC)	1.62
Xanthan	2.65

2.5.2 Oil-based fracturing fluid

The most common oil-based fracturing gel available today is a reaction product of aluminum phosphate ester and a base, typically sodium aluminate.

2.5.3 Fracturing-fluid additives

The fracturing-fluid additives are:

* **Biocides**: biocides are used to eliminate surface degradation of the polymers in the tanks.

- **Breakers**: a breaker is an additive that enables a viscous fracturing fluid to be degraded controllably to a thin fluid that can be produced back out of the fracture.
- **Buffers**: common buffering agents are used in fracturing fluids to control the pH for specific crosslinkers and crosslink times.
- **Surfactants and nonemulsifiers**: surfactants are used to prevent to treat near-wellbore water blocks. Surfactants lower the surface tension of the water and reduce capillary pressure.

2.5.4 Fluid-loss additives

The fluid-loss additives are:

- **Foamers**: they are now available for virtually any base fluid from fresh water to high-brine fluids contaminated with large amounts of hydrocarbons to water/alcohol mixtures varying from 0 to 100% methanol.
- **Friction reducers**: the most efficient and cost-effective friction reducers used for fracturing fluids are low concentrations of polymers and copolymers and of acrylamide.
- **Temperature stabilizers**: a basic use for temperature stabilizing is to remove free oxygen from the system. A temperature stabilizer commonly used for this purpose is sodium thiosulfate.
- **Diverting agents**: a diverting agent is typically a graded material that is insoluble in fracturing fluids but soluble in formation fluids. Also included are slurries of resins, viscous fluids, and crosslinked fluids.

2.5.5 Applications

Table J2 introduces hydraulic fracturing applications (*source*: Halliburton).

Table J2   Hydraulic fracturing applications [9].

Symptom or typical problem	Fracturing service or fluid	Application	Properties
Low permeability, limits production or injection rates.	Fracturing process which uses a thickened water formed by addition of a gelling agent.	Oil, gas, or injection wells – sandstone or limestone formations.	Specifically designed to match well characteristics. Fluid usually more available than crude. Additives help protect formation.
Production potential limited by permeability. Wells on wide spacing or in thick zones.	Water base gel with apparent viscosity in excess of 20,000 cp and friction loss properties comparable to that of water.	Oil or gas wells – sandstone and some limestone formations. BHT to 270°F (130°C).	Crosslinked. Low fluid loss. Excellent proppant transport capability. Water base gel with friction properties comparable to water.

(to be continued)

Table J2 *(cont'd)* Hydraulic fracturing applications [9].

Symptom or typical problem	Fracturing service or fluid	Application	Properties
Deep, high temperature, low permeability zones where small tubing limits effective treating rates.	Process including a special gel formula: the viscosity of the gel actually increases during a fracturing treatement, then decreases downhole.	Oil or gas wells – deep, high temperature (to 450°F or 230°C) formations.	Can be formulated to meet wide range of time or temperature requirements. Superior bottom hole viscosity at high temperatures results in wider fractures with better proppant distribution and fewer screen-outs.
Low temperature, low permeability or damaged zones, Need for extended drainge area.	Water base system thickened with a combination of two completely soluble agents.	Oil, gas, waterflood and supply, salt water disposal, Irrigation wells (60-220°F or 15-105°C).	Water clear no-residue gel. Excellent proppant transport qualities. Low viscosity in pipe yet high viscosity in fracture. Especially adapted to water injection wells where recovery of broken gel is not possible. Not affected by high salt concentrations.
Low pressure, low permeability zones where economics will not permit conventional fracturing due to extensive zone thickness. Zone may be lenticular in nature.	Fluid including an exclusive cross-linking agent.	Oil and gas wells.	Viscous water gel, crosslinked for maximum viscosity yield, yet breaks back to thin fluid with internal breaker system. Friction loss properties less than water lowers hydraulic horsepower requirements.
Slow cleanup following treatment, or gas permeability blocked by fluid (water) saturation especially in low pressure, low permeability formations.	Combination of methyl alcohol and treated water.	Low pressure gas wells with low permeability and high fluid saturation.	Viscous alcohol-water base gel with excellent proppant transport properties. Lower surface tension and easy gelation properties of alcohol, yet treated water reduces cost. Friction pressures approximately 50% of water. Gel may be broken to a thin fluid with no residue.
Large vertical zone thickness or water blockage in low pressure, low permeability formations.	Combination of methyl alcohol and treated water.	Gas wells – formations susceptible to water blockage.	Same base composition as above but more stable (to 280°F or 140°C). Able to place higher proppant concentration.

(to be continued)

Table J2 *(cont'd)* Hydraulic fracturing applications [9].

Symptom or typical problem	Fracturing service or fluid	Application	Properties
Low permeability formations or damage at the well bore.	Fracturing process using thickened crude oil or refined oils as the fracturing fluid.	Oil wells – formations contain water sensitive clays.	Thickened crude kerosene or diesel fuel with internal breakers. When diluted by formation crude, will break back to base viscosity.
Low permeability or damaged, water sensitive zones producing some oil or gas condensate.	Viscous oil fracturing fluid.	Oil and gas wells – higher temperature (240°F or 115°C).	Viscous oil gel, up to 80% friction reduction compared to base fluid. May be prepared from many lease crudes or condensates. Controlled viscosity reduction allows wells to be placed on production quicker.
High permeability, high fluid loss zones producing some oil or gas condensate.	Service based on a technique that lubricates the pumped fracturing fluids down the well bore on an outer ring of water.	Oil well – high permeability, high fluid loss formations.	Thick fracturing fluid pumped down the well bore with an outer ring of water at pressures less than water alone. Low fluid loss characteristics aid proppant placement.
Low effective permeability formations or well-bore damage.	It is a polyphase emulsion embodying characteristics not encountered in other fracturing fluid.	Oil and gas wells – tight gas sands with BHT to 350°F (or 175°C).	Highly viscous emulsion with good proppant transport characteristics. Less water contacts formation.

2.6 Proppants

Some of the successful and more commonly used propping agents today include:
(a) Sand
(b) Resin-coated sand
(c) Intermediate-strength proppant (ISP) ceramics
(d) High-strength proppants (sintered bauxite, zirconium oxide, etc.).

2.6.1 Size of proppants

Table J3

Mesh size	Diameter	
	(in.)	(mm)
4	0.187	4.789
6	0.132	3.353
8	0.094	2.387
10	0.079	2.007
12	0.066	1.676
16	0.047	1.194
20	0.033	0.838
40	0.017	0.432
60	0.01	0.254

2.6.2 Typical proppants and their characteristics [19]

The characteristics of some typical proppants are given in Table J4.

Table J4

Proppant	Mesh size	Specific gravity	Porosity (%)
Northern White sand	12/20	2.65	38
	16/30	2.65	39
	20/40	2.65	40
Texas Brown sand	12/20	2.65	39
	16/30	2.65	40
	20/40	2.65	42
Curable resin-coated sand	12/20	2.55	43
	16/30	2.55	43
	20/40	2.55	41
Precured resin-coated sand	12/20	2.55	38
	16/30	2.55	37
	20/40	2.55	37
ISP	12/20	3.17	42
	20/40	3.24	42
ISP-lightweight	20/40	2.63	40
Sintered bauxite	16/20	3.70	43
	20/40	3.70	42
	40/70	3.70	42
Zirconium oxide	20/40	3.16	42

2.6.3 Mechanical properties of proppants

Table J5

Proppant	Specific gravity	Maximum closure stress	
		(psi)	(MPa)
Sand	2.65	6 000	42
Resin-coated sand	2.55		
ISP ceramics	2.7 to 3.3	5 000 to 10 000	35 to 70
High-strength proppants	3.4	>10 000	>70

2.6.4 Propped fracture conductivity [16]

The fracture conductivity *FC* is given as follows:

$$FC = \overline{w}_f k_f \tag{J24}$$

where

$\overline{w}_f$ final average fracture width

k_f permeability of proppant-packed fracture.

Figure J7 shows the permeability of various sand sizes as a function of closure stress.

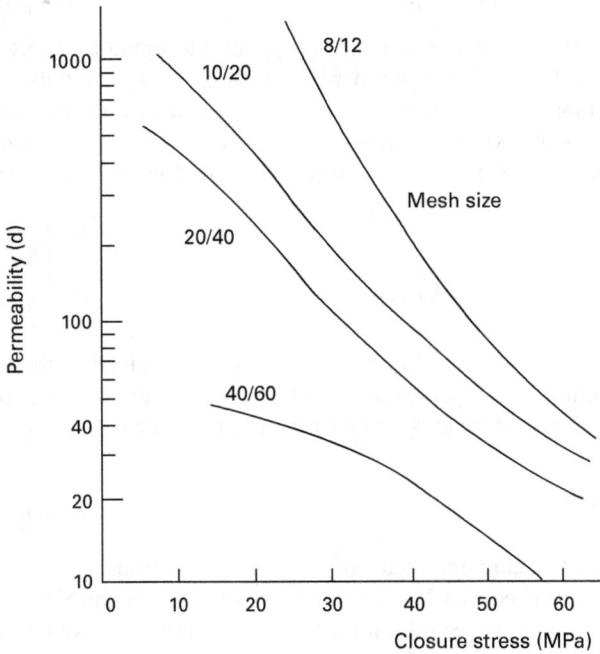

Figure J7 Permeability of various sand sizes vs. closure stress [16].

2.6.5 Fracture permeability

Final fracture permeability is strictly a function of the diameter of the proppant particles used in the treatment:

$$k_f = \frac{d_p^2 \Phi_f^3}{150(1 - \Phi_f)^2} \qquad \text{(J25)}$$

where
  d_p diameter of the proppant particle
  Φ_f porosity of the packed, multilayer bed of proppant particles ($\approx 0.32 - 0.38$).

2.6.6 Fracture width

The final fracture width $\overline{w}_f$ is strictly related to the concentration of proppant in the fracture when it closes:

$$\overline{w}_f = \frac{\overline{w}(t_f)}{\rho_p (1 - \Phi_f)} \left(\frac{m}{1 + m_i / \rho_p} \right) \qquad \text{(J26)}$$

where
  $\overline{w}(t_f)$ average dynamic fracture width at the end of pumping
  ρ_p density of proppant
  $m_i / (1 + m_i / \rho_p)$ mass of proppant per total volume, including both proppant and fluid.

Normally, the injected concentration of proppant will range from 0.12 kg of proppant per liter of fluid to 0.80 kg of proppant per liter of fluid (1 lb/gal to 6 lb/gal) although even larger concentrations have been reported when cross-linked fluids are used. Proppant concentrations in excess of 1 kg/liter of fluid should be used with care, because it may be difficult to get all of the proppant into the fracture if the fluid loss is somewhat more than anticipated.

J3 MATRIX ACIDIZING [16]

In this section, matrix acidizing of both sandstone and carbonate formations is described. This treatment method is defined as the injection of acid into the formation porosity at a pressure less than the pressure at which a fracture can be opened.

3.1 Acid systems

- **Mineral acids**. Most acid treatments of carbonate formations employ hydrochloric acid (HCl). Usually it is used as a 15 wt% solution of hydrogen chloride gas in water. With the development of improved inhibitors, high concentrations have become practical and in some cases concentrations to 30 wt% are used. For sandstone, mixtures of hydrochloric and fluorhydric acids are applied in almost all situations.

- **Organic acids**. The principal virtues of the organic acids are their lower corrosivity and easier inhibition at high temperatures. Only two, acetic and formic, are used to any great extent in well stimulation.

  Mixtures of organic acid and hydrochloric acid are also used. They generally have been designed to exploit the dissolving power economics of HCl while attaining the lower corrosivity (especially at high temperatures) of the organic acids.

- **Powdered acids**. Sulfamic and chloroacetic acids have only limited use in well stimulation, most of which is associated with their portability to remote locations in powdered form. They are white crystalline powders that are readily soluble in water. Generally, they are mixed with water at or near the wellsite.

- **Retarded acids**. The acid reaction rate can be slowed or retarded in a number of ways. The viscosity of the acid can be increased by gelling, thereby slowing the diffusion of acid to the rock surface. Acids in emulsion are of wide use in high temperature reservoirs.

3.2 Stoichiometry of acid–carbonate reactions [16]

3.2.1 Typical reactions

Typical reactions are:

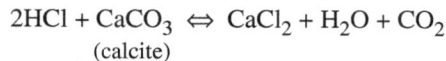

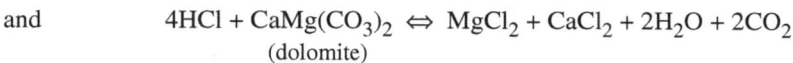

$$2HCl + CaCO_3 \Leftrightarrow CaCl_2 + H_2O + CO_2$$
$$\text{(calcite)}$$

and
$$4HCl + CaMg(CO_3)_2 \Leftrightarrow MgCl_2 + CaCl_2 + 2H_2O + 2CO_2$$
$$\text{(dolomite)}$$

3.2.2 Acid characteristics

Table J6 gives the characteristics of acids used in carbonate.

Table J6 Acid used in carbonate [16].

Category	Molecular weight
Mineral acids	
Hydrochloric (HCl)	36.47
Organic acids	
Formic (HCOOH)	46.03
Acetic (CH$_3$COOH)	60.05

3.2.3 Gravimetric dissolving power

The gravimetric dissolving power β is given by the equation:

$$\beta = \left[\frac{1 \text{ mole } CaCO_3}{2 \text{ moles } HCl} \right] \left[\frac{\text{molecular weight } CaCO_3}{\text{molecular weight } HCl} \right] \left[\frac{\text{mass } HCl}{\text{mass acid solution}} \right] \quad (J27)$$

▼ **Example**

Calculate β for the dissolution of dolomite using a 30 wt% solution of HCl:

$$\beta = \frac{1(184.3)30}{4(36.47)100} = 0.379 \text{ mass of dolomite/mass of acid}$$

▲

3.2.4 Specific gravity of aqueous hydrochloric acid solutions

The specific gravity of aqueous hydrochloric acid solutions are given in Table J7.

Table J7 Specific gravity of aqueous
hydrochloric acid solutions (at 20°C) [16].

Percent HCl	Specific gravity
1	1.0032
2	1.0082
4	1.0181
6	1.0279
8	1.0376
10	1.0474
12	1.0574
14	1.0675
16	1.0776
18	1.0878
20	1.0980
22	1.1083
24	1.1187
26	1.1290
28	1.1392
30	1.1493
32	1.1593
34	1.1691
36	1.1789
38	1.1885
40	1.1980

3.2.5 Volumetric dissolving power

The volumetric dissolving power X is defined as the volume of rock dissolved per volume of acid reacted:

$$X = \beta \frac{\rho_{\text{acid}}}{\rho_{\text{rock}}} \tag{J28}$$

3.2.6 Dissolving power of various acids

Table J8 gives the values of the dissolving power X for the organic acetic and formic acids. The gravimetric dissolving power, β_{100}, refers to a value for an acid having 100% strength. To find for lesser strength solutions, one need only multiply by the weight fraction of acid in the solution.

Table J8 Dissolving power of various acids [16].

Formation	Acid	β_{100}	X			
			5%	10%	15%	30%
Limestone: CaCO$_3$ ρCaCO$_3$ = 2.71 g/cm^3	Hydrochloric (HCl)	1.37	0.026	0.053	0.082	0.175
	Formic (HCOOH)	1.09	0.020	0.041	0.062	0.129
	Acetic (CH$_3$COOH)	0.83	0.016	0.031	0.047	0.096
Dolomite: CaMg(CO$_3$)$_2$ ρCaMg(CO$_3$)$_2$ = 2.87 g/cm^3	Hydrochloric	1.27	0.023	0.046	0.071	0.152
	Formic	1.00	0.018	0.036	0.054	0.112
	Acetic	0.77	0.014	0.027	0.041	0.083

▼ **Example**

Calculate the volume of 10 wt% formic acid required to increase the permeability of a limestone formation 10 m thick by a factor of 10 in a zone 1.5 m in radius around the wellbore. The wellbore radius is 0.2 m and the permeability response of the limestone is given by:

$$\frac{k}{k_0} = \left(\frac{\Phi}{\Phi_0}\right)^{10}$$

where k_0 and Φ_0 are the original permeability and porosity, respectively.
In this example $\Phi_0 = 0.1$.

Solution

Since
$$\frac{k}{k_0} = 10 = \left(\frac{\Phi}{\Phi_0}\right)^{10} = \left(\frac{\Phi}{0.1}\right)^{10}$$

we find $\Phi = 0.1266$.
Thus, to increase the porosity of the rock from $\Phi_0 = 0.1$ to $\Phi_0 = 0.126$, the volume of rock to be dissolved is:

$$V = \pi\left[(1.5)^2 - (0.2)^2\right](0.126 - 0.1)(10) \quad \text{or} \quad V = 1.80 \text{ m}^3 \text{ of rock}$$

From Table J8 we see that the volumetric dissolving power is:

$$X = 0.041 \text{ m}^3 \text{ of rock/m}^3 \text{ of acid}$$

Therefore: acid volume = 1.80/0.041 = 44 m^3 ▲

3.3 Stoichiometry of acid–sandstone reactions [16]

Acidizing treatments in sandstone formations normally employ a mixture of HCl and HF. Equations describing the reaction of HF with SiO_2 are as follows:

$$SiO_2 + 4HF \Leftrightarrow SiF_4 + 2H_2O$$

and:

$$SiF_4 + 2HF \Leftrightarrow H_2SiF_6$$

J

▼ Example

Calculate the volumetric dissolving power of a 3 wt% HF and 12 wt% HCl solution, assuming that the formation is composed of quartz. Take the predominant reaction product to be SiF_4.

Solution

The gravimetric dissolving power is given by:

$$\beta_{100} = \left[\frac{1 \, \text{mole} \, SiO_2}{4 \, \text{moles HF}}\right]\left[\frac{1 \, \text{mole HF}}{20 \, \text{kg HF}}\right]\left[\frac{60 \, \text{kg} \, SiO_2}{\text{mole} \, SiO_2}\right]$$

giving:

$$\beta_{100} = 0.75 \, \text{kg} \, SiO_2/\text{kg HF}$$

Taking $\rho SiO_2 = 2\,420$ kg/m^3 and $\rho HF = 1\,070$ kg/m^3 (same as 15 wt% HCl, see Table J7), then:

$$X = (0.75)\left(\frac{3}{100}\right)\left(\frac{1\,070}{2\,420}\right) = 0.010 \, \text{m}^3 \, \text{of rock/m}^3 \, \text{of acid}$$

▲

3.4 Well preparation

It is advisable to remember that before a fluid enters the formation, it is pumped from surface containers and through tubing. Unless all the materials that come into contact with the acid are thoroughly cleaned, acid-insoluble solids deposited on the tubing walls (oxide scale, pipe dope, paraffin, asphaltenes, etc.) will all be carried into the formation creating further damage. It is highly recommended that, to wash the tubing, acid be pumped down the tubing near the bottom and subsequently produced back into the waste pit before injecting into the formation. If the reservoir pressure is too low to lift the acid out of the tubing, a foamed acid should be used. Futhermore, the proper spotting of the acid is crucial. To be successful, the acid must be diverted into the damaged zone.

3.5 Additives

To ensure that the acid is diverted into the damaged zones, a diverting agent consisting of particulate matter to be subsequently removed is often added to the acid. These agents may consist of benzoic acid powder or oil-soluble resins. Both of these components are soluble in crude oils and will presumably dissolve when the well is put in production.

3.6 Which acid to apply

In a matrix, acidizing various acids may be appropriate depending on the particular situation. Table J9 has been prepared as a guide for acid selection.

Table J9 Acid use guidelines [16].

Situation	Acid to apply
Carbonate acidizing	
Perforating fluid	5% acetic acid
Damaged perforations	(a) 9% formic acid
	(b) 10% acetic acid
	(c) 15% HCl
Deep wellbore damage	(a) 15% HCl
	(b) 28% HCl
	(c) emulsified HCl
Sandstone acidizing	
HCl solubility > 20%	Use HCl only
High permeability (100 md plus)	
high quartz (80%), low clay (< 50%)	10% HCl – 3% HF[a]
high feldspar (> 20%)	13.5% HCl – 1.5% HF[a]
high clay (> 10%)	6.5% HCl – 1% HF[b]
high iron chlorite clay	3% HCl – 0.5% HF[b]
Low permeability (10 md or less)	
low clay (< 5%)	6% HCl – 1.5 HF[c]
high chlorite	3% HCl – 0.5% HF[d]

HF[a]: preflush with 15% HCl
HF[b]: preflush with sequestered 5% HCl
HF[c]: preflush with 7.5% HCl or 10% acetic acid
HF[d]: preflush with 5% acetic acid.

J4 ACID FRACTURING [5]

Acid fracturing is a well stimulation process in which acid, usually hydrochloric acid (HCl), is injected into a carbonate formation at a pressure sufficient to fracture the formation or to open existing natural fractures. As the acid flows along the fracture, portions of the fracture face are dissolved. Since flowing acid tends to etch in a nonuniform manner, conductive channels are created which usually remain when the fracture closes. The effective length of the fracture is determined by the volume of acid used, its reaction rate, and the acid fluid loss from the fracture into the formation. The effectiveness of the acid fracturing treatment is largely determined by the length of the etched fracture.

Acid fracturing is normally limited to limestone or dolomite formations. However, treatments have been successful in some sandstone formations containing carbonate-filled natural fractures. Fluid loss is a greater problem when using acid and is very difficult to control.

4.1 Factors controlling the effectiveness of acid fracturing treatments

The two major factors controlling the effectiveness of acid fracturing treatments are the resulting fracture length w_a and conductivity.

The effective fracture length is controlled by the acid fluid-loss characteristics, the acid reaction rate, and by the acid flow rate in the fracture.

Ideal conductivity can be estimated by the following equations:

$$wk_{f\max} = 7.8 \times 10^{12} \left(\frac{w_a}{12} \right)^3 \tag{J29}$$

where w_a is in inches and $wk_{f\max}$ is in millidarcy-ft, or:

$$wk_{f\max} = 8.40 \times 10^{10} w_a^3 \tag{J30}$$

where w_a is in meters and $wk_{f\max}$ is in darcy-meters.

4.2 Treatment design

Most of the treatement designs use acid gelled with polymer or surfactant. The products have the advantage of limiting the fluid loss from the fracture walls and of decreasing the dissolution reaction rates.

Table J10 shows a comparison of actual reaction data, measured under flowing conditions, for ungelled acid vs. acid gelled with polymer or surfactant gelling agents.

Table J10 Effect of acid viscosity on reaction rate of flowing acid [5].

Gelling agent	Acid concentration (C) (gmole/1 liter)		ΔC	Velocity	Change in reaction rate vs. ungelled acid
type	initial	final		(cm/s)	(%)
None	0.165	0.150	0.015	1.20	+ 33
Polymer	0.165	0.147	0.018	1.33	
None	0.319	0.282	0.037	1.14	+ 14
Polymer	0.319	0.283	0.036	1.33	+ 16
Surfactant	0.315	0.279	0.036	1.32	
None	0.980	0.898	0.082	1.08	+ 3
Polymer	0.980	0.906	0.070	1.30	− 8
Surfactant	0.976	0.911	0.065	1.26	

Note: Reaction rate comparisons are adjusted for differences in reaction time resulting from variations in acid flow rates.

J5 FRAC PACKS [11, 15]

A frac pack is a hydraulically induced propped fracture that is followed, without interruption, by a circulating gravel pack to bypass the near-wellbore damage caused by operational procedures.

5.1 Producing mechanism [15]

A frac pack also creates a large amount of surface area outside of the well bore. The large surface area of the propped fracture compared to perforation tunnels results in a much lower flow velocity at the interface. The typical dimensions of these fractures are about 30 m (100 ft) high and 15 m (50 ft) long, which reduces fine particle movement and minimizes turbulence by providing sufficient surface area at the fracture face for very low fluid velocity into the fracture.

The short fractures of frac packed wells result in low drawdown, with fluid flow into the well bore distributed throughout the entire pay interval. The reduced near-wellbore drawdown during a well's production is a key element to improving production efficiencies. As drawdown—which is the driving force for flow into the well bore—increases from higher production rates or depletion, formation instability can cause fines and sand to migrate into the wellbore region. Fracturing beyond the wellbore region bypasses the damaged zone, increasing the effective radius of the wellbore and enabling higher flow rates with lower drawdown pressures. Consequently, the reservoir energy is used more efficiently because the conductive proppant bed bypasses the near-wellbore region. Operators have reported not only accelerated production rates, but also decreased abandonment pressures as a result of the low drawdown associated with frac packs.

In addition, hydraulically fractured wells exhibit linear flow characteristics rather than the typical radial flow. At comparable flow rates, the pressure drop for linear flow is less than the radial flow, reducing turbulence and improving recovery. When the dominant flow regime change from radial to linear in the near wellbore region, fines migration into the screen is also mitigated.

5.2 Frac pack applications [11]

Table J11 Completion type selection guidelines [11].

	Well condition
Reservoir conductivity	Average reservoir permeablity < 10 millidarcy for gas well or < 50 millidarcy for oil well
Fracturing constraints	Individual sand interval, k_h > 12 darcy-meters w/initial skin > 10 k_h > 12 darcy-meters and high-viscosity crude ($\approx$ 20° API or less) k_h for any sand interval from 5 to 12 darcy-meters w/skin > 5
Fracture growth	Thinly bedded sand/shale when shale barrier between pay and water sand > 6 meters thick
Operational constraints	Severely overpressured Severely underpressured and damaged Severely underpressured and low initial skin

5.3 Guidelines for screenless frac pack completions

The role of the proppant pack has changed from providing stimulation and reduce drawdown to supporting the perforations and formation in the near-wellbore region as stresses caused by depletion increase. Stability of the perforations can be maximized by the following:

- Orienting perforations in the direction of maximum principal stress to maximize the stability of the perforation tunnels.
- Stabilizing the proppant pack, by use of deep-penetrating charges that result in less damage at the perforation face, a more stable perforation tunnel, and a smaller diameter hole.
- Restricting perforation intervals to competent rock with unconfined compressive strengths of 15 MPa (2 200 psi) or greater, then fracturing to establish communication with high-permeability, weak formation intervals.
- Consolidating the near-wellbore region before stimulation.

The proppant pack stability can be improved by the following:

- Creating a high-strength proppant pack that will withstand stresses imposed during depletion and provide continuous support for the perforation tunnel and fracture.

- Performing a squeeze job with liquid-resin-coating (LRC) treated proppant after the main fracture treatment by use of a pinpoint injector tool to ensure all perforations are filled and packed tightly.
- Minimizing formation drawdown by fracturing to bypass near-wellbore damage.
- Bringing the well on production slowly during cleanup to reduce initial stress and allow the formation sand to form stable bridges.

J6 PERFORATING REQUIREMENTS FOR FRACTURE STIMULATIONS [1]

6.1 Penetration depth

- Perforations need to penetrate only 4 to 6 inches into the formation.
- The minimum casing-hole diameter should be 8 to 10 times the proppant diameter.

6.2 Perforated interval

The perforated-interval length should be limited even when the perforated portion of the well is nominally with the preferred fracture plane.

6.3 Shot density and hole diameter

The number of perforations in contact with the fracture determines the average injection rate per perforation.

- For 0 to 180° phased guns, all perforations should contribute to the fracture.
- Only two-thirds of the perforations from a 120° phased gun are likely to communicate with the fracture.
- And only one-third of the perforations from a 60° phased gun are likely to be effective.

6.4 Frac packs

- A gun with shots phased at 12, 16, and 21 shots per foot should be used.
- The frac-packed interval should not exceed approximately 50 ft to achieve a minimum injection rate per perforation.

REFERENCES

1   Behrmann LA, Nolte KG (1998) Perforating Requirements for Fracture Stimulations. *Paper SPE 39453* presented at the SPE International Symposium on Formation Damage Control held in Lafayette, Louisiana

2   Chambre Syndicale de la Recherche et de la Production du Pétrole et du Gaz Naturel. Comité des techniciens (1983) *Manuel d'acidification des réservoirs*. Editions Technip, Paris

3   Charlez PA (1997) *Rock Mechanics*, vol. 2, *Petroleum Applications*. Editions Technip, Paris

4   Dowell Schlumberger (1982) *Field Data Handbook*

5   Economides MJ, Nolte KG (1987) *Reservoir Stimulation*. Schlumberger Educational Services

6   Gay L, Hentz A (1970) *Formulaire du producteur*. Editions Technip, Paris

7   Gay L, Sarda JP, Roque C (1984) *Séminaire Fracturation et stimulation des puits*. BEICIP

8   Gidley JL, Stephen SA, Nierode DE, Veatch RW Jr. (1989) *Recent Advances in Hydraulic Fracturing*. Monograph Series, SPE, Dallas, Texas.

9   Halliburton Cie, *Technical Data Sheet*

10 Le Tirant P, Gay L (1972) *Manuel de fracturation hydraulique*. Editions Technip, Paris

11 Mathis SP, Saucier RJ (1997) Water-fracturing vs. Frac-packing: Well Performance Comparison and Completion Type Selection Criteria. *Paper SPE 38593,* Ann. SPE Tech Conf.

12 Nordgren RP (1972) *Propagation of a Vertical Hydraulic Fracture*. SPEJ (August 1972) 306-314

13 Nguyen PD, Dusterhoft RG, Dewprashad BT, Weaver JD (1998) New Guidelines for Applying Curable Resin-Coated Proppants. *Paper SPE 39582*, SPE, Lafayette, Louisiana

14 Perrin D, Caron M, Gaillot G (1995) *La production fond*. Editions Technip, Paris

15 Frack Packs: A Specialty Option or Primary Completion Technique? (1997) *Petro Eng Int,* V.70, No. 3 (Suppl), March 1997

16 Schechter RS (1992) *Oil Well Stimulation*. Prentice Hall Inc, New Jersey

17 Valko P, Economides MJ (1995) *Hydraulic Fracture Mechanics*. John Wiley & Sons

18 Williams BB, Gidley JL, Schechter RS (1979) *Acidizing Fundamentals*. Monograph Series, SPE, Dallas, Texas

19 Economides MJ, Hill AD, Ehlig-Economides C (1994) *Petroleum Production Systems*. PTR, Prentice Hall Inc., Englewood Cliffs, New Jersey

20 Lietard 0, Ayoub J, Pearson A (1996) Hydraulic Fracturing of Horizontal Wells: An Update of Design and Execution Guidelines. *Paper SPE 37122* presented at the 2nd International and Exhibition on Horizontal Well Technology held in Calgary, Alberta, Canada.

K

Horizontal and Multilateral Wells

K

Horizontal and Multilateral Wells

K1 HORIZONTAL WELL PRODUCTIVITY [1, 2, 3]

1.1 Presentation

A horizontal well of length L penetrating a reservoir with horizontal permeability k_h and vertical permeability k_v creates a drainage pattern that is different from that a vertical well. Fig. K1 presents this drainage pattern, together with important variables affecting well performance. The drainage shape formed is ellipsoidal, with the large half-axis of the drainage ellipsoid, r_{eh}, related to the length of the horizontal well (Fig. K2).

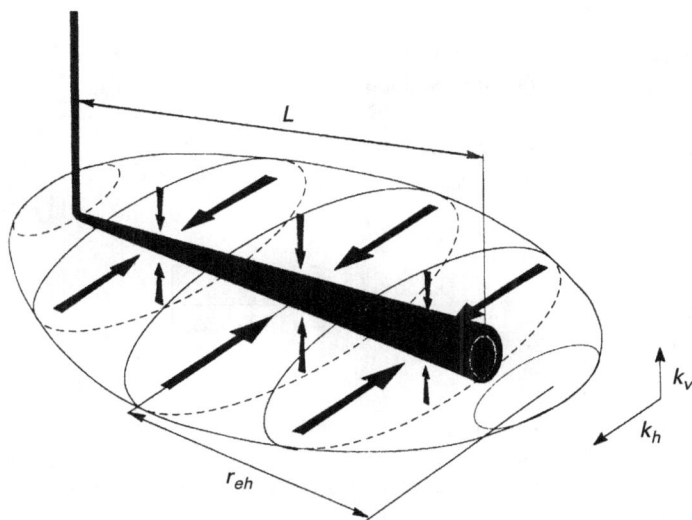

Figure K1 Drainage pattern formed around a horizontal well [3].

217

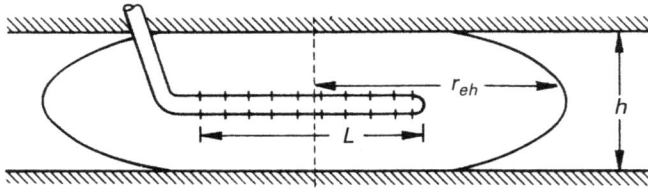

Figure K2 A schematic of horizontal well drainage area [2].

1.2 Steady-state productivity [2]

Generally the length L of a horizontal well is significantly longer than the reservoir thickness h, i.e., $L \gg h$, and, in generalized units, the flow rate q_h is given by:

$$q_h = \frac{2\pi k_h h \Delta p}{B\mu} \left[\frac{1}{\ln(4r_{eh} / L)} \right] \tag{K1}$$

- *In practical U.S. field units*

$$q_h = \frac{0.007078 k_h h \Delta p}{B\mu} \left[\frac{1}{\ln(4r_{eh} / L)} \right] \tag{K2}$$

where
  q_h flow rate (st.bbl/D)
  k_h horizontal permeability (md)
  h reservoir thickness (ft)
  Δp pressure drop from the drainage radius to the wellbore (psi)
  B formation volume factor (res.bbl/st.bbl)
  μ oil viscosity (cp)
  r_{eh} horizontal well drainage radius (ft)
  L horizontal well length (ft).

- *In practical field metric units*

$$q_h = \frac{0.536 k_h h \Delta p}{B\mu} \left[\frac{1}{\ln(4r_{eh} / L)} \right] \tag{K3}$$

where
  q_h flow rate (m³/d)
  k_h horizontal permeability (10^{-3} μm² ≈ md)
  h reservoir thickness (m)
  Δp pressure drop from the drainage radius to the wellbore (MPa)
  B formation volume factor (m³/m³)
  μ oil viscosity (mPa.s)
  r_{eh} horizontal well drainage radius (m)
  L horizontal well length (m).

K

▼ **Example**

A 1 000 ft-long horizontal well is drilled in a reservoir with the following characteristics:

k_h = 75 md
B = 1.34 res.bbl/st.bbl
h = 30 ft
μ = 0.62 cp
r_{eh} = 1 053 ft.

The productivity index J_h can be obtained by dividing q_h by Δp:
$J_h = q_h/\Delta p$

$$J_h = \frac{0.007078 \times 75 \times 30}{1.34 \times 0.62}\left[\frac{1}{\ln(4 \times 1053/1000)}\right]$$

J_h = 13.3 st.bbl/(D-psi)
J_h = 336.3 m³/(d.MPa). ▲

1.3 Pseudosteady-state productivity [5]

There is an easy-to-use equation for calculating the productivity of a horizontal well in a pseudosteady-state regime. It preserves the form of the most familiar flow equation for a vertical well.

1.3.1 Physical model

The physical model in a rectangular porous medium is shown in Fig. K3.

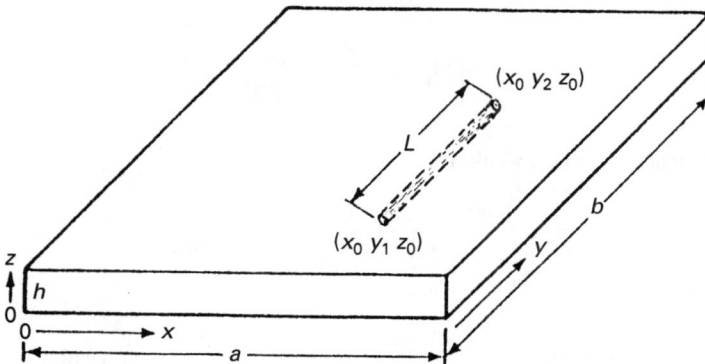

Figure K3 Physical model [5].

The following nomenclature is used in Fig. K3:

r_w radius of well
L horizontal length of well
h thickness of the drainage volume
a length of the drainage volume (x direction)
b width of the drainage volume (y direction).

1.3.2 Pseudosteady-state flow equation

To preserve the form of the most familiar flow equation for a vertical well[1], an easy-to-use
equation is chosen for calculating the productivity of a horizontal well:

$$q = \frac{0.007078b\sqrt{k_xk_z}\left(\bar{p}_R - p_{wf}\right)}{B\mu\left[\ln\dfrac{\sqrt{A}}{r_w} + \ln C_H - 0.75 + s_R\right]} \tag{K4}$$

where

q constant rate (uniform flux) (st.bbl/D)
b extension of drainage volume of horizontal well in y direction (ft)
k_x permeability in x direction (md)
k_z permeability in z direction (md)
$\bar{p}_R$ average pressure in drainage volume of well (psi)
p_{wf} average flowing bottomhole pressure (psi)
B formation volume factor (res.bbl/st.bbl)
μ viscosity (cp)
A vertical section of drainage area of horizontal well, $a \times h$ (ft^2)
C_H geometric factor defined by Eq. K7
s_R skin resulting from partial penetration.

The productivity index J (st.bbl/D/psi) results from Eq. K4:

$$J = \frac{0.007078b\sqrt{k_xk_z}}{B\mu\left[\ln\left(C_H\dfrac{\sqrt{A}}{r_w}\right) - 0.75 + s_R\right]} \tag{K5}$$

---

1. Productivity index of vertical wells [16]:

$$J_v = \frac{0.007078h\sqrt{k_xk_y}}{B\mu\left[\ln\dfrac{0.565\sqrt{ab}}{r_w} - 0.75\right]} \quad \text{(st.bbl/D/psi)} \tag{K6}$$

where

h	formation thickness (ft)	a	reservoir length (ft)
k_x	permeability in x direction (md)	b	wellbore width (ft)
k_y	permeability in y direction (md)	r_w	wellbore radius (ft).

Calculation of $\ln C_H$

C_H is a shape factor ; it is a function of the scaled aspect ratio $a\sqrt{k_z}/h\sqrt{k_x}$, of the values of k_x, k_y, and of k_z, and of the well location, and results from the following equation:

$$\ln C_H = 6.28\frac{a}{h}\sqrt{k_z/k_x}\left[\frac{1}{3} - \frac{x_0}{a} + \left(\frac{x_0}{a}\right)^2\right]$$

$$- \ln\left(\sin\frac{180°z_0}{h}\right) - 0.5\ln\left[\frac{a}{h}\sqrt{k_z/k_x}\right] - 1.088 \tag{K7}$$

where x_0 and z_0 are the coordinates of the center of the well in the vertical plane (Fig. K3).

Values of s_R

The skin, s_R, is a function of k_y. Thus, k_y affects J through s_R.
- If $L/b = 1$, $s_R = 0$ and k_y has no effect.
- There is no bottom water and gas cap.
- As L/b decreases, the effect of k_y increases. For example, a decrease in k_y by a factor of 4 decreases J by a factor of 1.8 for $L/b = 0.5$, and by a factor of 2.8 for $L/b = 0.25$.

K2 PRESSURE DROPS IN HORIZONTAL WELLS [7]

In a single phase flow, simple formulas can be derived under the assumption that inflow is uniform along the well. This is appropriate in cases in which:
- Friction is not severe.
- There is no bottom water and gas cap.
- The reservoir is homogeneous or nearly.
- The entire horizontal section is open to the formation.

2.1 Laminar flow

With uniform inflow of liquid, the pressure drop in horizontal wells is:

$$\Delta p_w = 1.920 \times 10^{-10}\frac{\mu_{bh}BQ_{sc}L}{d^4} \tag{K8}$$

where
Δp_w well drawdown (psi)
μ_{bh} bottomhole viscosity (cp)
B formation volume factor (res.bbl/st.bbl)
Q_{sc} production rate in standard conditions (st.bbl/D)
L length of well (ft)
d diameter of well (in.).

2.2 Turbulent flow

In the case of liquids, the pressure drop in horizontal wells is:

$$\Delta p_w = \frac{6.158 \times 10^{-14}}{\log^2(\varepsilon/3.7d)} \frac{\rho_{bh} B^2 Q_{sc}^2 L}{d^5} \tag{K9}$$

where
- ρ_{bh} bottomhole density (lbm/ft^3)
- ε/d normalized wall roughness.

2.3 Application to field data

Figure K4 gives the curves that bound regions in which friction is negligible in oil recovery.

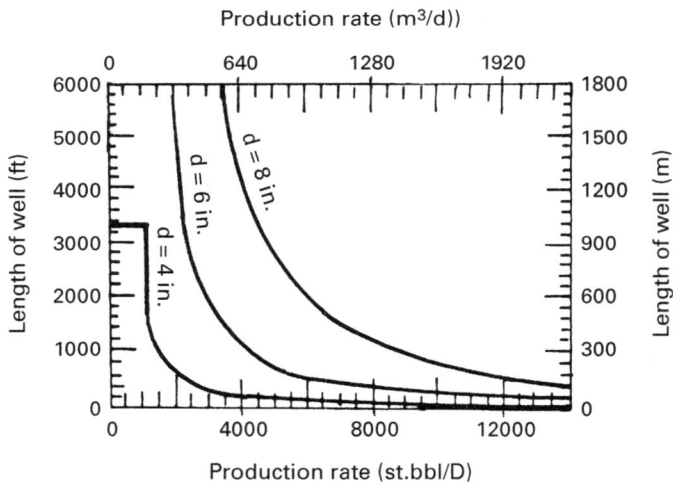

Figure K4 Regions in which friction is negligible in oil recovery [7].

A key factor is the ratio of wellbore pressure drop to drawdown at the producing end. A conservative rule-of-thumb is that, should this ratio exceed 10–15%, then wellbore friction can reduce productivity by 10% or more. Susceptible to this loss are oil wells that produce more than 1 500 st.bbl/D (240 m^3/d) and gas wells that produce more than 2 MMscf/D (50 000 m^3/d).

K3 FLOW PATTERNS IN HORIZONTAL WELLS [4]

3.1 Classification

Figure K5 represents the flow patterns in horizontal flow.

3.1.1 Stratified flow

Liquid flows along the bottom of the pipe with gas along the top. Both phases are continuous in the axial direction. Two sub-patterns are defined: stratified smooth and stratified wavy.

3.1.2 Intermittent flow

Plugs or slugs of liquid are separated by gas zones which overlay a stratified liquid layer flowing along the bottom of the pipe. The intermittent pattern is sometimes subdivided into slug and elongated bubble patterns. The elongated bubble pattern should be considered the limiting case of intermittent flow when the liquid slug is free of entrained gas bubbles.

K

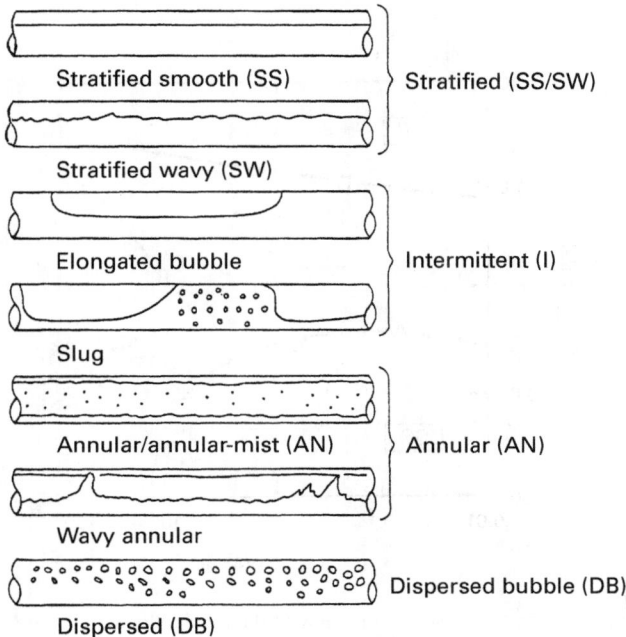

Stratified smooth (SS)   ⎫
                          ⎬ Stratified (SS/SW)
Stratified wavy (SW)   ⎭

Elongated bubble   ⎫
                          ⎬ Intermittent (I)
Slug   ⎭

Annular/annular-mist (AN) ⎫
                          ⎬ Annular (AN)
Wavy annular   ⎭

Dispersed (DB)   — Dispersed bubble (DB)

Figure K5 Flow patterns in horizontal flow [4].

3.1.3 Annular flow

The liquid exists as a continuous film around the perimeter of the pipe and is also continuous axially thus forming an annulus. Two patterns exist:

- **Annular or annular-mist flow pattern**: condition where the film thickness at the top is fairly steady with time.

- **Wavy-annular flow pattern**: large aerated waves moving along the bottom of the pipe are high enough to wet the top surface.

3.1.4 Dispersed bubble flow

The gas phase is distributed as discreet bubbles in an axially continuous liquid phase. The concentration of bubbles is higher near the top of the pipe, but as the liquid rate increases, the bubbles are dispersed more uniformly.

K 3.2 Location of transition mechanisms

Figure K6 displays the resulting generalized pattern map. Once the physical properties and tube diameter are specified, the only remaining variables are the two superficial velocities and the transitions can be mapped by U_{LS} and U_{GS} coordinates.

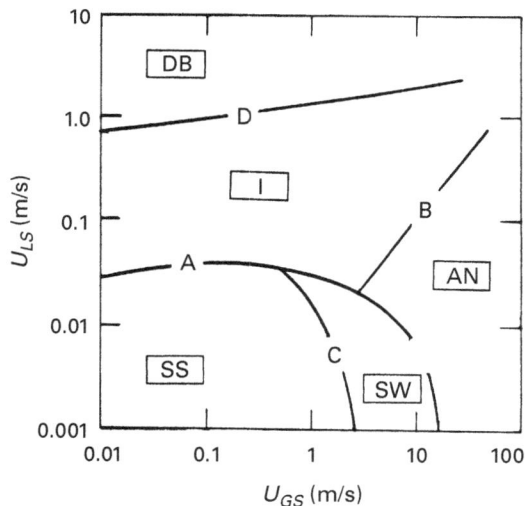

Figure K6 A schematic view of the relative location of transition mechanisms [4].

The following nomenclature is used in Fig. K6:

U_{GS} superficial gas velocity (m/s)
U_{LS} superficial liquid velocity (m/s)
SS   stratified smooth
SW stratified wavy
I   intermittent
AN annular
DB dispersed bubble

with the transition boundary as shown in the table below.

Transition boundary	Patterns	Mechanism
A	Stratified to non stratified	
B	Intermittent to annular	h/D ≥ 0.35
C	Stratified smooth to wavy	Wind-wave interactions
D	Intermittent to dispersed bubble	Turbulent fluctuations vs. buoyancy forces

K4 MULTILATERAL WELLS [11–14]

4.1 Introduction

With the wide range of multilateral (ML) complex wells drilled, development of a common classification system has considerable added value during the planning phase of a ML well. The main benefits are:

- **Determination of functional requirements.** It is agreed that the determination of functional requirements of a proposed ML well is one of the key success factors in delivering a well that meets its objectives. A classification system provides a "road map" which allows well and petroleum engineers to efficiently achieve this.
- **Utilization of the most appropriate system.** With functional determined requirements, a classification "code" enable the comparison of well requirements and capabilities of various systems on the market.
- **Transfer of learning.** A classification code is relevant with a comparison of case histories and performance indicators.

4.2 Multilateral well classification [11, 12, 19]

Figure K7 proposes a multilateral well classification where six levels are defined.

The following demonstrates the progression of multilateral completion support complexity [19].

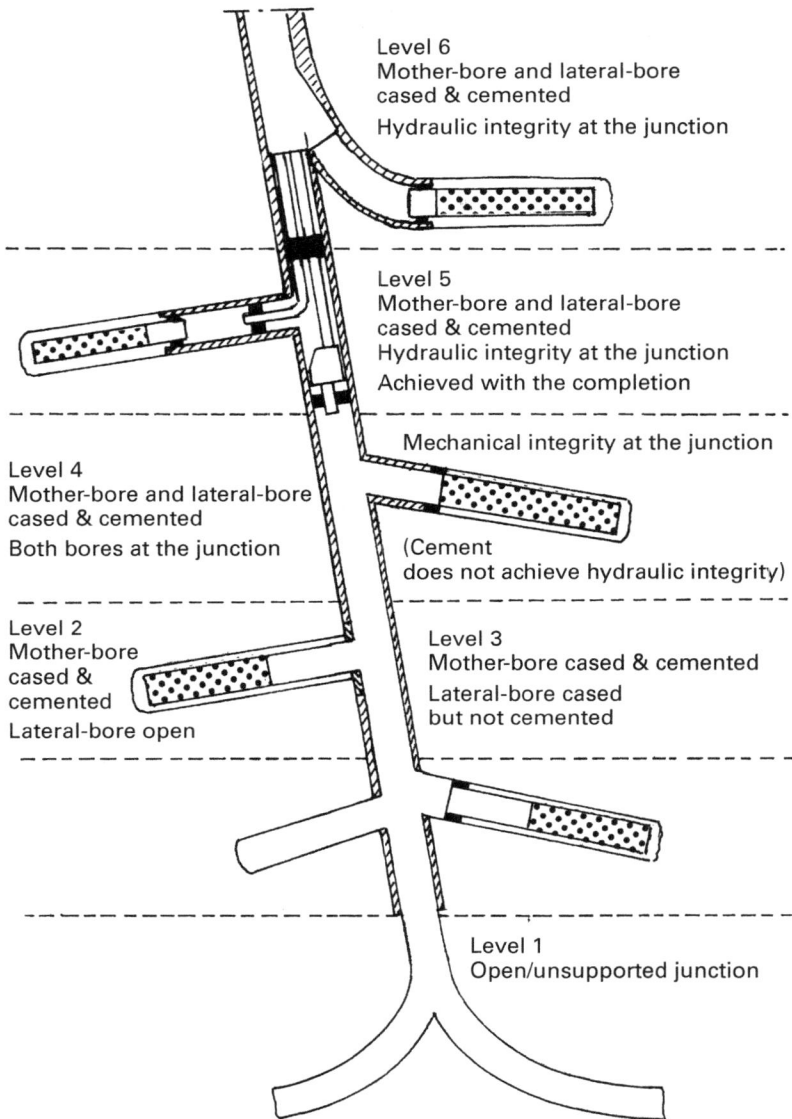

Figure K7 Multilateral well classification [12].

- **Level 1.** Openhole laterals from an openhole mother bore require no mechanical or hydraulic junction. Thus, they are usually carried out in consolidated formations as barefoot completions.

- **Level 2.** Multilaterals in which the main bore is cased and cemented with the lateral open represent a significant step in complexity. The completion is economical, allows selective production and can be carried out in standard casing sizes.

- **Level 3.** Main bore is cased and cemented and lateral bore is cased but not cemented. Once again, continual improvement has led to the development of new liner hangers and whipstock system modifications which have, in turn, improved the final completion.

- **Level 4.** The main bore and lateral are cased and cemented to provide mechanical junction integrity. These system can be simple, or they can be the basis for more complex systems, such as dual packer completions, single string selective re-entries and single strings with lateral entry nipples. The main wellbore is drilled and cased; a window is milled; the lateral is drilled and the whipstock is recovered.

- **Level 5.** When possible, junction kick-off points for multilateral wells should be located in a strong, competent, consolidated formation. However, economic, geologic or drilling conditions often preclude this ideal scenario. In these cases, pressure integrity is necessary to prevent junction collapse due to drawdown. In a level 5 multilateral, full hydraulic and mechanical pressure integrity at the junction is achieved with the completion.

- **Level 6.** The completion system used for this application achieves junction pressure integrity with casing and is developed to address needs raised through use of level 5 system, most notably:
  - Elimination of debris
  - Risk reduction
  - Simpler installation
  - Top-down construction.

4.3   Gas and water cresting situations

4.3.1 Gas cresting situations [13]

Multilateral wells can be envisioned as a good production tool to recover oil in reservoirs in the presence of a gas cap. The gain in oil recovery using a multilateral well instead of a pattern of several parallel horizontal wells can be very significant with recovery ratio increasing as the well length exposed to the reservoir is increased.

4.3.2 Water cresting situations [14]

Multilateral wells can be envisioned as a good production tool to recover in reservoirs that experience bottom water coning phenomenon. The gain in oil recovery using a multilateral well instead of a pattern of several parallel horizontal wells can be very significant with recovery ratio greater than 2. This recovery ratio increases as the reservoir depth increases.

Interferences between the laterals and the main hole have a very slight impact on the performance of a multilateral well.

4.4 Downhole intelligent oil water separation in multilateral applications [17, 18]

Multilateral wells permit to reinject produced water without transporting it to the surface. Fig. K8 illustrates this ability.

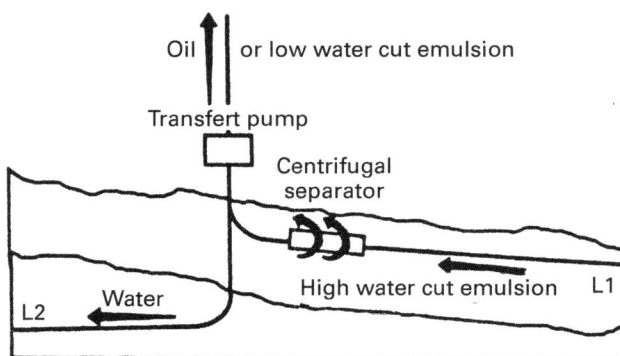

Figure K8 Multilateral well completed to accomodate reinjection of produced water [17]. (© John Wiley & Sons Limited. Reproduced with permission).

Lateral L1 is drilled into the pay zone and L2 is drilled into the water zone. Oil and water will be produced from L1 and will be fed into a downhole separator. The oil (or low watercut emulsion) is then directed to the surface from the separator, and the water is directed down to L2 for subsequent reinjection. Advantages to a design such as this are listed as follows:

- Upstream corrosion and scaling are limited.
- Oil does not have to be separated as completely from the water as it would if the water were to be disposed of (depending of the disposal formation characteristics).
- There is no pressure reduction resulting from increased hydrostatic pressure because no water column is formed in the production string.
- Because the water is pushed down rater than up, the strain on the downhole pump is minimized and its life span lengthened.

Completion and production equipment is becoming intelligent with the addition of electronics and software for data acquisition, data processing, communications, and control of electromechanical and hydraulic devices from the surface (Fig. K9). These modules provide the resources required to perform the following control functions:

- Optimize the production of hydrocarbons from a geological formation.
- Produce simultaneously from several laterals in multilateral well applications.
- Slow the flow of water from non-producing into producing zones.

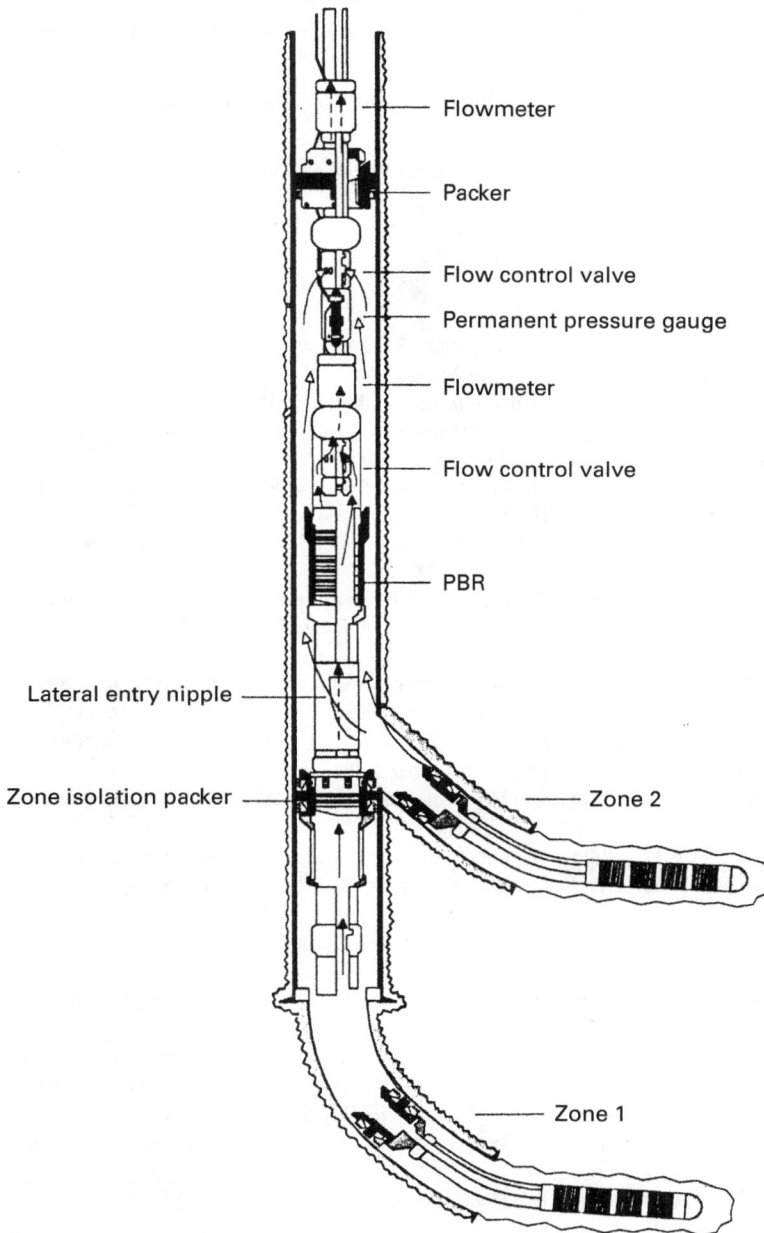

Figure K9 Intelligent system in a multilateral well [18].

- Equalize the main bore pressure in multilateral applications to prevent cross flow.
- Control the amount of fluid which is produced from the formations into the downhole oil/water separator.
- Control the back pressure to the injection pump to optimize injection and pump performance.
- Separate and re-inject the produced water inside the wellbore.

REFERENCES

1 Butler RM (1994) *Horizontal Wells for the Recovery of Oil, Gas and Bitumen.* Petroleum Society monograph of Canada Institute of Mining, No. 2

2 Joshi SD (1991) *Horizontal Well Technology*, Ch. 2 to 7. PennWell Books

3 Economides MJ, Hill AD, Ehilig-Economides C (1994) *Petroleum Production Systems*, Ch. 2, 18. Prentice-Hall Inc., Englewood Cliffs, New Jersey

4 Dukler AE (1992) *Flow Pattern Transitions in Gas-Liquid Systems Measurements and Modelling.* University of Houston, Houston, TX

5 Babu DK, Odeh AS (1989) Productivity of a horizontal well. *SPE Reservoir Engineering*, Vol. 4, No. 4, November 1989

6 Giger FM, Renard G (1987) Low permeability reservoirs development using horizontal wells. *Paper SPE 16406* presented at the 1987 SPE/DOE Low Permeability Reservoirs Symposium, Denver, Colorado

7 Novy RA (1992) Pressure drops in horizontal wells: when can they be ignored? *Paper SPE 24941* presented at the 67th SPE Annual Technical Conference, Washington, DC

8 Dikken BJ (1990) Pressure drop in horizontal wells and its effect on production performance. *Journal of Petroleum Technology,* November 1990

9 de Montigny O, Combe J (1998) Hole benefits, reservoir types key to profit. *Oil and Gas Journal*, April 1988

10 Burnett DB (1998) Wellbore cleanup in horizontal wells: an industry funded study of drill-in fluids and cleanup methods. *Paper SPE 39473* presented at the SPE Int. Symp. on Formation Damage Control held in Lafayette, Louisiana

11 Diggins E (1997) A proposed multi-lateral well classification matrix. *World Oil*, November 1997

12 Goffart A (1998) *La complétion et la production en drains horizontaux.* AFTP/SPE France Conference, June 1998

13 Renard G, Gadelle C, Dupuy JM, Alfonso H (1997) Potential of multilateral wells in gas coning situations. *Paper SPE 38760,* SPE Annual Technical Conference, San Antonio, TX

14 Renard G, Gadelle C, Dupuy JM, Alfonso H (1997) Potential of multilateral wells in water coning situations. *Paper SPE 39071* presented at the Fifth Latin American and Caribbean Conference, Rio de Janeiro, Brazil

15 Renard G, Dupuy JM (1990) Influence of formation damage on the flow efficiency of horizontal wells. *Paper SPE 19414* presented at the SPE Formation Damage Control Symposium, Lafayette, Louisiana

16 Aguilera R, Artindale JS, Cordell G, Ng MC, Nicholl GW, Runions GA (1991) *Horizontal Wells.* Gulf Publishing Company, 158–209

17 Hardy M, Lockhart T (1998) *Water Control. Petroleum Well Construction.* John Wiley & Sons, Ltd, Ch. 20, 571–591

18 Tubel P, Herbert RP (1998) Intelligent System for monitoring and control of downhole oil water separation. *Paper SPE 49186* presented at the SPE Annual Technical Conference and Exhibition, New Orleans, Louisiana

19 MacKenzie A, Hogg C (1999) Multilateral classification system with example applications. *World Oil,* January 1999, 55–61

20 Hogg C (1997) Comparison of multilateral completion scenarios and their application. *Paper SPE 38493* presented at the 1997 Offshore Europe Conference held in Aberdeen, Scotland.

K

L

Water Management

L

Water Management

L1 BASIC WATER–OIL FLOW PROPERTIES OF RESERVOIR ROCK [1]

They consist of two main types:

- Properties of the rock skeleton alone, such as porosity, permeability, pore size distribution and surface area.
- Combined rock–fluid properties such as capillary pressure (static) characteristics and relative permeability (flow) characteristics.

Some basic definitions:

Absolute permeability. Permeability of rock saturated completely with one fluid.

Effective permeability. Permeability of rock relative to one fluid, the rock being only partially saturated with that fluid.

Relative permeability. Ratio of effective permeability to some base value.

Porosity. Portion of rock bulk volume composed of interconnected pores.

1.1 Rock wettability

It is defined as the tendency of a fluid to spread on or adhere to a solid surface in the presence of other immiscible fluids.

The contact angle, θ_c, has achieved significance as a measure of wettability. As shown in Fig. L1, the value of the contact angle can range from zero to 180° as limits. Contact angles of less than 90°, measured through the water phase, indicate preferentially water-wet conditions, whereas contact angles greater than 90° indicate preferentially oil-wet conditions.

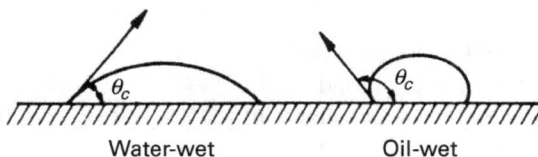

Figure L1 Wettability of oil-water-solid system [1].

235

1.2 Capillary pressure

The water-oil capillary pressure is the pressure in the oil phase minus the pressure in the water phase, or:

$$P_c = p_o - p_w$$

1.3 Relative permeability

The relative permeability characteristics are a direct measure of the ability of the porous system to conduct one fluid when one or more fluids are present. These flow properties are the composite effect of pore geometry, wettability, fluid distribution, and saturation history.

The differences in the flow properties that indicate the different wettability preferences can be illustrated as follows on Table L1.

Table L1 [1].

Flow parameters	Water-wet	Oil-wet
Connate water saturation	>20 to 25% pore volume	Generally, <15% pore volume Frequently, <10% pore volume
Saturation at which oil and water relative permeabilities are equal	>50% water saturation	<50% water saturation
Relative permeability to water at maximum water saturation, i.e., floodout	Generally <30%	>50% and approaching 100%

L2 WATERFLOODING [1, 2]

2.1 Mobility ratio

It is defined as:

$$M = \frac{k_w}{\mu_w} \frac{\mu_o}{k_o} = \frac{k_{rw}}{\mu_w} \frac{\mu_o}{k_{ro}}$$

where
- k_w effective permeability to water (md)
- k_o effective permeability to oil (md)
- k_{rw} relative permeability to water (fraction)
- k_{ro} relative permeability to oil (fraction)
- μ_w water viscosity (cp)
- μ_o oil viscosity (cp).

By conventional use, mobility ratios less than unity are termed "favorable", and those greater than unity are "unfavorable". One way to reduce the mobility ratio is to thicken the water, for example, by addition of polymers.

2.2 Efficiency of oil displacement by water

The fractional flow f_w equation is, in a simplified form:

$$f_w = \frac{1}{1 + \dfrac{\mu_w}{\mu_o}\dfrac{k_{ro}}{k_{rw}}}$$

2.3 Areal sweep efficiency

In waterflooding, water is injected into some wells and produced from other wells.

A wide variety of injection–production well arrangements have received attention in the literature. These are shown in Table L2 and Fig. L2.

Table L2 [1].

Pattern	Ratio of producing wells to injection wells	Drilling pattern required
Four-spot	2	Equilateral triangle
Skewed four-spot	2	Square
Five-spot	1	Square
Seven-spot	1/2	Equilateral triangle
Inverted seven-spot (single inj. well)	2	Equilateral triangle
Nine-spot	1/3	Square
Inverted nine-spot (single inj. well)	3	Square
Direct line drive	1	Rectangle
Staggered line drive	1	Offset lines of wells

2.4 Factors affecting selection of waterflood pattern

The proposed waterflood pattern should fulfill the following:
1. Provide desired oil production capacity.
2. Provide sufficient water injection rate to yield desired oil productivity.
3. Maximize oil recovery with a minimum of water production.

4. Take advantage of known reservoir nonuniformities, i.e. directional permeability, regional permeability differences, formation fractures, dip, etc.
5. Be compatible with the existing well pattern and require a minimum of new wells.
6. Be compatible with flooding operations of other operators on adjacent leases.

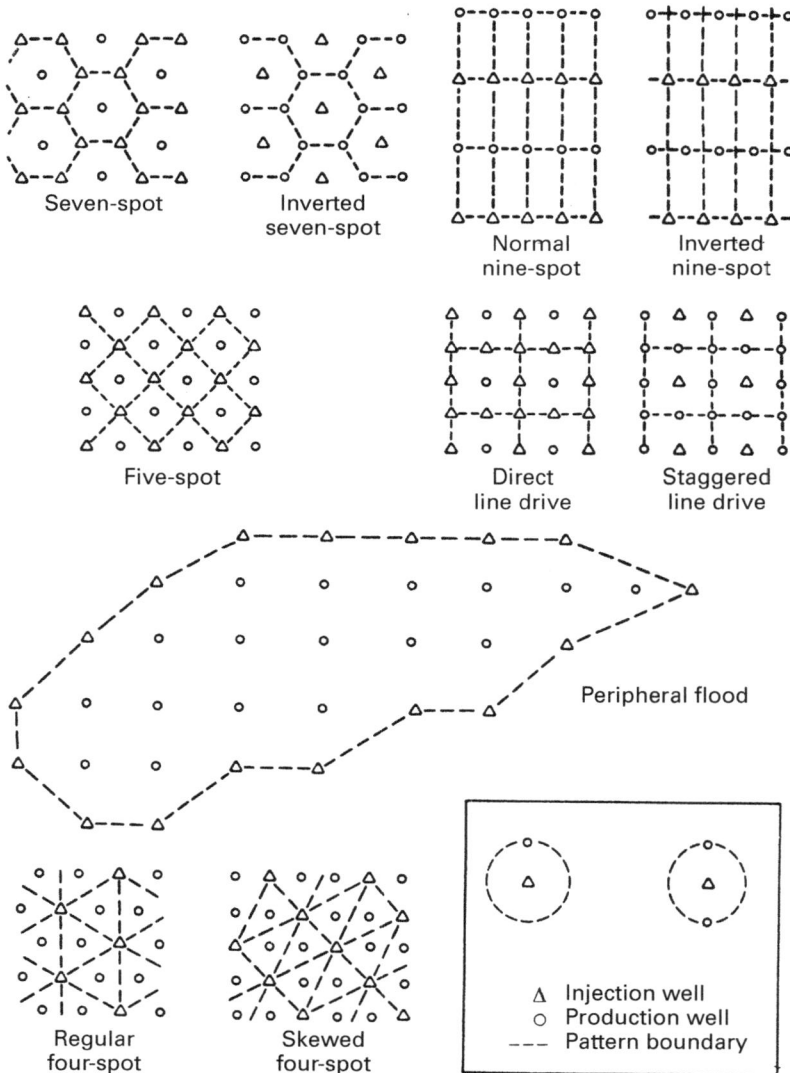

Figure L2 Flooding patterns [1, 2].

2.5 Factors affecting waterflood oil recovery performance

Any waterflood performance prediction method requires a description of the reservoir. An engineer should attempt to obtain answers to these questions:

1. Is the reservoir likely to perform as a series of independent layers, or as zones of differing permeability with fluid crossflow?
2. Are there zones of high gas saturation or high water saturation that could serve as channels for bypassing water?
3. Does the reservoir contain long natural fractures or directional permeability that could cause preferential areal movement in some direction?
4. Are there areas of high and of low permeability that might cause unbalanced flood performance?
5. Is crossbedding present to the degree that fluid communication between injection and producing wells might be impaired?
6. Is the reservoir likely to contain planes of weakness or closed natural fractures that would open at bottomhole injection pressures?
7. Water injection quality (total dissolved solids, injectivity impairment).

2.6 Profile correction

In the case of strong water channeling through high permeability streaks, gel treatment of injection wells may be a suitable way to improve sweep efficiency.

Water channeling problems may be detected by tracer injection (radioactive, iodide…) or interference pressure test.

L3 WATER CONING

Correlations programmed on a personal computer allow rapid analysis of water coning. Compared to complicated numerical models, these correlations are particularly useful when detailed reservoir data are not available, or when decision time and project cost are limited.

3.1 Schols critical production rate [3]

Schols empirical formula offers a quick method for calculating critical rate:

$$q_c = \left(\frac{(\rho_w - \rho_o)k(h^2 - D^2)}{2049\mu_o B_o} \right) \left(0.432 + \frac{\pi}{\ln(r_e / r_w)} \right) \left(\frac{h}{r_e} \right)^{0.14} \tag{L1}$$

where

q_c	critical production rate (st bbl/D) (stock-tank barrels per day)
$\rho_w - \rho_o$	density difference between water and oil (g/cc)
h	oil zone thickness (ft)
D	penetration (ft) (from the top of the sand)
k	permeability (md)
μ_o	oil viscosity (cp)
B_o	oil formation volume factor (res-bbl/st-bbl) (reservoir oil barrels/stock-tank oil barrels)
r_e	drainage radius (ft)
r_w	well bore radius (ft).

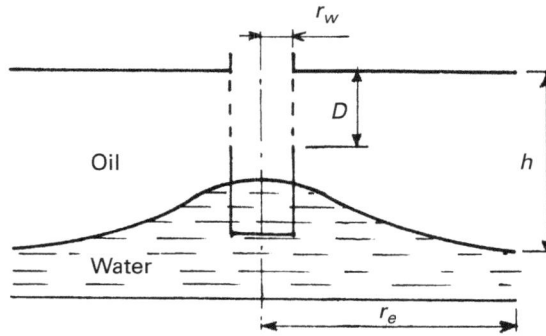

Figure L3　Flow configuration for critical rate.

3.2　Sobocinski critical production rate [4–6]

Sobocinski's formula is derived simply from the expression of the forces in action (Fig. L3):

$$q_c = 0.878 \times 10^{-3} \frac{(\rho_w - \rho_o)kh(h - D)}{B_o \mu_o} \qquad \text{(L2)}$$

This formula is applicable when the water–oil mobility ratio M is below 10 ($M < 10$), where

$$M = \frac{k_w}{k_o} \times \frac{\mu_o}{\mu_w}$$

▼ Example

Calculate the critical production rate for water coning at initial conditions for the reservoir and fluid properties listed below:

$\rho_w = 1$

$\rho_o = 0.85$

h = 100 ft (30.5 m)
D = 60 ft (18.3 m)
k = 100 md
μ_o = 2 cp
B_o = 1.2
r_e = 2 000 ft (610 m)
r_w = 0.3 ft (0.10 m).

Schols method

From Eq. L1:

$$q_c = \frac{(1-0.85)(100)\left(100^2 - 60^2\right)}{2049(2)(1.2)} \times \left(0.432 + \frac{\pi}{\ln(2000/0.3)}\right) \times \left(\frac{100}{2000}\right)^{0.14}$$

q_c = 10.15 st bbl/D   or   1.6 m³/d

Sobocinski method

From Eq. L2:

$$q_c = 0.878 \times 10^{-3} \frac{(1-0.85)(100)(100)(100-60))}{(1.2)(2)}$$

q_c = 21.95 st bbl/D   or   3.5 m³/d   ▲

This example shows the low critical rate related to the initial conditions.

L4   WATER CONTROL IN PRODUCTION WELLS [7–12]

Almost all oil and gas reservoirs produce water. Since nature does not like vacuum, water usually replaces oil as hydrocarbon reserves decline in the field. Usually, water production increases as oil or gas decline. In mature or old reservoirs, most of fluid production is water, with oil or gas representing a few percent of total production. Moreover, many reservoirs are submitted to water injection, that provides pressures and improves sweep efficiency. In such cases, large quantities of injected water are produced in the field. A continuous increase in water production is thus a normal behavior in the lifetime of a field.

4.1   Causes of excessive water production

4.1.1   Completion accidents

- Tubing/casing/packer leaks (easiest problem).
- Flow behind pipe.

4.1.2 Reservoir problems

- Layered reservoirs with vertical flow barriers.
- Individual fractures between injectors and producers.
- "2-D coning" through fractures.
- Channeling through naturally fractured reservoirs.
- 3-D coning or cusping.
- Layered reservoirs without vertical flow barriers (trickest problem).
- Wormhole development (heavy oil in consolidated sands).

4.2 Key ideas

To deal with excessive water production, a solution is the injection of chemicals into the formation.

Two kinds of injection are applied:

- Permanent barriers for strong treatments.
- Selective barriers for soft treatments.

Their characteristics are specified in Table L3.

Table L3

Permanent barriers	Selective barriers

Types:
  – Cement squeeze
  – Resins
  – Strong gels.
	Types:
	  – Hydrosoluble polymers
	  – Weak gels.

Characteristics:
  • Good results when clearly separated oil and water zones.
  • Risky otherwise with possible plugging of formation.
  • Workover always required.
	Characteristics:
	  • Applies to multilayered reservoirs and leaves k_{ro} and k_{rg} almost unaffected.
	  • Reversible process, i.e. chemicals can be removed.
	  • Workover sometimes required. Most often injection through existing completion.
	  • Wag (water and gas).
	  • Wax or deal oil squeeze.

k_{ro} Relative permeability to oil.
k_{rg} Relative permeability to gas.

4.3 Principle of water shutoff treatment [7, 8]

Layered reservoirs with vertical flow barriers constitute a case frequently encountered in field situations (Fig. L4).

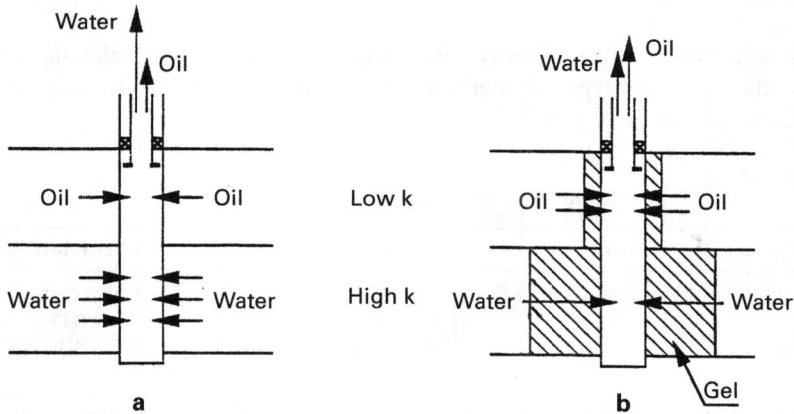

Figure L4 Profile correction by polymers in heterogeneous reservoirs [7].
a. Before treatment. **b.** After gel treatment.

4.4 Polymer/gel placement around the wellbore

- The target is to:
  - Maximize penetration depth and permeability reduction in the high k-watered out zone
  - Minimize penetration depth and permeability reduction in the low k-oil bearing zone.
- The placement is a key factor of success (even more important than the relative permeability modification).
- Because of the difficulty of placement control, most people recommend, whenever possible, to use mechanical zone isolation to place the polymer/gel at the right place.
- Bullhead treatments require better placement control via diversion techniques.

4.5 Symbols of polymer/gel

Different polymers are used:

PAM	Nonionic polyacrylamide
HMTA	Hexamethyl tetramine
PATBA	Acrylamide tertio butyl acrylate copolymer
PVA	Polyvinyl alcohol
CPAM	Cationic polyacrylamide
HPAM	Hydrolyzed polyacrylamide or acrylamide-acrylic acid copolymer
HEA	2-hydroxyethylacrylate.

4.6 Relative permeability modifiers

These modifiers are: bullheaded polymers or weak gels without zone isolation. In Table L4 are described the vinyl type polymers and copolymers, in Table L5 the polysaccharides, and in Table L6 the weak gels.

Table L4 Vinyl type polymers and copolymers.

Polymer	Chemical name	Manufacturer	Characteristics
PAM HPAM CPAM	See paragraph L4.5.	Floerger Allied Colloïds and others	Are limited in temperature at 60°-70°C and their adsorbability on reservoir rock decreases in the order: CPAM > PAM > HPAM.
AM-N-VP (HE polymers)	Acrylamide-N-vinylpyrrolidone nonionic copolymers	Drilling specialities for Phillips Petroleum	Are characterized by: – high thermal stability, – high adsorbability on reservoir rock, but due to their low molecular weight, their properties are improved by using a crosslinker.
AM-AMPS	Acrylamide-acrylamido methyl propane sulfonic acid copolymers	Floerger	Compared to polyacrylamides they have improved thermal stability (up to 100°C) but their adsorbability on reservoir rock decreases with AMPS content.
AM-N-VA-AMPS	Acrylamide-N-vinyl amide-AMPS terpolymers.	Hœchst	Similar to the previous AM-AMPS copolymers with somewhat improved thermal stability.

Table L5 Polysaccharides

Polymer	Chemical name	Manufacturer	Characteristics
Scleroglucan	Nonionic polysaccharide	Sanofi	High shear resistance. Thermal stability up to 120°C. High adsorbability on reservoir rock.
HEC	Hydroxy-ethyl-cellulose	Hercules for Total	Used for injection wells.

Table L6   Weak gels.

Polymer	Trade name	Licensee	Characteristics
Amphoteric polymer	Worcon (Water oil ratio control)	Halliburton	Temperature limit: 150°C (or 300°F).
Amphoteric polymer	Aquatrol	B.J.	Temperature limit: 80°C (or 175°F).
PAM-Glyoxal	IFPOL 300	IFP	Low temperature applications (up to 60°C).
CPAM-Glyoxal	FLOPERM 500	OFPG	Low temperature applications (up to 60°C).
Sceroglucan-Zr(IV) lactate	IFPOL 200	IFP	High temperature applications (up to 120°C).
HPAM/CPAM + Al(III) citrate	CAT-AN	Tiorco	Sequential injection, or dilute mixture.
HPAM-AMPS copolymer	IFPOL 400	IFP	High temperature applications (up to 100°C).

4.7 Strong gels

These chemical systems are preferably used for zonal isolation. In Table L7 are described ionic type gels, in Table L8 organic gels, and in Table L9 in-situ polymerisation-type gels.

Table L7   Ionic gels.

Polymer	Trade name	Licensee	Characteristics
HPAM + Cr(VI) + reducer	Channelblock	Dowell	Reducer: thiosulfate, thiourea. Environmental restrictions.
HPAM + Al(III)	Zonetrol	Dowell	
HPAM + Sodium aluminate		Unocal	
HPAM + Cr(III) acetate	Marcit (h.m.w.)*	Marathon Oil Co	Temperature limit: 120°C. Fault blocking system.
	Maraseal (l.m.w.)**	Marathon Oil Co	Matrix blocking chemical.
HPAM + Zr(IV) lactate		Phillips Petroleum	
Xanthan gum + Cr(III)		OFPG	Only injector treatments.

*   (h.m.v.): high molecular weight.
** (l.m.v.): low molecular weight.

Table L8 Organic gels.

Polymer	Trade name	Licensee	Characteristics
PAM + Phenol-formaldehyde		Phillips Petroleum	Environmental restrictions. High temperature applications.
PAM + HMTA + Resorcinol or salicyclic alcohol		Unocal	High temperature applications.
PATBA + Polyethylene imine	H_2Zero	Halliburton	High temperature applications.
PVA + Formaldehyde		Unocal	High temperature applications.

Table L9 In situ polymerisation gels.

Polymer	Trade name	Licensee	Characteristics
Acrylamid monomer + Catalyst	K-Trol	Halliburton	Low temperature applications (up to 60-70°C).
HEA + Azo compound		British Petroleum	High temperature applications (up to 147°C).

4.8 Other systems

4.8.1 Inorganic gels

- Silicates
- Aluminates.

4.8.2 Resins

- Phenol-formaldehyde (used for sand control also)
- Melamine-formaldehyde
- Epoxy.

4.8.3 Cements

- Ultra fine cement (preferred) for fault blocking systems.

4.8.4 Quick-set formation treating methods [4]

A method for fluid control or plugging of a well is to inject a mixture of an acid polymerizable resin, a polar organic diluent, and an acid catalyst, and later injecting an acidic fluid to quick-set a portion of the resin and hold it in place while the pre-mixed catalyst sets the resin. In an alternate sand consolidation embodiment, a fluid slug is injected between the resin and acidic fluid injections to create permeability in the resin saturated area of the formation prior to final set of the resin (Trade name: TexPlug).

4.9 Mechanical ways of placement

Let us consider a two-layer well with one layer producing mainly water (thus to be plugged) and the other producing oil (to be protected from gelant invasion) (Fig. L4). Two possibilities may occur.

4.9.1 Case 1: The upper layer is the target

The goal is to protect the bottom layer from gelant injection. This could be achieved by setting a packer between the two layers or filling the bottom part of the wellbore with a fine $CaCO_3$ powder while the gelant is injected into the upper layer. To spot the gelant at the right place and avoid the filling up of the wellbore, the coiled tubing could be equipped with a stradle packer which could be set on top of the upper layer.

After gelant squeeze and sufficient time given for gelation, the packer placed at the intermediate level can be removed, and the wellbore cleaned (gel and $CaCO_3$) by water/acid recirculation from the coiled tubing.

4.9.2 Case 2: The bottom layer is the target

The best way is to use a coiled tubing with a straddle packer, set between the two layers. In such a way, a dual injection can be performed, with the gelant in the coiled tubing and the confined fluid (water or diesel) in the annulus. The role of the confining fluid is dual:
  (a) To protect the upper zone from gelant invasion
  (b) To counterbalance the pressure build-up during gelant squeeze into the formation.

4.10 Water shutoff using an inflatable composite sleeve polymerized in-situ

4.10.1 Principle [18]

The principle of the technology is to use an Inflatable Setting Element (ISE) to convey a composite sleeve in the well. The composite sleeve is manufactured using thermosetting resins and carbon fibre, so it will be soft and deformable when running in hole. Once the tool is opposite the zone to be treated the ISE can be inflated to push the composite into place against the inside of the casing, where it is heated to polymerize the resins. The ISE is then deflated and extracted to leave the composite sleeve as a hard pressure resistant lining inside the casing.

4.10.2 Applications [19]

The ISE is particularly suitable for:
* Sealing perforations for water shutoff, gas shutoff, selective profile modification and zone isolation
* Repairing damaged tubing and/or casing with minimal diameter loss
* Setting in openhole for zone isolation
* Use as a temporary casing to cure total loss zones while drilling
* Use in re-entry wells for sealing between a side-track liner and the main bore casing.

4.10.3 Advantages [19]

- Running through tubing to expand into casing
- No need to kill the well
- Minimum diameter loss
- Can be re-perforated
- Can be run through a previous patch (see below).

4.10.4 Limitations [18]

- The length is limited to 16 m
- The temperature limitation of the resins is 110°C
- The smallest possible run-in diameter of 3.6 in. for setting in 7-in. liners only gives access to wells with 4 1/2-in. tubing or larger.

4.10.5 Setting procedure [18, 19]

Figure L5 describes the successive phases of the setting procedure.

L

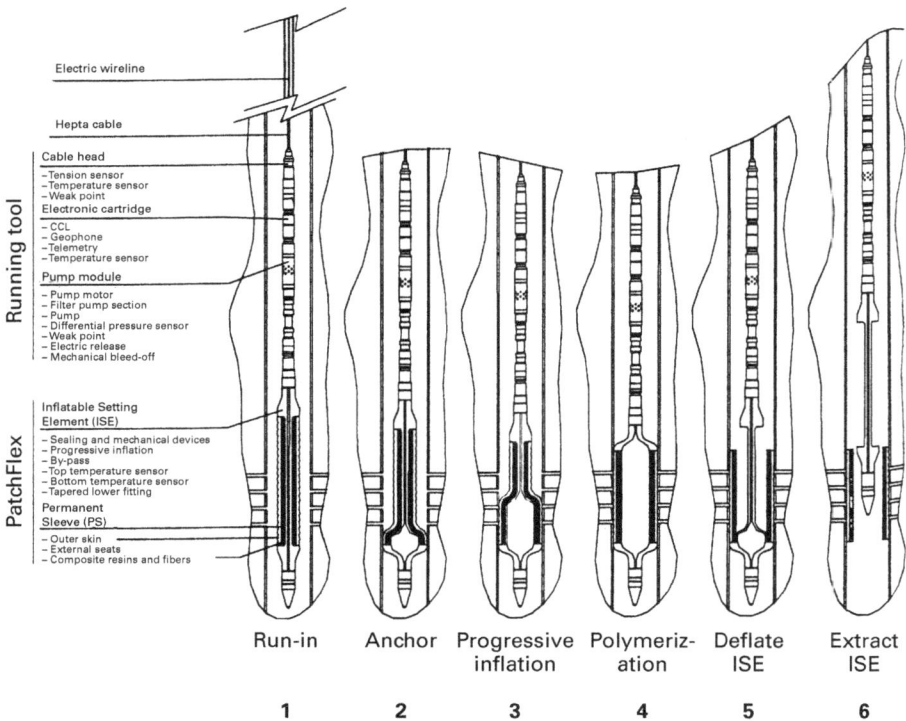

Figure L5 ISE setting procedure [18, 19].

1. **Run in.** When the assembly is in place, a surface control module is used to activate a pump in the downhole running tool and begin expansion of the ISE.
2. **Anchor.** The lower end inflates first to anchor the composite sleeve in place.
3. **Progressive inflation.** The inflation then progressively moves upwards so no well fluid can become trapped behind the permanent sleeve and create a hydraulic lock.
4. **Polymerisation.** Heating begins when the product is fully inflated.
5. **Deflate ISE.** The pump is reversed to deflate the ISE.
6. **Extract ISE.** When the pressure has been bled-off, the ISE is pulled through the permanent sleeve and recovered at the surface with the running tool.

4.11 Selection of candidate wells

To evaluate the feasibility of a water control treatment, it is necessary to collect several data. Ideally, the information needed concerns the following:
- Reservoir lithology and stratification, existence of natural fractures, presence of clay lenses and their extension, directional permeability trends.
- Well pay-zone average thickness, number and size of perforated intervals.
- Average permeability and anisotropy.
- Complete completion diagram.
- Well production history, well logs.
- Analysis of cores and of produced fluids.
- Production mode and constraints, field facilities, availability and analysis of injection water.

Once all information is collected, the choice of the candidate well is based on the following criteria:
- **Heterogeneity.** A certain degree of vertical heterogeneity is required. Those wells which have to be treated should be perforated in either two or more independent reservoirs or else in a single reservoir exhibiting several inter-bedding of sizable lateral extension.
- **Sweeping efficiency.** Water injection within the reservoir should not have been initiated too long ago (about 10 years is a reasonable figure). A favorable criterion is a recent perforation of a reservoir layer, either in a producer or in an injector.
- **Production.** The water-cut should be in the range 70–90%. Moreover, in order to obtain a reasonable pay-out, the fluid rate should exceed 50 m³/d (or 300 bbl/D).

L5 DOWNHOLE OIL WATER SEPARATION (DHOWS) [13–16]

Original ways are proposed for the improvement of management techniques when facing large quantities of water production.

5.1 Principle

Figures L6 and L7 illustrate the principle of the Downhole Oil Water Separator (DOWS). A set-up made of one hydrocyclon and dual stream pumps is placed down the wellbore. The fluid produced from the target formation is separated continuously in the hydrocyclon into a water phase and an oil-rich phase. The oil-rich phase is produced at the wellhead, whereas the water phase is pumped into an aquifer located either above or below the productive interval.

The dual stream pumps could be rod pumps, electric submersible pumps, progressing cavity pumps or even gas lift systems.

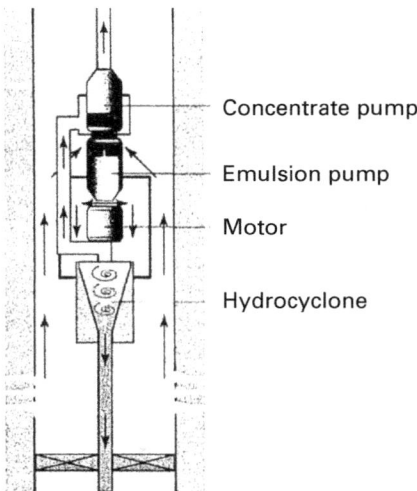

Figure L6 C FER project oil-water system concept [14].

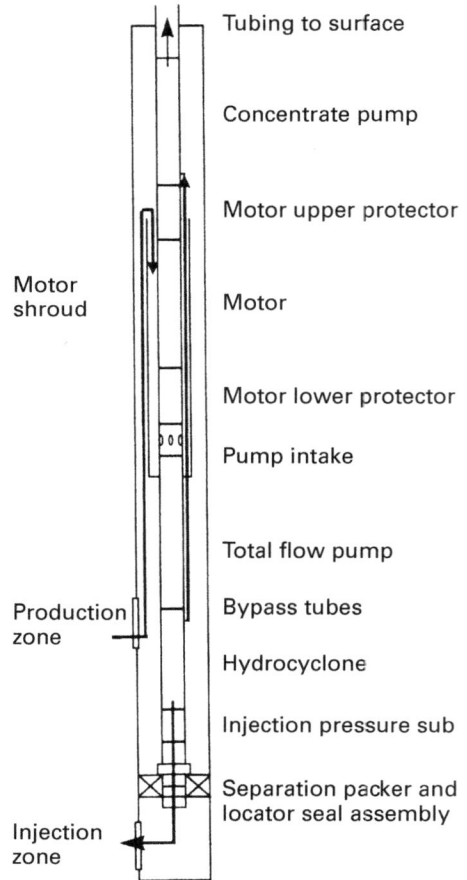

Figure L7 Downhole equipment line-up in well completion and flow paths [16].

5.2 Advantages of this technology [13, 15]

1. **Reduction in operating expense** (OPEX), through less use of treatment chemicals and reduced plant maintenance. Chemical treatment inside the wellbore will increase the system reliability and improve the performance of the equipment.
2. **Increase in net present value** (NPV). By leaving water downhole, the well inflow can be maximized without regard to process capacity constraints.
3. **Increase in ultimate recovery of reserves**. By injecting the water to a target zone, pressure maintenance and sweep efficiencies can be improved in areas of the reservoir that may otherwise have been poorly serviced. Wells will be able to stay in production longer.
4. **Reduction of the frequency of intervention and possibility of planned interventions** by use of intelligent separator system completions.

L6 KEY MESSAGES [17]

- **Inject water in the right place (fracture management).**
- **Much produced water (bad water) is not working for you. Keep it in the ground.**
- **Make the water work in the reservoir to maximize pressure across oil.**

REFERENCES

1 Craig FF, Jr. (1993) *The Reservoir Engineering Aspects of Waterflooding*. Monograph Volume 3 of the Henry L. Doherty series, Chapters 1 to 8. Society of Petroleum Engineers of AIME, New-York, Dallas

2 Rose SC, Buckwalter JF, Woodhall RJ (1989) *The Design Engineering Aspects of Waterflooding*. Monograph of the Henry L. Doherty series, Chapter 3. Society of Petroleum Engineers of AIME, Richardson, TX

3 Kuo MCT and DesBrisay CL (1983) A simplified method for water coning predictions. *Paper SPE 12067,* presented at the 58th ATCE held in San Francisco, CA

4 Surles BW, Fader PD, Pardo CW (1995) Quick-set formation treating methods. *US Patent N° 5,423,381*

5 Sobocinski DP, Cornelius AJ (1965) A correlation for predicting water coning time. *Journal of Petroleum Technology*, May 1965, 594–600

6 Cossé R (1993) *Basics of Reservoir Engineering*. Editions Technip, Paris

7 Zaitoun A, Kohler N, Bossie-Codreanu D (1998) Water control in hydrocarbon reservoirs and storages: A literature review. *IFP Report 45 076* (prepared for the Workshop on Water Control in Oil and Gas Production, Gas Storage State of the Art, Leipzig, June 8, 1998). European Commission Directorate-General for Energy, Contrat Thermie, No. DIS-1203-97-DE

8 Zaitoun A (1998) *Water control by polymer/gels*. Seminar presentation

9   Hardy M, Botermans CW (1992) New organically crosslinked polymer system provides competent propagation at high temperature organic gels for water control. *Paper SPE 39690,* SPE Symposium on IOR, Tulsa, OK

10   Hardy M, Lockhart T (1998) Water control. *Petroleum Well Construction.* John Wiley & Sons, Chapter 20, 571–591

11   Seright RS (1991) Effect of rheology on gel placement. *SPE Reservoir Engineering,* 6(2), 212–18

12   Liang JT, Lee RL, Seright RS (1993) Gel placement in production wells. *SPE Production Facilities,* 8(4), 276–84

13   Peachy BR, Matthews CM (1994) Downhole oil/water separator development. *Journal of Canadian Petroleum Technology,* September 1994, 33.7, 17–21

14   Foulser B (1998) Water management Down-hole oil-water separation. *SPE Review,* 105, June 1998, 18–21

15   Tubel P, Herbert RP (1998) Intelligent system for monitoring and control of downhole oil-water separation. *Paper SPE 49186,* presented at the SPE Annual Technical Conference and Exhibition, New-Orleans, Louisiana

16   Verbeek PHJ, Smeenk RG, Jacobs D (1998) Downhole separator produces less water and more oil. *Paper SPE 50617,* presented at the SPE European Petroleum Conference held in The Hague, The Netherlands

17   Stevens DG (1998) *Water Management and Control.* SPE Conference, Paris

18   Leighton J, Saltel JL, Morrison J, Welch R, Pilla (1998) Water shut-off using an inflatable composite sleeve polymerised in-situ. A case history on forties delta. *Paper SPE 50620,* presented at the SPE European Petroleum Conference held in The Hague, The Netherlands

19   Drillflex (1998) *Patchflex.* Document Drillflex, Schlumberger

20   Stebinger LA, Elphingstone GM, Jr. (1998) Multipurpose wells: downhole oil-water separation in your future. *Paper SPE 49050,* presented at the SPE Annual Technical Conference and Exhibition, New-Orleans, Louisiana

21   Shaw C, Fox M (1998) Economics of downhole oil-water separation. A case history and implications for the North Sea. *Paper SPE 50618,* presented at the SPE European Petroleum Conference held in The Hague, The Netherlands.

M

Heavy Oil Production
Enhanced Oil Recovery

M

Heavy Oil Production
Enhanced Oil Recovery

M1 HEAVY OIL PROPERTIES

1.1 Definitions [1]

What are heavy crude and tar sand oils? In an effort to set agreed meanings, UNITAR, in a meeting in 1982 in Venezuela, proposed the following definitions (see also Table M1):

1. Heavy crude oil and tar sand oil are petroleum or petroleum-like liquids or semi-solids naturally occurring in porous media. The porous media are sands, sandstones, and carbonate rocks.

2. These oils will be characterized by viscosity and density. Viscosity will be used to define heavy crude oil and tar sand oil, and density (°API) will be used when viscosity measurements are not available.

3. Heavy crude oil has a gas-free viscosity of 100 to 10 000 cp (or mPa.s) at reservoir temperatures, or a density of 943 kg/m^3 (20°API) to 1 000 kg/m^3 (10°API) at 15.6°C and at atmospheric pressure.

4. Tar sand oil has a gas-free viscosity greater than 10 000 cp (or mPa.s) at reservoir temperatures, or a density greater than 1 000 kg/m^3 (less than 10°API) at 15.6°C and at atmospheric pressure.

5. Heavy crude oils generally contain 3% weight or more of sulfur, and as much as 2 000 ppm of vanadium. Nickel and molybdenum are also frequently encountered.

Table M1 UNITAR definitions of heavy crude and tar sand oils [1].

Classification	Viscosity (cp or mPa.s)	Density at 15.6°C (kg/m^3)	°API at 15.6°C
Heavy crude	100–10 000	943–1 000	20–10
Tar sand oil	>10 000	>1 000	<10

1.2 Temperature effect on oil viscosity [5]

Figure M1 shows the relationship between viscosity and temperature for oils of various gravities. It can be seen that orders of magnitude changes in viscosity can be achieved by variations of temperature.

When the viscosity of a crude oil is known at a given temperature (point A), it is possible to obtain an approximative value of viscosity at another temperature by drawing a line parallel to the slanted lines (point B).

However, the oil viscosity measured in laboratory at a given temperature is not the viscosity of the oil into the formation. It generally contains free and dissolved gas, contributing to a much lower visible oil viscosity. This apparent viscosity should then be measured in laboratory, in order to evaluate more precisely the frictional pressure drops.

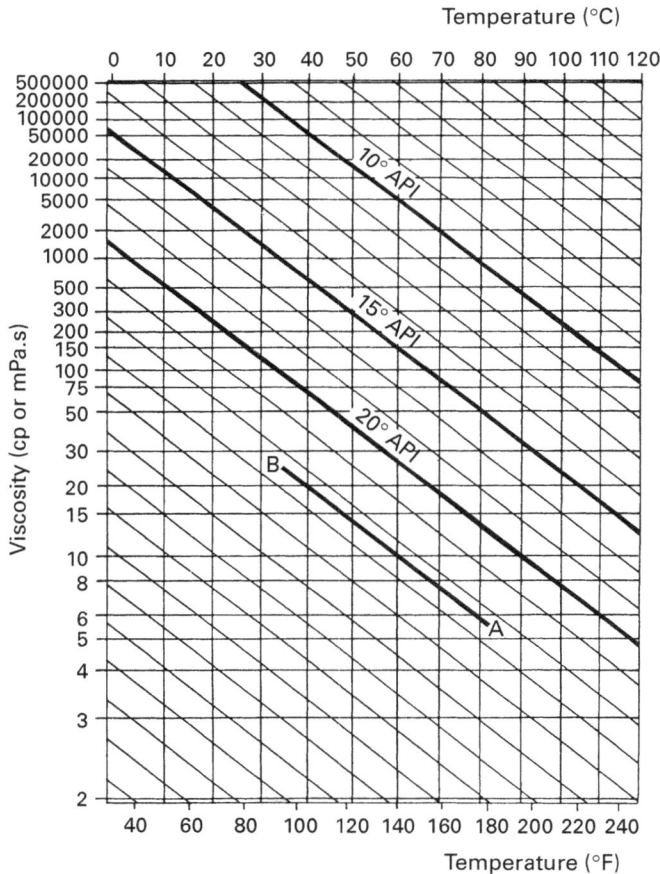

M

Figure M1 Temperature effect on oil viscosity [after Owens and Souler].

1.3 Frictional pressure drop [6]

Heavy and viscous oils lead to frictional pressure drop in tubing and line pipe, hence a decrease in the performance of the pumping system as far as the fluid height to be lifted is concerned. So, these pressure drops have to be evaluated and a higher head rating pump should be chosen.

Also, the viscosity increases the value of the resistant torque of the drive strings in the case of rotating pumping.

The discharge of high viscosity fluids through the production tubing can generate a significant pressure drop which is proportional to the oil viscosity.

$$\Delta P_f = \frac{705 \times 10^{-4}}{(D+d)(D-d)^3} \times Q \times \mu_f \times L \times \frac{1}{\ln \frac{\mu_s}{\mu_f}} \left(\frac{\mu_s}{\mu_f} - 1 \right) \tag{M1}$$

where

ΔP_f	pressure drop due to friction (kPa)
D	inside diameter of the tubing (cm)
d	drive string diameter (cm)
Q	pumped flow rate (m³/d)
μ_f	viscosity of the effluent at the inlet temperature (cp or mPa.s)
μ_s	viscosity of the effluent at the surface (cp or mPa.s)
L	length of the tubing (m).

1.4 Resistant torque generated by the viscosity [6]

In rotating pumping, heavy oils generate a torque increase proportional to the viscosity:

$$\Gamma_v = 0.165 \times 10^{-8} \times \mu_f \times L \times N \times \frac{d^3}{D-d} \times \frac{1}{\ln \frac{\mu_s}{\mu_f}} \left(\frac{\mu_s}{\mu_f} - 1 \right) \tag{M2}$$

where

Γ_v	resistant torque (m.daN)
μ_f	viscosity of the effluent at the inlet temperature (cp or mPa.s)
μ_s	viscosity of the effluent at the surface (cp or mPa.s)
L	length of the tubing (m)
N	rotating speed (rpm)
d	drive string diameter (cm)
D	inside diameter of the tubing (cm).

The evaluation of the pressure drops and of the resistant torque due to the viscosity is then very important for selecting:

(a) The pumping system (admissible head rating)
(b) Drive strings (admissible torque)
(c) The motor (required power).

These relations point out the influence of viscosity and temperature. As a matter of fact, if the formation oil is at the bottomhole temperature, at the pump inlet this temperature will decrease according to the geothermal gradient. The viscosity increases perceptibly as the pressure drops along the tubing. The above Fig. M1, which represents the viscosity variation according to the temperature, shows that a variation of temperature from 60° to 40°C increases the viscosity, for a 10°API crude oil, from 5 000 cp to about 50 000 cp, or 10 times. A pressure variation generated by an increase or a decrease of pressure drops will be expressed by output power variations of the drive motor.

1.5 Solutions for reducing the pressure drop [6]

There are several solutions for reducing pressure drops generated by the oil viscosity, when it is possible to use a diluent as it is shown in the next paragraph:
- Use tubing with larger diameter. The above relations show that pressure drops are varying with the power 4 of the diameter. Nevertheless, the tubing diameter is limited by the casing diameter, and a flow speed which will have to take into account the sand sedimentation possibly transported. However, when the oil is very viscous, the sedimentation speed of the sand is low.
- Use continuous drive strings, without sub-coupling which create restrictions.
- Insulate tubing near the wellhead in cold areas, or during the cold seasons of the year.

To conclude, it is better to use high capacity pumps running at low rotating speed, minimize restrictions, and reduce the formation oil viscosity with the injection of diluent.

1.6 Diluent effect on oil viscosity [6]

Figure M2 shows that the viscosity can be reduced by injection of diluent (light oil as kerosene, water with additives). It can be seen that an injection of 20% kerosene reduces the oil formation viscosity by 50, as well as for the frictional pressure drops.

The injection should be made at the pumping system inlet in order to help the mixing of the various oil specific gravities.

M2 HEAVY OIL PRODUCTION PROBLEMS [15]

In shallow, low pressure heavy oil wells, production is generally dominated by viscosity. Some types of damage will affect well production in most completions and can be categorized into three types.

2.1 Emulsions

Near wellbore emulsions can severely affect flow paths. Drilling and cementing deposit filter cake, drilling fines, cement particles, and fluid invasion in the pore spaces resulting in an area of reduced permeability.

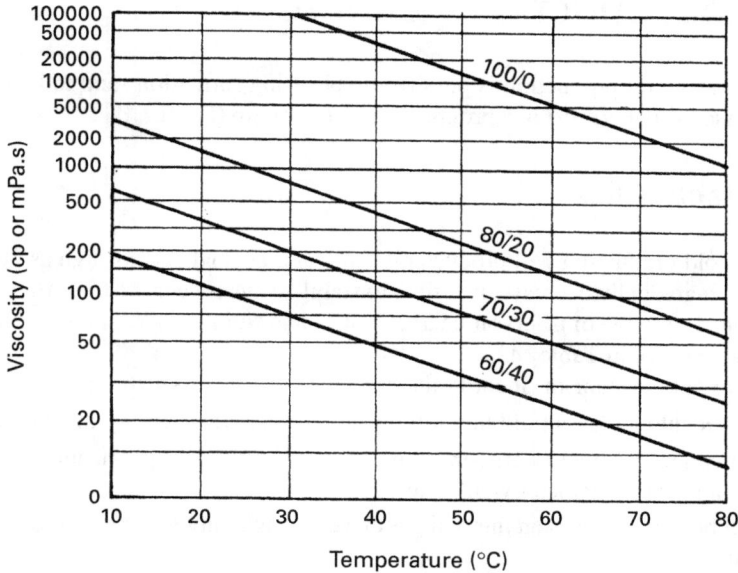

Figure M2 Heavy oil/kerosene dilution [6]. Viscosity of heavy oil versus temperature and heavy oil/kerosene ratio. (80/20: read 80% heavy oil volume/20% kerosene volume).

M

During perforating, large shock loads are exerted on the formation rock resulting in a crush zone with pulverization of the rock grains. The large surface area of this "rock flour" material when mixed with oil can form a viscous plug.

2.2 Visco-skin effect

The pressure drop across the sand face leads to evolution of solution gas from the oil in the near wellbore area resulting in a region of higher viscosity oil. This higher viscosity oil impedes the flow of the low viscosity oil immediately behind it.

2.3 Sand arches

Sand arches are formed by migration of sand particles during production. Under certain conditions adjacent sand grains can form a stable arch that will prevent the movement of the grains and oil behind the arch due to gravity or flow drag forces. These arches may be formed in both the near wellbore area and at a significant distance from the wellbore.

A related phenomenon known as "wormholes" has also been shown to exist. Wormholes can be defined as a circumstance where the majority of production for a given well may come from an isolated area of an otherwise homogeneous reservoir. This area may or may not be near the wellbore.

M3 COLD PRODUCTION TECHNOLOGY [4, 11, 12]

The primary heavy oil production is possible by allowing formation sand to be produced along with reservoir fluids using a progressing cavity pump (PCP).

3.1 Characteristics

Typically, a cold production will produce oil and water at sand cuts as high as 30 to 40% initially, which gradually decrease over time to stabilize at between 1 to 5% after one year of production. The range of reservoir characteristics and fluid properties amenable to cold production are shown in Table M2.

Four important mechanisms identify the dynamics of cold production:

- Sand movement increases fluid mobility by an intrinsic permeability increase effect.
- Sand production generates a growing zone of greatly enhanced permeability, therefore an expanding "large diameter well" effect occurs.
- Foamy oil behaviour gives an internal gas drive through bubble expansion as the slurry approaches the wellbore.
- Continuously moving sand means that fines blockage, gas bubble blockage, and asphaltene precipitation near the borehole are eliminated, thus "skin" effects are absent.

M

Table M2 Cold production application ranges to date [4].

Reservoir characteristics	U.S. units	Metric units
Depth	1 300–2 500 ft	400–760 m
Oil saturation	67 to 87%	67 to 87%
Porosity	30 to 34%	30 to 34%
Net pay	13 to 80 ft	4 to 25 m
Permeability	500 to 10 000 md	0.5 to 10 μm^2
Oil gravity	11 to 14°API	0.990 to 0.970
Reservoir pressure	400 to 848 psi	2.8 to 6 MPa
Reservoir temperature	61 to 70°F	16 to 21°C
Oil viscosity at reservoir temperature	600 to 160 000 cp	600 to 160 000 mPa.s
Gas oil ratio (GOR)	0–56 scf/bbl	0–10 m^3/m^3

3.2 Other factors contributing to cold production

- Higher PC pumps allow co-production of the sand with the heavy oil.
- Sand production is encouraged through large diameter perforations in vertical wells and wide slots in horizontal well liners.
- Perforating with large perforation density.

M4   GENERAL CLASSIFICATION OF ENHANCED OIL RECOVERY (EOR) PROCESSES [1, 2, 3, 5]

4.1   Thermal methods

Thermal methods are steam injection and in-situ combustion. In these methods the viscosity reduction in crude oil is brought about by the application of heat into the reservoir.

4.1.1   Steam injection

Steam is generated at the surface in an electric, gas or oil fired boiler and injected into the reservoir through insulated tubing, as shown in Fig. M3.

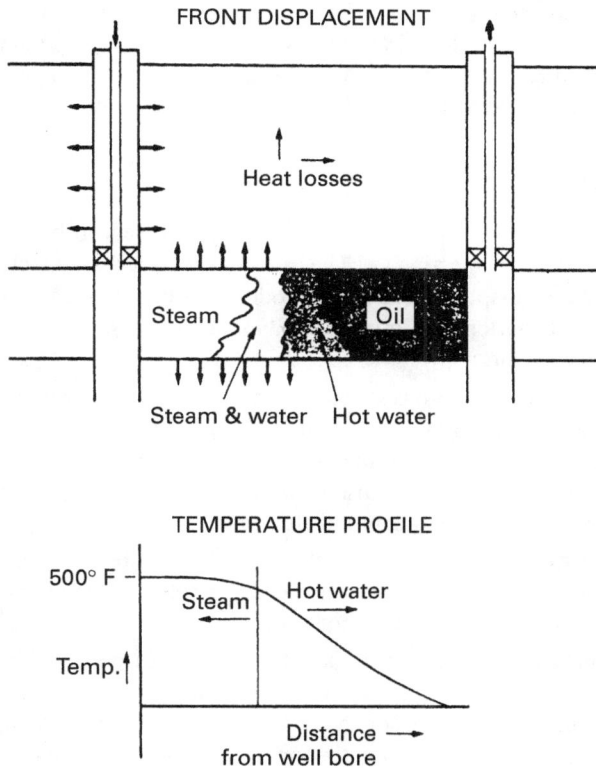

Figure M3   Steam injection [5].

Field tested processes are:

- **Cyclic steam stimulation** (huff and puff). It is a single well process where steam is injected for a period of time before the well is put on production, either immediately or after a soak period.

- **Fracture-assisted steam stimulation**. Steam is injected into the reservoir at pressures above the parting pressure for the formation. Steam injection continues for about 30 days after which the well is put on production and large amounts of bitumen can be produced at high rates and a steam/oil ratio (SOR) around 3.

- **Continuous steam drive**. It is a multi-well process in which steam is injected in one or more wells and water and oil are produced from nearby wells.

- **Heated annulus steam drive**. A long, unperforated horizontal well is placed near the base of the reservoir. Steam is circulated through the well to heat the surrounding reservoir and mobilize the oil. Oil is recovered by injecting high pressure steam into a vertical well landed in the heated zone with recovery from one or more offsetting vertical wells at intervals along the heated horizontal well.

- **Steam assisted gravity drainage (SAGD)**. The process involves forming a steam chamber above a horizontal well at the base of the reservoir, injecting steam into the chamber and producing oil and condensed steam that flow to the horizontal well by gravity drainage. More details about this process may be found in Paragraph M5.

4.1.2 In-situ combustion

In the in-situ combustion process, heat is generated in the reservoir by igniting the oil and maintaining combustion by injecting air. The heat generated serves to increase the mobility of the oil, permitting its displacement to the producing wells. Fig. M4 illustrates the temperature and fluid distribution in a forward combustion process.

Field tested processes are:

- **Forward combustion**. The fire is started downhole at the point where air is injected. A combustion front is therefore located in the bed, ahead of which a hot zone causes the vaporization and cracking of the molecules in place. In front of this hot zone is a cold zone where the vapors are recondensed. As the front advances, it finds only coke to burn. In the combustion zone, the temperature may exceed 600°C.

- **Reverse combustion**. Air is injected until it reaches the production well. The fire is ignited by an electrical system downhole. Hence, a countercurrent displacement of the combustion front and of the hydrocarbons will occur.

- **Top down combustion**. The conceptual strategy of this process involves the stable propagation of a high temperature combustion front from the top to the bottom of a heavy oil or bitumen containing reservoir. Combustion is initiated and maintained by injecting an oxygen containing gas at the top of the reservoir, with mobilized oil draining to a lower horizontal production well.

- **Combustion override split-production horizontal-well (COSH)**. This new process is designed to overcome the early oxygen breakthrough, sanding and pump gas locking problems typically associated with in-situ combustion. In a heavy oil reservoir, vertical wells are drilled in rows and perforated near the top of the pay zone. Vertical or horizontal gas producers are drilled between the injector rows. These wells are also completed near the top of the pay zone. If vertical wells are used they may be recompleted in the lower part of the reservoir once combustion is well established. A horizontal production well is drilled near the base of pay zone. Air is injected at sub-fracture pressure and the oil ignited. The combustion gases override the oil, form a high gas saturation layer at the top of the pay zone and are eventually produced at the gas production wells. A counter-current flow system is formed near the top of the pay zone once combustion is established. Oil and water, heated by the combustion are drained along the gas/liquid interface and are eventually produced at the horizontal production well.

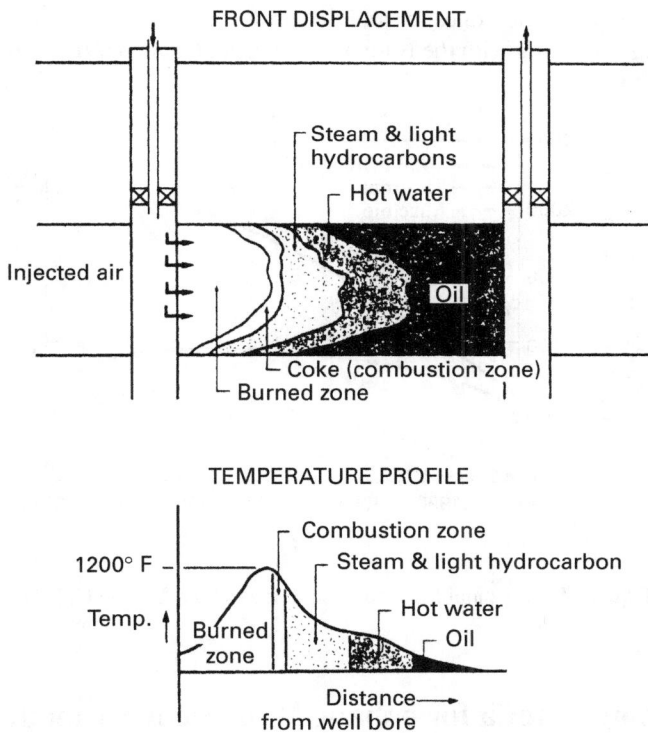

Figure M4 In-situ combustion [5].

4.2 Chemical flooding

Chemical flooding is accomplished by injecting chemicals in order to lower the interfacial tension between oil and water which mobilizes by-passed oil in the reservoir, and/or improves mobility ratio.

4.3 Miscible flooding

Miscible flooding is a process where total or partial miscibility of CO_2, hydrocarbon gases, N_2 or flue gas with crude oil, is accomplished by injecting them into the reservoir under optimum conditions.

Moreover, CO_2, which dissolves in the oil, increases the volume of the oil considerably (20 to 100%) and significantly lowers its viscosity. This is valid for heavy oils, even if CO_2 is not miscible with them.

4.4 Potential of the different processes [14]

Actual cumulative production rates for all EOR projects in the U.S. for the years 1984-95 are shown in Fig. M5 along with the National Petroleum Council (NPC) projections.

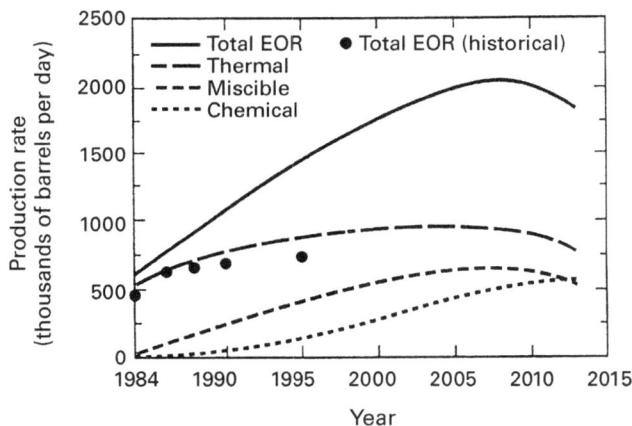

Figure M5 Potential EOR production rates (*from* U.S. NPC) [14].

4.5 Screening criteria for enhanced oil recovery methods [14]

Screening criteria for the different EOR methods are detailed in the following tables. In Table M3 are treated the gas injection methods, in Table M4 the chemical flooding methods, and in Table M5 the thermal methods.

Table M3 Gas injection methods.

EOR method	Oil properties				Reservoir characteristics				
	°API	Viscosity (cp)	Composition	Oil saturation (%PV)	Formation type	Net thickness (ft)	Average permeability (md)	Depth (ft)	Temperature (°F)
Nitrogen (& flue gas)	>35→48	<0.4→0.2	High % of C_1–C_2	>40→75	Sandstone or Carbonate	Thin unless dipping	N.C.	>6 000	N.C.
Hydrocarbon	>20→41	<3→0.5	High % of C_2–C_7	>30→80	Sandstone or Carbonate	Thin unless dipping	N.C.	>4 000	N.C.
Carbon dioxide	>22→36	<10→1.5	High % of C_5–C_{12}	>20→55	Sandstone or Carbonate	Wide range	N.C.	>2 500	N.C.

Underlined values represent the approximate mean or average for current field projects.
N.C.: not critical.

Table M4 Chemical methods.

EOR method	Oil properties				Reservoir characteristics				
	°API	Viscosity (cp)	Composition	Oil saturation (%PV)	Formation type	Net thickness (ft)	Average permeability (md)	Depth (ft)	Temperature (°F)
Micellar/ polymer, Allkaline/ polymer (ASP), and Allkaline flooding	>20→35	<35→13	Light, intermedi­ate Some organic acids for alkaline floods	>35→53	Sandstone preferred	N.C.	>10→450 →3 250	<9 000	<200→80
Polymer flooding	>15→40	<150,>10	N.C.	>70→80	Sandstone preferred	N.C.	>10^3 →800	<9 000	<200 →140

Underlined values represent the approximate mean or average for current field projects.
N.C.: not critical.

Table M5 Thermal methods.

EOR method	Oil properties				Reservoir characteristics				
	°API	Viscosity (cp)	Composition	Oil saturation (%PV)	Formation type	Net thickness (ft)	Average permeability (md)	Depth (ft)	Temperature (°F)
Combustion	>10→16	<5 000 →1 200	Some asphaltic compo­nents	>50→72	High porosity sand/ sandstone	>10	>50^4	<11 500 →3 500	>100 →80
Steam flooding	>8-13.5 →?	<200 000 →4 700	N.C.	>40→66	High porosity sand/ sandstone	>10	>200^5	<4 500 →1 500	N.C.

Underlined values represent the approximate mean or average for current field projects.
N.C.: not critical.

M

265

M5 STEAM ASSISTED GRAVITY DRAINAGE AND VAPOR EXTRACTION PROCESSES [2, 7, 8]

5.1 Steam assisted gravity drainage (SAGD)

The SAGD technique appears to be most promising for the exploitation of heavy oil and bitumen. In this process, steam is injected through a horizontal well and forms a steam chamber in the reservoir sand. Steam condenses at the bitumen interface and heats the oil. The hot oil, being less viscous, drains by gravity to the horizontal production well. The concept of this process is shown in Fig. M6

Characteristics

* Steam condenses at the interface.
* Oil and water drain to the production well at the base of the reservoir.
* Flow is caused by gravity.
* Chamber grows upwards and sideways.
* Mostly governed by thermal conduction.
* Large oil productivity.
* Large recovery (50 to 60%).
* Low Steam/Oil Ratio (SOR).
* Production: 0.25 to 0.5 m^3/day/m of horizontal well.

M

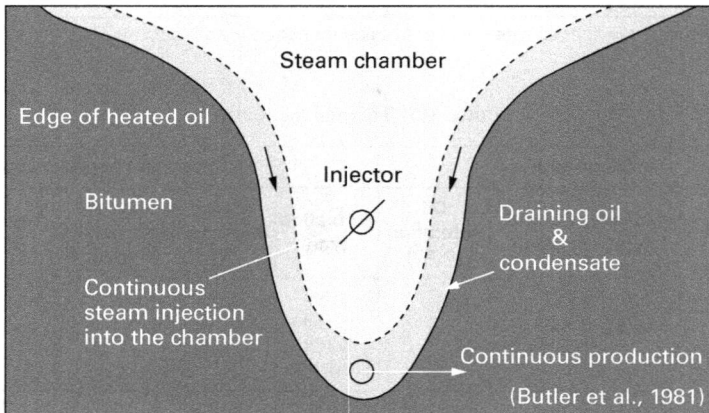

Figure M6 The concept of steam assisted gravity drainage process [7, 8].

266

5.2 Vapor extraction process (VAPEX)

VAPEX is an in-situ recovery process for bitumen and heavy oil that utilizes a vaporized solvent such as propane. It operates at reservoir temperatures and utilizes gravity to achieve contact of the solvent vapor with the crude oil and to transport the solution to the base of the reservoir.

5.3 Comparison of SAGD and VAPEX processes [8]

Table M6 shows screening criteria for application of these processes in heavy oil and bitumen reservoirs.

Table M6

	SAGD	VAPEX
Temperature and pressure requirements	Operating temperature is guided by the injection pressure which depends on the reservoir pressure. The problem may be less severe when there is a lack of confinement of the vapor chamber.	Operating may be close to the reservoir pressure. Due to the low temperature operation, the energy dissipation will be negligible. It is not necessary to have a lack of confinement of the vapor chamber.
Operational constraints	SOR = 2.5 - 3, i.e. about 3 volumes of steam as condensed water equivalent is used per 1 volume of produced oil. Process is not recommended for formations with a swelling clay content of more than 10%.	Less than a volume of solvent (liquid equivalent) under reservoir condition is injected per volume of oil produced.
Aquifer constraints	If a steam chamber contacts an aquifer, some of the injected steam is llikely to be lost there due to the high relative permeability to water. Oil lost in aquifer.	The solvents are virtually insoluble in water therefore losses are minimal.
Effect of gas cap	A thin gas cap may not be a counter-indicator for the application.	Solvent losses to the gas cap may adversely affect the process economics.
Upgrading potential	No significant upgrading except some thermal degradation of the asphaltic components.	It is possible to de-asphalt the crude to the desired extent by controlling the solvent pressure.
Oil nature (°API)	Not critical	Important

5.4 Common features of SAGD and VAPEX processes [8]

SAGD and VAPEX present the following same characteristics:
- They are applicable for the extraction of high viscosity crudes.
- Gravity drainage is the drive mechanism.

- A pair of horizontal wells is the preferred configuration.
- Confinement of the vapor chamber is an essential criterion for the economic success.

5.5 Factors controlling drainage rate for an established vapor chamber [8]

- **Viscosity of oil within and around the vapor chamber**. Rate becomes lower with an increase in oil viscosity. However, the decrease in production rate is much slower than the increase in viscosity.
- **Reservoir heterogeneity**. It controls the shape of the vapor/steam chambers, and may alter the performance significantly.
- **Chamber dimension**. The production rate is proportional to the square root of the height of the chamber.
- **Diffusivity** (thermal as well as molecular). Convection and turbulence at a microscopic level may enhance the heat and mass transfer yielding a higher rate of extraction.
- **Interfacial tension within the vapor chamber**. Lower interfacial tension alters the capillary gravity equilibrium reducing the residual oil saturation.

M

5.6 Factors promoting an economic project [8]

1. Unbroken pay zone thickness >10 m.
2. Gas caps/bottom water thinner than 1 m. Thicker gas cap/bottom water associated with pay zones thicker than 20 m may also yield an economic production rate.
3. Permeabilities in the range of 1 darcy or higher are desirable.
4. Mobile oil content per volume of the reservoir should be >500 bbl/acre-ft ($0.25 \ m^3/m^3$) of the reservoir.
5. The in-situ viscosity may be a guiding factor for optimizing the well configuration.
6. The geological factors favouring large confined steam/vapor chambers, and the recovery of a significant portion of the injected vapor during the blow down phase, are of obvious advantage.

REFERENCES

1 Okandan E (1984) *Heavy Crude Oil Recovery*. Martinus Nijhoff Publishers, V–VII
2 Townson DE (1997) Canada's heavy oil industry: A technological revolution. *Paper SPE 37972*, presented at SPE International Thermal Operations and Heavy Oil Symposium, Bakersfield, CA
3 Cossé R (1993) *Basics of Reservoir Engineering*, Chapter 8. Editions Technip, Paris
4 Bill Huang WS, Marcum BE, Chase MR and Yu CL (1997) Cold production of heavy oil from horizontal wells in the Frog Lake field. *Paper SPE 37545*, presented at the SPE International Thermal Operations and Heavy Oil Symposium, Bakersfield, CA

5   Ryalls P (1985) Reservoir problems associated with heavy oil, presented at *Seminar Problems Associated with the Production of Heavy Oil*, 18th March 1985, London, organised by Oyez Scientific & Technical Sevices Ltd

6   Cholet H (1997) *Progressing Cavity Pumps*, Chapter 3. Editions Technip, Paris

7   Butler RM, McNab GS, Lo HY (1981) Theoretical studies: The gravity drainage of heavy oil during steam heating. *Canadian Journal of Chemical Engineering*, Vol. 59, Aug. 1981, 455–460

8   Singhal AK, Das SK, Leggitt SM, Kasraie M and Ito Y (1996) Screening of reservoirs for exploitation by application of steam assisted gravity drainage/Vapex processes. *Paper SPE 37144,* presented at the 1996 SPE International Conference on Horizontal Well Technology, Calgary, Alberta, Canada

9   Das SK (1998) Vapex: an efficient process for the recovery of heavy oil and bitumen. *SPE Journal*, September 1998, 232–237

10   Edmunds N (1999) On the difficult birth of SAGD. *Journal of Canadian Petroleum Technology*, January 1999, 14–17

11   Deruyter C, Moulu JC, Nauroy JF, Renard G, Sarda JP (1998*) Production sans réchauffement des huiles visqueuses,* Bibliography. IFP internal report

12   Dusseault MB (1990) Canadian heavy oil production experience using cold production. *Paper SPE,* introduced to the 12th SPE Trinidad and Tobogo Section Biennal Technical Conference and Exposition, March 1998

13   Bavière M et al (1996) Basic concepts in enhanced oil recovery processes. *Critical Reports on Applied Chemistry*, Volume 33. Published for SCI by Elsevier Applied Science, London and New York

14   Green DW, Willhite GP (1998) *Enhanced Oil Recovery*. Henry L. Doherty Memorial Fund of AIME, Society of Petroleum Engineers, Richardson, Texas

15   Haney BL, Cuthill DA (1997) The application of an optimized propellant stimulation technique in heavy oil wells. *Paper SPE 37531,* presented at the SPE International Thermal Operations and Heavy Oil Symposium, Bakersfield, CA

16   Huc AY, Ed. (to be published) *Heavy Oils: Production and Upgrading*. Editions Technip, Paris.

M

N

Artificial Lift

N

Artificial Lift

N1 COMMON METHODS

Artificial lift is used when the pressures in the oil reservoir have failed to the point where a well will not produce at its most economical rate by natural energy. The most common methods of artificial lift are:

- Plunger lift and sucker rod pumping (Chapter O)
- Gas-lift (Chapter P)
- Electric submersible pumping (ESP) (Chapter Q)
- Progressing cavity pumping (PCP) (Chapter R)
- Hydraulic pumping (Chapter S).

1.1 Artificial lift efficiency [1]

Table N1

System	Efficiency (%)
Rod pump	30–40
Gas lift	25–32
Electric submersible pump	50–60
Progressing cavity pump	60–80
Jet pumping	10–25
Hydraulic pump	30–40

1.2 Artificial lift limitations [1, 2]

Depending on application, solid handling capabilities, bottomhole temperature, and volume capabilities, Table N2 gives indications and limitations to optimize the choice of an artificial lift method.

1.3 Adapting well activation processes to operating conditions [2]

Many criteria take into account the production constraints. They are reported in Table N3: environment, positioning in the well of the artificial lift system, characteristics of the reservoir fluid, type of completion.

	Rod pump	Gas lift	ESP	PCP	Hydraulic piston pump	Hydraulic jet pump
Offshore application	Good if space for power system	Excellent	Good Little surface equipment needed	Good if pulling unit available	Good if space for power system	Good Can use prod. water as power fluid
Solid handling capabilities	Prod. less than 200 ppm Power less than 10 ppm	No problem	Prod. less than 200 ppm	No problem up to 5%	Prod. less than 200 ppm Power less than 10 ppm	Prod. up to 3% Power up to 200 ppm
Maximal temperature	260°C	175°C	160°C	120°C	260°C	260°C
Volume capabilities (turndown)	less than 100 m^3/d	8–8 000 m^3/d (good)	40–8 000 m^3/d (fair)	5–800 m^3/d (good)	15–800 m^3/d (good)	15–2 500 m^3/d (fair)

Table N3

Systems	Rod pump	Gas lift	ESP	PCP	Hydraulic piston pump	Hydraulic jet pump
Environment						
Offshore	*	**	**	**	***	***
Desert	**	**	*	**	**	***
Urban area	**	**	**	***	***	***
Insulated well	***	–	*	**	***	***
Many wells	*	***	**	***	***	***
Positioning						
Very deep	*	**	–	*	**	***
Low pressure	***	*	**	***	*	*
High temperature	**	***	–	*	***	***
Fluids						
Viscous	*	*	–	***	**	**
Corrosive	*	**	–	**	***	***
Abrasive	*	**	–	**	*	*
Depositing	*	–	*	*	**	**
Emulsifiable	***	*	*	**	***	–
High GLR	*	**	–	**	*	**
Completion						
Multilayer	*	**	–	*	**	**
Very slanted	*	*	*	***	***	***
TFL	–	**	–	–	–	**
Dogleg	–	***	*	**	****	***

*** Very good   * Acceptable
** Good   – Unsuitable except for a specific solution.

N2 PRODUCTION CRITERIA

The selection of the pump and its positioning in the well depend on:
- The flowrate
- The dynamic level and bubble point
- The admissible head rating of the pump (column height to be discharged, pressure drops due to friction generated by the effluent viscosity, and the wellhead pressure)
- The abrasion of the pumped product (sand).

2.1 Determination of the positioning level

2.1.1 Position according to the dynamic level (or submergence level)

In spite of its self-priming characteristic, a submergence of about 100 m (300 ft) is sufficient.

2.1.2 Position with respect of the bubble point level

Figure N1 is a representation of the evolution of the pumping conditions according to the GOR, and the influence of the pump position according to the dynamic level and bubble point.

For example, at GOR = 0, the oil flow rate is about 7.2 m^3/d. But, if the pump is positioned at a higher level (GOR = 0.13, as indicated on the figure), it is then necessary to run it at a higher speed in order to generate a flow rate of oil + gas = 8.3 m^3/d, in order to keep the same oil production of 7.2 m^3/d.

2.2 Evaluation of the minimum head rating of the pump

It is the sum of:
- Well pressure
- Pressure generated by the column height to be discharged from the dynamic level to surface
- Pressure drop generated by the viscosity of the effluent:

$$\Delta P_f = \frac{7.05 \times 10^{-5}}{(D+d)(D-d)^3} \times Q \times \mu_f \times L \times \frac{1}{\ln \dfrac{\mu_s}{\mu_f}} \times \left(\frac{\mu_s}{\mu_f} - 1 \right) \tag{N1}$$

where
ΔP_f pressure drop due to friction (MPa)
D inside diameter of the tubing (cm)
d rod string diameter (cm)
Q pumped flow rate (m^3/d)
μ_f viscosity of the effluent at the inlet temperature (cp or mPa.s)
μ_s viscosity of the effluent at the surface (cp or mPa.s)
L length of the tubing (m).

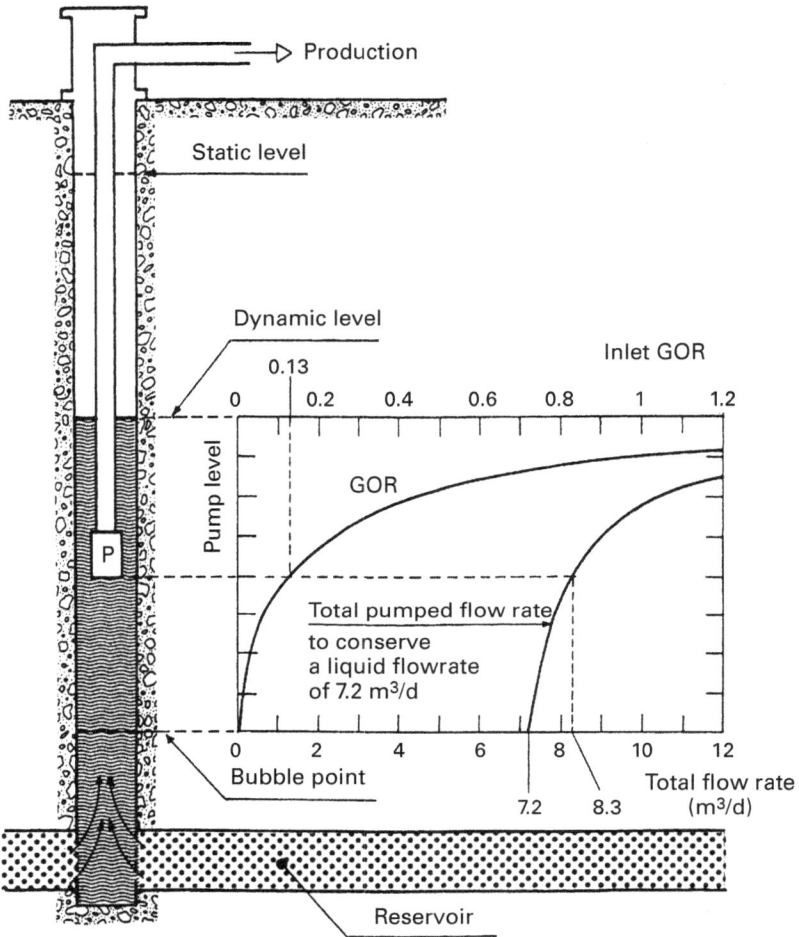

Figure N1 Incidence on the flowrate of the pump position in an oil well containing gas [4].

N3 GOR CALCULATION AT THE PUMP INLET [3, 4]

3.1 Solution gas/oil ratio

• *In metric units*

$$R_s = 0.342Y_g \times \left[\frac{P_b}{10^{0.0091 \times (1.8T+32)}} \times 10^{0.0125 \times °\text{API}} \right]^{1.2048} \tag{N2}$$

- *In U.S. units*

$$R_s = Y_g \times \left[\frac{P_b}{10^{0.0091 \times (1.8T+32)}} \times \frac{10^{0.0125 \times °API}}{18} \right]^{1.2048} \tag{N3}$$

where

Y_g specific gravity of gas (air = 1)
P_b bubble point pressure (bar or psi)
T temperature at pump inlet (°C or °F).

R_s must be corrected "R_s(corr.)" when the submergence level is below the bubble point:

$$R_s(\text{corr.}) = R_s \times \text{factor}$$

The factor is taken from Fig. N2, by locating the ratio of "submergence pressure/bubble point pressure" on the graph and determining the corresponding ratio of "GOR remaining in solution/solution GOR at bubble point", which is then multiplied by the calculated R_s to get R_s (corr.).

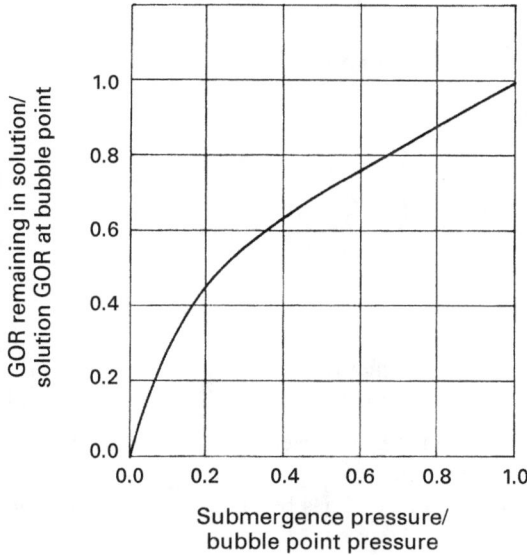

Figure N2 Non-dimensional gas liberation curve [5].

3.2 Gas volume factor

- *In metric units*

$$B_g = 0.00378 \times \frac{ZT}{P} \tag{N4}$$

- *In U.S. units*

$$B_g = 5.05 \times \frac{ZT}{P} \qquad \text{(N5)}$$

where

Z   gas compressibility factor (0.81 to 0.91)
T   bottomhole temperature (°K = 273 + °C)
P   submergence pressure (bar).

The gas volume factor B_g is expressed in m³/m³ or in bbl/mil st cu ft gas.

3.3 Formation volume factor

$$B_o = 0.972 + 0.000147. \times F^{1.175} \qquad \text{(N6)}$$

where

- *In metric units*

$$F = 5.61 \times R_s(\text{corr.}) \times \sqrt{\frac{Y_g}{Y_o}} + 1.25(1.8T + 32) \qquad \text{(N7)}$$

- *In U.S. units*

$$F = R_s(\text{corr.}) \times \sqrt{\frac{Y_g}{Y_o}} + 1.25 \qquad \text{(N8)}$$

where

Y_g specific gravity of gas
Y_o specific gravity of oil
T temperature at the pump inlet (°C or °F)
$R_s(\text{corr.})$ R_s(corrected) is defined from Fig. N2 (Paragraph 3.1 above).

The formation volume factor B_o represents the ratio of an oil volume in the formation compared to a stock tank volume.

From R_s, B_g, B_o, the percentage of gas by volume can be calculated.

3.4 Total volume of fluids

- **Total gas volume (free and in solution)** V_{gt}:

$$V_{gt} = \text{stock oil volume} \times \text{GOR}_{\text{standard}} \qquad \text{(N9)}$$

$\text{GOR}_{\text{standard}}$ is the gas/oil ratio measured at the surface.

- **Gas volume in solution at submergence pressure** V_{gs}:

$$V_{gs} = \text{stock oil volume} \times R_s(\text{corr.}) \qquad \text{(N10)}$$

- **Free gas volume** V_{gl}:

$$V_{gl} = V_{gt} - V_{gs} \tag{N11}$$

- **Oil volume in the formation** V_o:

$$V_o = \text{stock oil volume} \times B_o \tag{N12}$$

- **Gas volume in the formation** V_g:

$$V_g = V_{gl} \times B_g \tag{N13}$$

- **Water volume into the formation:** V_w

- **Total volume** V_t:

$$V_t = V_o + V_g + V_w \tag{N14}$$

Finally, for a given oil volume measured in the stock tank, the pump selection criteria must be the calculated volume V_t.

REFERENCES

1 Gray F (1995) *Petroleum Production in Nontechnical Language*. PennWell Books
2 Corteville J, Hoffmann F, Valentin E (1986) Activation des puits : Critères de selection des procédés. *Revue de l'Institut Français du Pétrole*, Vol. 41, No. 6, November-December 1986
3 *Oilfield Centrilift-Hughes Submersible Pump Handbook*. Centrilift-Hugues, Inc
4 Cholet H (1998) *Progressing Cavity Pumps*. Editions Technip, Paris
5 Nabla corporation. *Principles of Rod and Submersible Pumping*, p. 26
6 Brown KE (1967) *Gas Lift Theory and Practice*. Petroleum Publishing Co
7 Frick TC (1962) *Petroleum Production Handbook*. Vol. 2. SPE of AIME.

N

O

Beam Pumping
and Other Reciprocating Rod Pumps

Beam Pumping
and Other Reciprocating Rod Pumps

O1 GENERAL CONSIDERATIONS [1, 2]

The beam pumping system consists essentially of five parts:
- The subsurface sucker rod driven pump
- The sucker-rod string which transmits the surface pumping motion and power to the subsurface pump
- The surface pumping equipment which changes the rotating motion of the prime mover into oscillating linear pumping motion
- The power transmission unit or speed reducer
- The prime mover which provides the necessary power to the system.

Figures O1 and O2 illustrate the various components of a complete beam pumping system.

The minimum amount of information which must be known, or assumed, to determine even approximate loads and pump displacements for sucker-rod pumping unit installation design must include:

1. Fluid level
2. Pump depth
3. Pumping speed, in stroke per minute
4. Length of surface stroke
5. Pump plunger diameter
6. Specific gravity of the fluid
7. The nominal tubing diameter and whether it is anchored or unanchored
8. Sucker-rod size and design
9. Unit geometry.

O

283

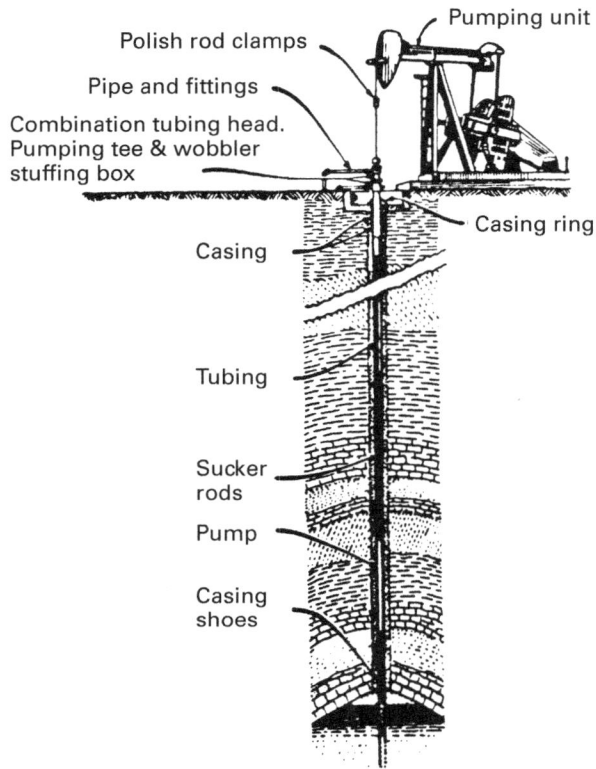

Figure O1 Beam pumping system [2].

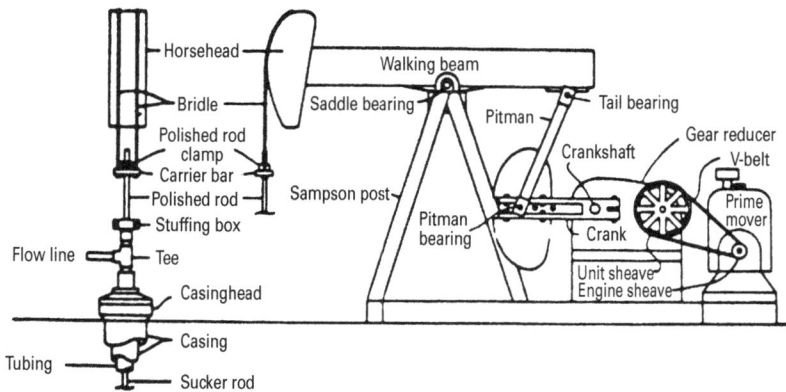

Figure O2 Surface equipment of a beam pumping unit [2].

With these factors, the designer should be able to calculate, with some degree of reliability, the following:

1. Plunger stroke
2. Pump displacement
3. Peak polished rod load
4. Mimimum polished rod load
5. Peak (crank) torque
6. Polished rod horsepower
7. Counterweight required.

O2 SUBSURFACE SUCKER-ROD PUMPS [1, 2]

2.1 Standard pump classifications

The standard designations for the various combinations are listed below. These classifications, taken from API Recommended Practice 11AR, are shown in Fig. O3.

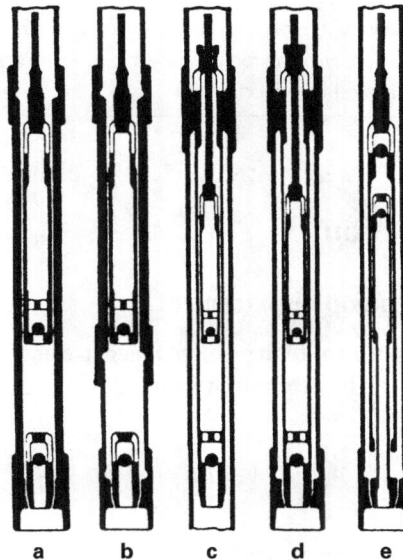

Figure O3 Diagrams illustrating standard pump classification [1].

The types of pumps are designed as indicated in Table O1.

Table O1 [1].

Ref	Type of pumps	Full barrel	Liner barrel
a	Tubing type, with shoe	TW	TL
b	Tubing type, with extension shoe and nipple	TWE	TLE
c	Rod type, stationary barrel, with top holddown	RWA	RLA
d	Rod type, stationary barrel, with bottom holddown	RWB	RLB
e	Rod type, traveling barrel	RWT	RLT

Definition of the symbols are given in Table O2.

Table O2 [1].

First letter	T	Tubing type; noninserted, run on tubing
	R	Rod or "inserted" type; run on rods; through tubing
Second letter	W	Full barrel
	L	Liner barrel
Third letter	E	Tubing pump with extension nipple and shoe
	A	Stationary-barrel rod pump with top holddown
	B	Stationary-barrel rod pump with bottom holddown
	T	Traveling-barrel rod pump

2.2 Pump size selection

2.2.1 Total theoretical pump displacement

The theoretical pump displacement for a given plunger size and for a given combination of pumping speed and stroke can be determined from:

- *In U.S. units*

$$V = KS_pN \tag{O1}$$

where
   V theoretical pump displacement (bbl/D)
   S_p effective plunger stroke (in.)
   N pump speed (spm).

- *In metric units*

$$V = 6.26\, KS_p N \tag{O2}$$

where
V theoretical pump displacement (m³/d)
S_p effective plunger stroke (m)
N pump speed (spm).

This applies if a pump constant is defined as:

$$K = 0.1484\, A_p \tag{O3}$$

where
A_p area of pump plunger (sq in.)
(see Table O3).

Table O3 Pump constant K [1, 2].

Plunger size (in.)	Constant K	Plunger size (in.)	Constant K
5/8	0.046	1 15/16	0.438
3/4	0.066	2	0.466
7/8	0.089	2 1/4	0.590
1	0.117	2 1/2	0.729
1 1/16	0.132	2 3/4	0.881
1 1/4	0.182	3	1.049
1 7/16	0.241	3 1/4	1.231
1 1/2	0.262	3 3/4	1.639
1 3/4	0.357	4 1/4	2.106
1 25/32	0.370	4 3/4	2.630

The effective plunger stroke is approximated at 80% of the surface stroke. Thus, the above equations can be written as:

- *In U.S. units*

$$V_{eff} = 0.8\, KSN$$

where S is the surface stroke in inches.

- *In metric units*

$$V_{eff} = 5\, KSN$$

where S is the surface stroke in meters.

2.2.2 Pump plunger size recommended

Table O4 can be used as a guide to determine pump plunger sizes for optimum conditions when surface stroke is less than 74 in. (or 1.90 m).

Table O4 Recommended pump plunger size [2].

Net lift of fluid (d=1)		Fluid production in bbl/D - 80% efficiency										Head rating	
(ft)	(m)	100	200	300	400	500	600	700	800	900	1000	(psi)	(MPa)
2 000	600	1 1/2	1 3/4	2	2 1/4	2 1/2	2 3/4	2 3/4	2 3/4	2 3/4	2 3/4	840	6
		1 1/4	1 1/2	1 3/4	2	2 1/4	2 1/2						
3 000	900	1 1/2	1 3/4	2	2 1/4	2 1/2	2 3/4	2 3/4	2 3/4	2 3/4	2 3/4	1 260	9
		1 1/4	1 1/2	1 3/4	2	2 1/4	2 1/4	2 1/2					
4 000	1 200	1 1/4	1 3/4	2	2 1/4	2 1/4	2 1/4	2 1/4	2 1/4			1 680	12
		1 1/2	1 3/4	2	2	2							
5 000	1 500	1 1/4	1 3/4	2	2							2 100	15
		1 1/2	1 3/4	1 3/4	2		2 1/4	2 1/4					
6 000	1 800	1 1/4	1 1/2	1 3/4	1 3/4							2 520	18
			1 1/4	1 1/2									
7 000	2 100	1 1/4	1 1/2									2 940	21
		1 1/8	1 1/4										
8 000	2 400	1 1/4										3 360	24
		1 1/8											
		16	32	48	64	80	96	112	128	144	160		
		Fluid production in m³/d - 80% efficiency											

▼ **Example**

A pump is to be set in a well at the working fluid level of 4 000 ft (1 200 m).

A value of 400 bbl/D (64 m³/d) of fluid at the surface is desired.

Thus, from Table O4, the suggested pump size to be installed is 2 in. or 2 1/4 in.

The exact size will depend upon other factors and other considerations to be discussed: pumping speed, effective plunger stroke...   ▲

2.2.3 Pumping speeds [1]

Maximum sucker-rod life comes from minimum pumping speeds because maximum load, range of load, and pumping speeds are directly proportional. Choose as long a stroke and slow a speed as the unit and prime mover will allow to make the desired production.

Figure O4 gives permissible speed and stroke based on 70% of maximum free-rod fall limit. For example, for a 54-in. (1.40 m) stroke, 26 spm is a maximum.

The product of speed and stroke giving 2 000 in./min (50 m/min) is the maximum rate. 70% of this is the maximum desirable. Vibratory problems also place a depth limit on pumping speeds.

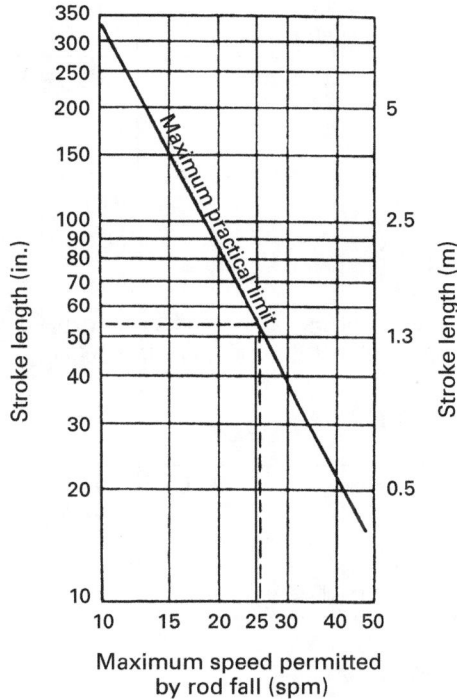

Figure O4 Permissible speed and stroke based on 70% of maximum free-rod fall limit [1].

O3 SUCKER RODS [1, 2, 8]

The sucker-rod string is a complex vibratory system that transmits energy from the surface equipment to the subsurface pump. Selection of a suitable sucker-rod string depends on the well depth and operating conditions. For well depths greater than 3 500 ft (1 000 m), it is common practice to use a tapered rod string that consists of different lengths of different rod sizes.

3.1 Mechanical properties

API classification of the mechanical properties are shown in Table O5.

Table O5 [8].

API grade	Minimum tensile strength		Maximum tensile strength	
	(psi)	(MPa)	(psi)	(MPa)
C	90 000	620	115 000	793
D	115 000	793	140 000	965

3.2 Service factors

The maximum allowable stress of the sucker-rod string depends on the service. Consequently, a service factor, defined in Table O6, must be applied.

Table O6 [2].

Service	API C	API D
Noncorrosive	1.00	1.00
Salt water	0.65	0.90
Hydrogen sulfide	0.50	0.70

3.3 Sucker-rod data

Sucker rods are furnished following the characteristics listed in Table O7.

Table O7 [8].

Rod size		Nominal diameter of pin		Metal area		Rod weight in air		Torque	
(in.)	(mm)	(in.)	(mm)	(sq in.)	(cm²)	(lb/ft)	(kg/m)	(lb-ft)	(daN.m)
1/2	12.7	3/4	19.1	0.196	1.27	0.72	0.993	90	12
5/8	15.9	15/16	23.8	0.307	1.98	1.13	1.66	180	24
3/4	19.1	1 1/16	27.0	0.442	2.85	1.63	2.37	300	41
7/8	22.2	1 3/16	30.2	0.601	3.88	2.22	3.17	450	61
1	25.4	1 3/8	34.9	0.785	5.07	2.90	4.20	675	91
1 1/8	28.6	1 9/16	39.7	0.994	6.41	3.67	5.36	900	122

3.4 Maximum rod loads [1]

By using the following formula, a reasonable estimate of the maximum well load may be obtained for average conditions.

- *In U.S. units*

$$P = W_r (1 + SN^2 / 70\,500) + W_f$$

- *In metric units*

$$P = W_r (1 + SN^2 / 1\,800) + W_f$$

where

P maximum polished-rod load (lbf or daN)
W_f weight of fluid based on net plunger area and head rating (lbm)
W_r weight of rods (lbm)
S polished-rod stroke (in. or m)
N strokes/min.

$(1 + SN^2 / 70\,500)$ in U.S. units or $(1 + SN^2 / 1\,800)$ in metric units is the acceleration factor (no-dimensional factor).

3.5 Minimum downstroke load [1]

Range of load, being the only undetermined factor of the three factors controlling sucker-rod life, must now be calculated. Assuming the simple harmonic motion and subtracting the same acceleration factor that was applied on the upstroke and the buoyant force on the rods, a rough approximation of minimum downstroke load may be evaluated as follows.

- *In U.S. units*

$$P_{min} = W_r (0.8725 - SN^2 / 70\,500)$$

- *In metric units*

$$P_{min} = W_r (0.8725 - SN^2 / 1\,800)$$

where

P_{min} minimum downstroke load (lbf)
W_r weight of rods (lbm)
S polished-rod stroke (in. or m)
N strokes/min.

3.6 Effects of maximum loads, minimum loads, and speed

The three controlling factors in sucker-rod life, related as they are, must be defined by limits to make all the foregoing calculations worthwhile. Figs. O4 and O5 give maximum pumping speeds, and the user is cautioned to select as slow a speed as is consistent with production volumes required, unit and prime-mover capacities.

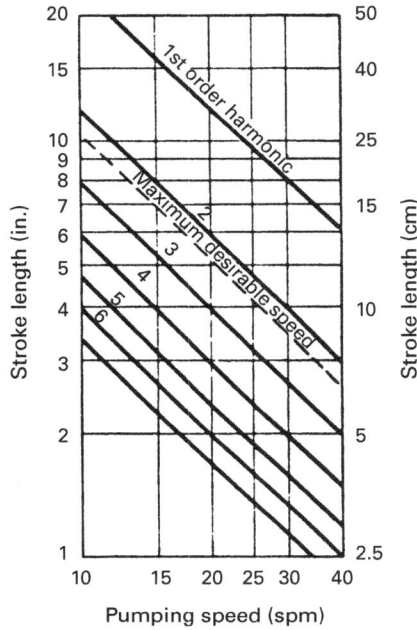

Figure O5 Synchronous speed curves [1].

Maximum load and range of load are linked in application although maximum-working-load values are based on average conditions and experience, which take into account an average amount of load range.

3.7 Joint make-up

The most important step in all rod handling is: **proper joint make-up**. API recommended practices include the torque value listed in Table O8.

Table O8 [1].

Rod size	<35 000 psi or <240 MPa rod strength		>35 000 psi or >240 MPa rod strength	
(in.)	(lbf-ft)	(N.m)	(lbf-ft)	(N.m)
1/2	110	150	120	160
5/8	220	300	240	330
3/4	350	470	385	520
7/8	520	700	570	770
1	800	1 080	880	1 190
1 1/8	1 100	1 490	1 210	1 640

O4 PUMPING UNITS [2]

4.1 Types

All beam type pumping unit geometrics fall into two distinct classes:
- The class I lever system which has its speed (gear) reducer-rear mounted with the fulcrum at mid beam, represented by the conventional unit (Fig. O6a). It is the most commun type of unit.
- The class III lever system, a push-up geometry with its speed reducer front-mounted, which is represented by the air balance (Fig. O6b), and Lufkin Mark II units, where the fulcrum is located at the rear of the beam (Fig. O6c).

Rod and structural ratings are expressed in terms of maximum polished rod loads, which can vary from 3 000 to over 42 000 lb (1.3 t to 20 t).

4.2 Qualitative performance characteristics of the different beam pumping unit types [2]

The performances of the three different types of beam pumping units may be classified as listed in Table O9.

4.3 Selection [1]

The selection of the proper size and type of pumping unit is based on well conditions and production requirements.

The size of a pumping unit is specified by the peak torque rating of the gear reducer, the length of stroke, and the polished-rod or beam capacity.

The following empirical formula will allow a rapid approximate evaluation of well conditions without the assistance of lengthy formulas.

Polished-rod load (PRL):

$$PRL = 1.25 \, (W_f + W_r) \ \text{(force = mass} \times \text{acceleration)}$$

where
- PRL polished-rod load (lbf)
- W_f weight of fluid based on net plunger area and head rating (lbm)
- W_r weight of rod string (lbm)
- 1.25   average acceleration factor (range 1.1 to 1.4).

Counterbalance (CB):

$$CB = W_r + W_f/2$$

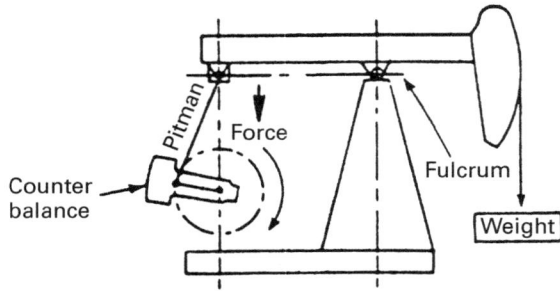

(a) CLASS I LEVER SYSTEM - CONVENTIONAL UNIT

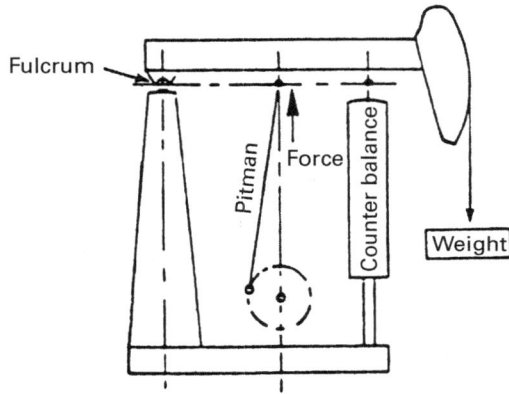

(b) CLASS III LEVER SYSTEM - AIR BALANCED SYSTEM

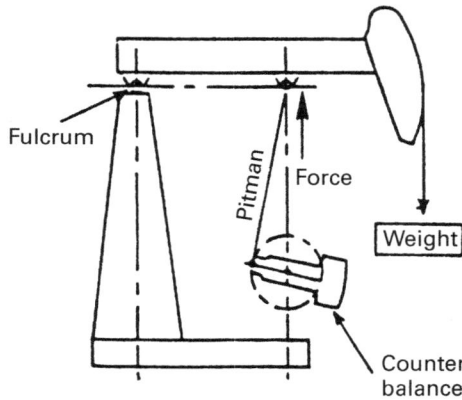

(c) CLASS III LEVER SYSTEM - LUFKIN MARK II

Figure O6 Geometries of a beam pumping unit [2].

Table O9 [2].

	Conventional unit	Air balanced unit	Mark II unit
1	Highly efficient	Normally somewhat less efficient than other unit types	Highly efficient
2	High reliability factor due to simplicity of design	Most complex of the unit types	Same reliability factor as conventional
3	Conventional unit cost used as reference	Occasionally higher cost than conventional	For comparable application, often about the same cost as conventional
4	Portability more limited than for air balance unit	Greatest ease of portability More compact	
5	Counterbalance more difficult to adjust	Counterbalance easily adjustable	Counterbalance more difficult to adjust
6	Generally widely-fluctuating torque loads	Torque peaks and range are often not as severe as on the conventional unit	Lowest and smoothest torque load with relative uniform torque system (Unitorque)
7	Impractical to install on two-point suspension for minimal foundation and base movement	Impractical to install on two-point suspension for minimal foundation and base movement	Can be mounted on two-point suspension for minimal foundation and movement
8	Relatively higher power cost and larger prime mover requirement	Relatively higher power cost and larger prime mover requirement	Generally, lowest power cost and smallest prime mover requirement because of uniform torque (Unitorque) system
9	Generally, higher rod and structural loads	Generally, lower rod and structural loads than conventional units	Generally, lowest rod and structural loads
10	Normally the highest (relative) maximum pumping speed	Often slightly reduced maximum pumping speed compared to conventional unit	Often slightly reduced maximum pumping speed compared to conventional unit
11	Less "fill-time" for subsurface pump barrel	Less "fill-time" for subsurface pump barrel	Greater "fill-time" for the subsurface pump barrel
12	Normally a lesser net plunger travel per stroke	Normally a lesser net plunger travel per stroke	Normally the greatest net plunger travel per stroke
13	Speed reducer maintenance nominal due to widely fluctuating torque loads and relatively small "flywheel" effect of cranks	Speed reducer maintenance may be less than on the conventional unit	Normally less speed reducer maintenance required than other unit types due to relative uniform torque system and substantial "flywheel" effect of cranks

Peak torque (PT):

$$PT = (PRL - CB) \times S/2$$

where

    PT   peak torque (lbf-in.).

Input power average:

- *In U.S. units*

$$HP_{avg} = 1.65 \times 10^{-5} D \times Q$$

- *In metric units*

$$WP_{avg} = 2.53 \times 10^{-4} D \times Q$$

where
HP	input power (hp)
WP	input power (kW)
D	well depth (ft or m)
Q	flow rate (bbl/D or m³/d).

Prime-mover input power (peak input power):

$$HP_{pm} = 1.5 HP_{avg} \quad \text{or} \quad WP_{pm} = 1.5 WP_{avg}$$

O5 LONG STROKE PUMPING UNIT [3]

The long stroke pumping unit (Fig. O7) is in the shape of a cylindrical column or a square tower equipped with two pulleys. The polished rod D is hung on two cables A. The cables are rolled up around the drum of a winch N and opposed to two other cables K which support the balancing counterweight B. This is all run by a motor that always goes in the same direction at a constant speed and drives a hydraulic motor P connected to the winch drum via a variable displacement pump. An inversion mechanism S driven by the winch automatically triggers the direction in which the winch rotates at the polished rod's top and bottom dead center. The pumping rate can range from 0 to 5 strokes per minute and is adjusted by changing the pump output. The stroke can go up to 10 m (or 394 in.), by regulating the inversion on the winch.

The ten-meter stroke allows the unit to pull its own rods out, thereby making further workover units unnecessary.

O6 JACK PUMPING UNIT [4]

The jack pumping unit (Fig. O8) uses a hydraulic jack to move the rod string. It consists of a jack (1) and its piston (2) connected to the polished rod and a stroke selector (3) in the axis of the well. The jack is connected by high pressure hoses to a hydraulic unit (4) with:
- (a) An electric motor or engine (electric, diesel, gas, etc.)
- (b) A hydraulic pump with a pumping rate selector
- (c) A pilot-operated slide valve (5) to transfer the oil from one side of the piston to the other

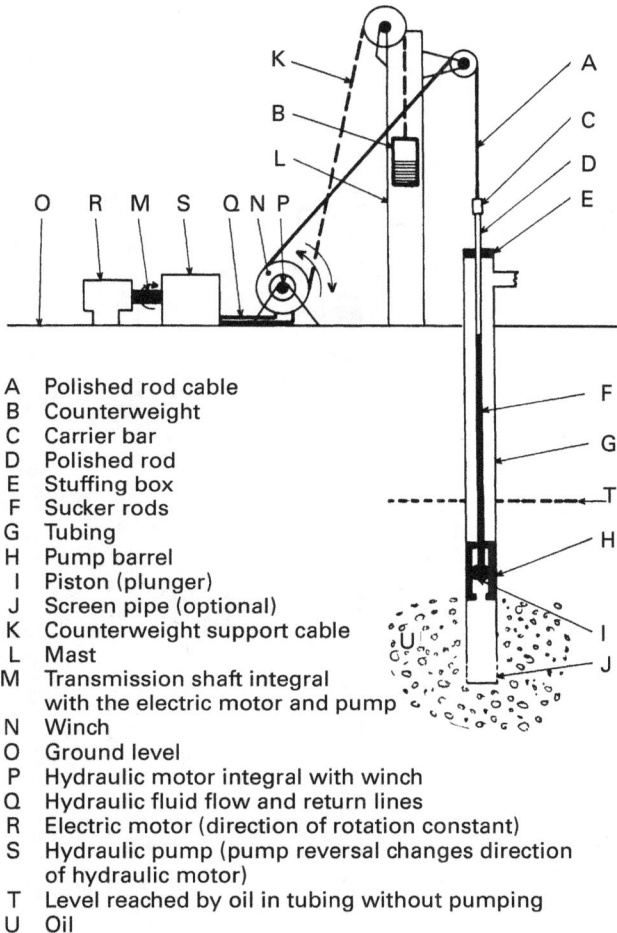

A Polished rod cable
B Counterweight
C Carrier bar
D Polished rod
E Stuffing box
F Sucker rods
G Tubing
H Pump barrel
I Piston (plunger)
J Screen pipe (optional)
K Counterweight support cable
L Mast
M Transmission shaft integral
   with the electric motor and pump
N Winch
O Ground level
P Hydraulic motor integral with winch
Q Hydraulic fluid flow and return lines
R Electric motor (direction of rotation constant)
S Hydraulic pump (pump reversal changes direction
   of hydraulic motor)
T Level reached by oil in tubing without pumping
U Oil

Figure O7 Long stroke pumping unit [3].

(d) An oil tank (6), accessible by a weighted valve for the return
(e) A pressure accumulator that acts as a counterweight: during the upstroke it adds its effect to the hydraulic pump and is inflated again by the pump during the down-stroke.

The manufacturer can provide hydraulic units that can activate one or two wells for cluster wells.

The stroke can go up to 9 m (360 in.).

In particular, this equipment uses a variable displacement pump that can reverse the flow gradually and smoothly, thereby lessening the fatigue and impacts on the system. Additionally, the energy efficiency of the installation is better.

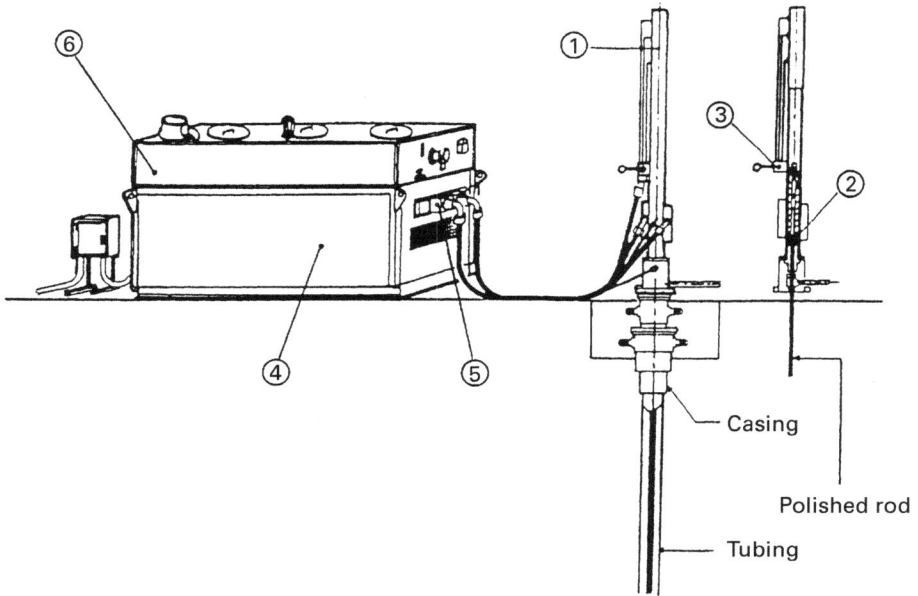

Figure O8 Jack pumping unit [4].

REFERENCES

1 Frick TC, Taylor BW (1962) *Petroleum Production Handbook*. SPE of AIME
2 Brown KE (1980) *The Technology of Artificial Lift Methods*, Vol. 2a. PennWell Books
3 MAPE, *Oil field product documentation*
4 Perrin D, Caron M, Gaillot G (1998) *Well Completion and Servicing*. Editions Technip, Paris
5 API RP 11L (1977, 1979) *Design Calculations for Sucker Rod Pumping Systems*
6 API RP 11AR (1968) *Recommended Practice for Care and Use of Subsurface Pumps*
7 API Spec 11E (1989) *Specification for Pumping Units*
8 API Spec 11B (1984) *Sucker Rods*
9 ISO 10428 (1998) *Sucker Rods (Pony Rods, Polished Rods, Couplings and Sub-couplings)*. Specifications
10 ISO 10431 (1998) *Pumping Units*. Specifications.

P

Gas Lift

Gas Lift

P1 TYPES OF GAS LIFT [1, 4, 5]

The gas can be injected either continuously or intermittently.

1.1 Continuous gas lift

Natural gas is injected at a given pressure and flow rate into the base of the production string or tubing.

This makes the density of the fluid in the production string lighter and allows the two-phase mixture to rise up to the surface, thereby making the well flow again.

1.2 Intermittent gas lift

A given volume of pressurized gas is injected at a high flow rate in the lower part of the production tubing. This gas flushes the volume of liquid it contains upward. As a result, the pressure on the pay zone decreases and it starts flowing again.

P

1.3 Limitations of continuous and intermittent gas lift

*   Continuous gas lift is suited to a liquid output of 30 to 3 000 m³/d.
*   Intermittent gas lift is used for outputs of less than 80 m³/d.
*   In the overlap between 30 and 80 m³/d, the preference goes to having the well produce by decreasing the diameter of the production tubing and prioritizing an increasing upward fluid velocity.
*   The intermittent method is the only gas lift technique possible in good producers that have a low downhole pressure because the reservoir pressure is initially low or has been depleted.
*   In general, 95% of gas lifted wells are produced by continuous gas lift.

Table P1 Limitations of continuous and intermittent gas lift [4].

	Continuous gas lift		Intermittent gas lift	
	U.S. units	metric units	U.S. units	metric units
Liquid output	200 to 20 000 bbl/D	30 to 3 000 m³/d	< 500 bbl/D	< 80 m³/d
Productivity index	> 0.45 bbl/D/psi	> 10 m³/d/MPa	< 0.45 bbl/D/psi	< 10 m³/d/MPa
Flowing pressure	> 0.08 psi/ft	> 1.8 kPa/m	> 145 psi	> 1 MPa
Injection GLR	50 to 250 ft³/bbl for 1 000 ft lifting	10 to 45 m³/m³ for 300 m lifting	250 to 300 ft³/bbl for 1 000 ft lifting	45 to 55 m³/m³ for 300 m lifting
Required inj. press.	> 100 psi for 1 000 ft lifting	> 700 Pa for 300 m lifting	< 100 psi for 1 000 ft lifting	< 700 Pa for 300 m lifting

P2 DESIGN OF A CONTINUOUS-FLOW GAS-LIFT SYSTEM [1, 4]

2.1 Introduction

In Fig. P1, depth is plotted vs. pressure and the pressure traverse of the fluids is shown as it starts from the flowing BHP at the bottom of the well and moves all the way to the surface.

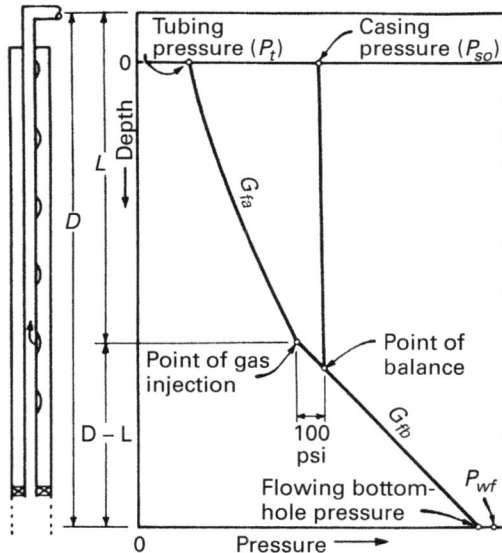

Figure P1 Continuous-flow well illustration [1].

If we assume an average flowing gradient beneath the point of gas injection and an average flowing gradient above the point of injection, we could start with the tubing pressure and calculate the flowing BHP as follows:

$$p_{wf} = p_t + G_{fa}L + G_{fb}(D - L) \tag{P1}$$

where

p_t tubing pressure
G_{fa} average flowing gradient above the point of injection
L vertical level of gas injection
G_{fb} average flowing gradient below the point of injection
D total vertical depth of well
p_{wf} flowing BHP.

2.2 Continuous-flow nomenclature

Figure P2 introduces the nomenclature used for continuous gas lift.

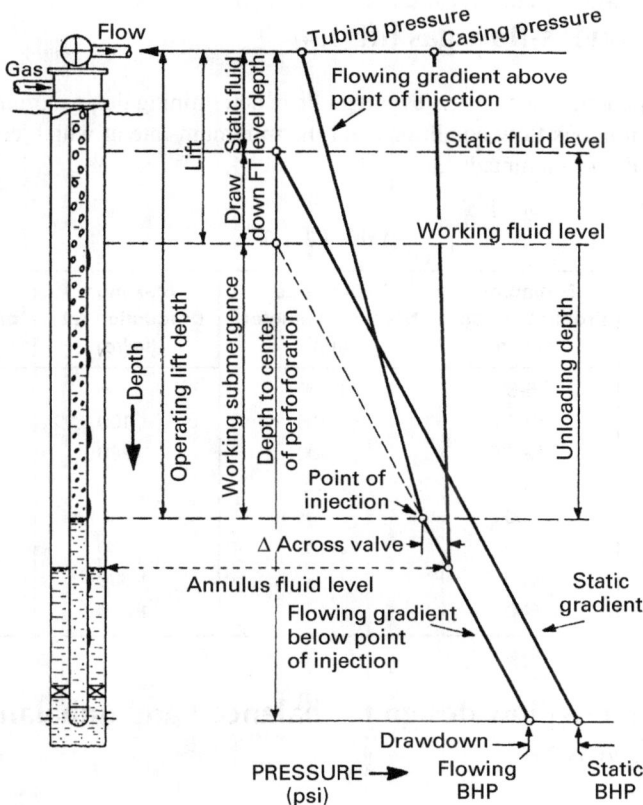

Figure P2 Continuous-flow nomenclature [1].

303

2.3 Approximations to be used in continuous-flow gas-lift installation

Table P2 specifies some of the constraints associated to continuous gas lift.

Table P2

Factors	U.S. units	Metric units
Pressure needed	100 psi min. per 1 000 ft of depth with a minimum of 300 psi	0.7 MPa min. per 300 m of depth with a minimum of 2.1 MPa
Gas volume needed	150–250 scf/bbl per 1 000 ft of lift	25–45 m³/m³ per 300 m of lift
Allowable continuous flow depth	$(p_c - p_t)/ 0.15$ = depth, in ft where p_c: casing pressure, in psi p_t: tubing pressure, in psi	$(p_c - p_t)/ 3.4$ = depth, in m where p_c: casing pressure, in kPa p_t: tubing pressure, in kPa

2.4 Tubing size and production rate in a continuous-flow gas lift [1, 4]

As an approximation, Table P3 serves as a guide in determining the maximum rate possible under good continuous flow conditions and the minimum rate at which continuous flow operations should be attempted.

Table P3 [1].

Maximum production rate (bbl/D)	Minimum production rate (bbl/D)	Tubing size Nominal diameter (in.)	Maximum production rate (m³/d)	Minimum production rate (m³/d)
350	25-50	1.050	55	4-8
600	50-75	1.315	100	8-12
1 800	75-125	1.900	280	12-20
3 000	200	2 3/8	480	30
4 000	250	2 7/8	640	40
6 000	300	3 1/2	950	50
8 000	400	4	1 300	65
10 000	600	4 1/2	1 600	100

2.5 Continuous-flow design for balanced and unbalanced gas-lift valves

- For balanced valves the surface opening pressures should be dropped by 25 psi per valve with a minimum of 15 psi per valve.

- For unbalanced valves that do not have to drop to the dome pressure to close (tapered stem) the valves should take a 10 psi drop per valve based on surface opening pressure (for continuous-flow operations it is better to work with surface opening pressures rather than surface closing pressures).
- For an unbalanced valve that exhibits spread and must drop to dome pressure to close, a minimum of 25 psi drop per valve in surface opening pressures should be taken.

2.6 Analytical solution to continuous-flow design

An analytical solution to the design of a continuous flow installation is recommended only when good BHP and PI information is not available on the well. It is common practice to select as the surface operating pressure of the uppermost valve a pressure equal to the gas pressure available less 50 psi.

To calculate valve installations analytically, the following formulas (in U.S. units) are used.

- **Valve 1:**

$$D_{v1} = \frac{(p_{ko} - 50) - p_{wh}}{G_s} \tag{P2}$$

where

D_{v1} depth from surface to valve 1 (ft)
p_{ko} available gas pressure for kick-off (psi)
p_{wh} surface back pressure on the tubing (psi)
G_s gradient of the fluid to be unloaded (psi/ft).

If balanced valves are used, a 25 psi drop per valve should be used, starting with the first valve being set with a surface operating pressure of 50 psi less than kick-off pressure.

- **Valve 2:**

$$D_{v2} = D_{v1} + \frac{\left[p_{so1} - (G_u \times D_{v1}) - p_{wh}\right]}{G_s} \tag{P3}$$

where

p_{so1} surface operating pressure of valve 1 (psi)
p_{so} normal gas pressure available at the well for lifting (psi)
p_{so2} $= p_{so} - 75$
p_{so3} $= p_{so} - 100$
p_{so4} $= p_{so} - 125$, etc.
G_u unloading gradient (from Figs. P3 and P4) with the preselected design daily production rate and tubing size (psi/ft).

- **Valve 3:**

$$D_{v3} = D_{v2} + \frac{\left[p_{so2} - (G_u \times D_{v2}) - p_{wh}\right]}{G_s} \tag{P4}$$

- **Valve 4:**

$$D_{v4} = D_{v3} + \frac{\left[p_{so3} - (G_u \times D_{v3}) - p_{wh}\right]}{G_s} \tag{P5}$$

and so on.

▼ **Example** (pages 306 to 310)

The following is an example of spacing model for balanced valves under continuous flow. Assume that only the following informations are known:
- Depth = 8 000 ft (or 2 440 m)
- Desired rate = 700 bbl/D (or 110 m³/d)
- Water = 95%
- Tubing size = 2 3/8-in. OD
- p_{wh} = 100 psi (or 700 kPa)
- Temperature at bottom = 210°F (or 100°C)
- Surface flowing temperature = 150°F (or 66°C)
- p_{so} available = 900 psi (operating), (or 6 300 kPa)
- p_{ko} = 950 psi (or 6 650 kPa)
- G_s ("kill" fluid gradient) = 0.50 psi/ft, (or 11.5 kPa/m).

Following the outlined procedure (Eq. P2):

$$D_{v1} = \frac{(p_{ko} - 50) - p_{wh}}{G_s}$$

where: $p_{wh} = 0$ for first valve only:

$$D_{v1} = \frac{(950 - 50) - 0}{0.50} = \frac{900}{0.50} = 1\,800 \text{ ft} \quad \text{(or } 550 \text{ m)}$$

Reference to Figs. P3 and P4 shows an unloading of 0.16 psi/ft (or 3.8 kPa/m) for 2 1/2-in. OD tubing and 1 000 bbl/D (or 160 m³/d). (Use a slightly higher design rate than the desired production of 700 bbl/D.) Spacing valves as previously outlined we have for the following valves (from Eqs. P2 to P5):

$$D_{v2} = 1\,800 + \frac{900 - (0.16 \times 1\,800) - 100}{0.50} = 2\,824 \text{ ft} \quad \text{(or } 860 \text{ m)}$$

$$D_{v3} = 2\,824 + \frac{875 - (0.16 \times 2\,824) - 100}{0.50} = 3\,470 \text{ ft} \quad \text{(or } 1\,058 \text{ m)}$$

$$D_{v4} = 3\,470 + \frac{850 - (0.16 \times 3\,470) - 100}{0.50} = 3\,860 \text{ ft} \quad \text{(or } 1\,176 \text{ m)}$$

$$D_{v5} = 3\,860 + \frac{825 - (0.16 \times 3\,860) - 100}{0.50} = 4\,070 \text{ ft} \quad \text{(or } 1\,240 \text{ m)}$$

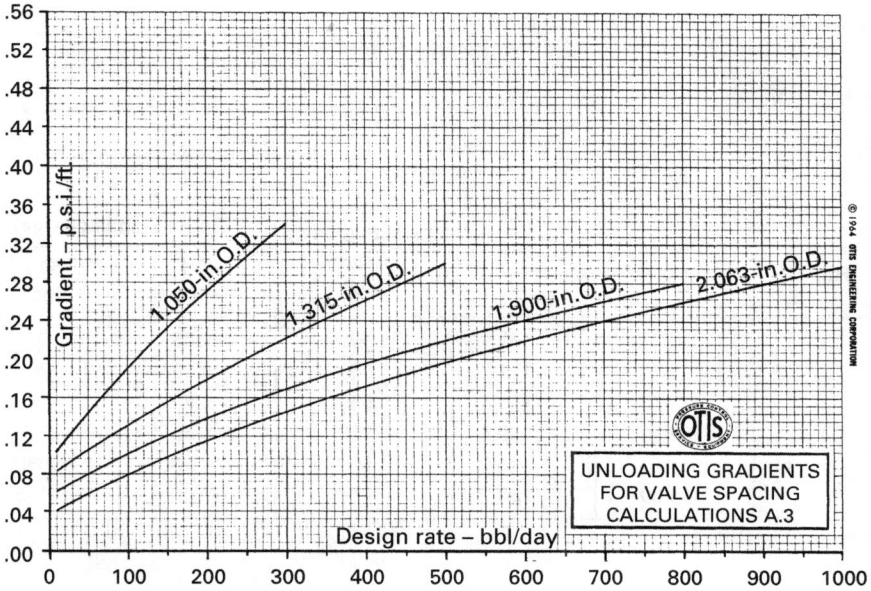

Figure P3 Unloading gradients for valve spacing calculations [1].

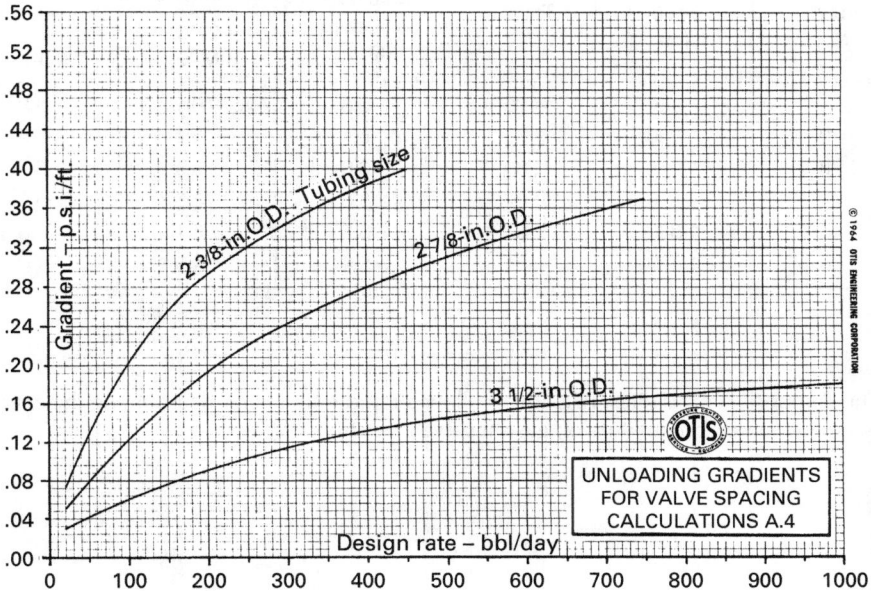

Figure P4 Unloading gradients for valve spacing calculations [1].

Since the spacing is less than 300 ft (or 90 m), place this valve at 4 160 ft (or 1 330 m). Place one more valve at 4 460 ft (or 1 420 m).

The following tables P4 and P5 show the final results.

- *In U.S. units*

Table P4

Valve No.	Depth (ft)	Temp. (°F)	Surface opening pressure (psi)	Settings (gas charge)	
				80°F (psi)	60°F (psi)
1	1 800	163	900	865	825
2	2 824	171	875	820	780
3	3 470	176	850	780	740
4	3 860	179	825	750	710
5	4 160	181	800	720	680
6	4 460	183	775	695	655

- *In metric units*

Table P5

Valve No.	Depth (m)	Temp. (°C)	Surface opening pressure (MPa)	Settings (gas charge)	
				27°C (MPa)	16°C (MPa)
1	550	73	6.3	6.1	5.8
2	860	77	6.1	5.7	5.5
3	1 058	80	6.0	5.5	5.2
4	1 176	82	5.8	5.3	5.0
5	1 330	83	5.6	5.1	4.8
6	1 420	84	5.4	4.9	4.6

The composite charts (Figs. P5a and b) allow a rapid estimate for determining the effect of gas gradient and/or temperature on gas-lift valve operation.

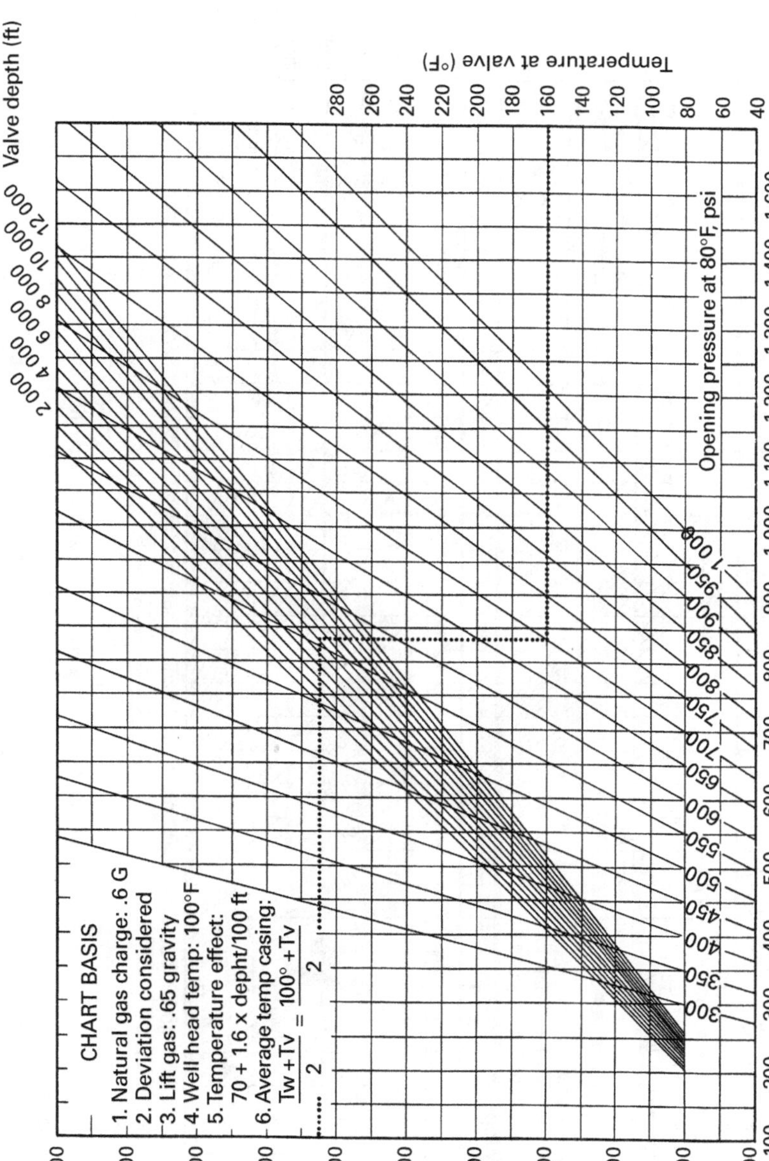

Figure P5a Determining operating pressures of balanced gas-lift valves [1].

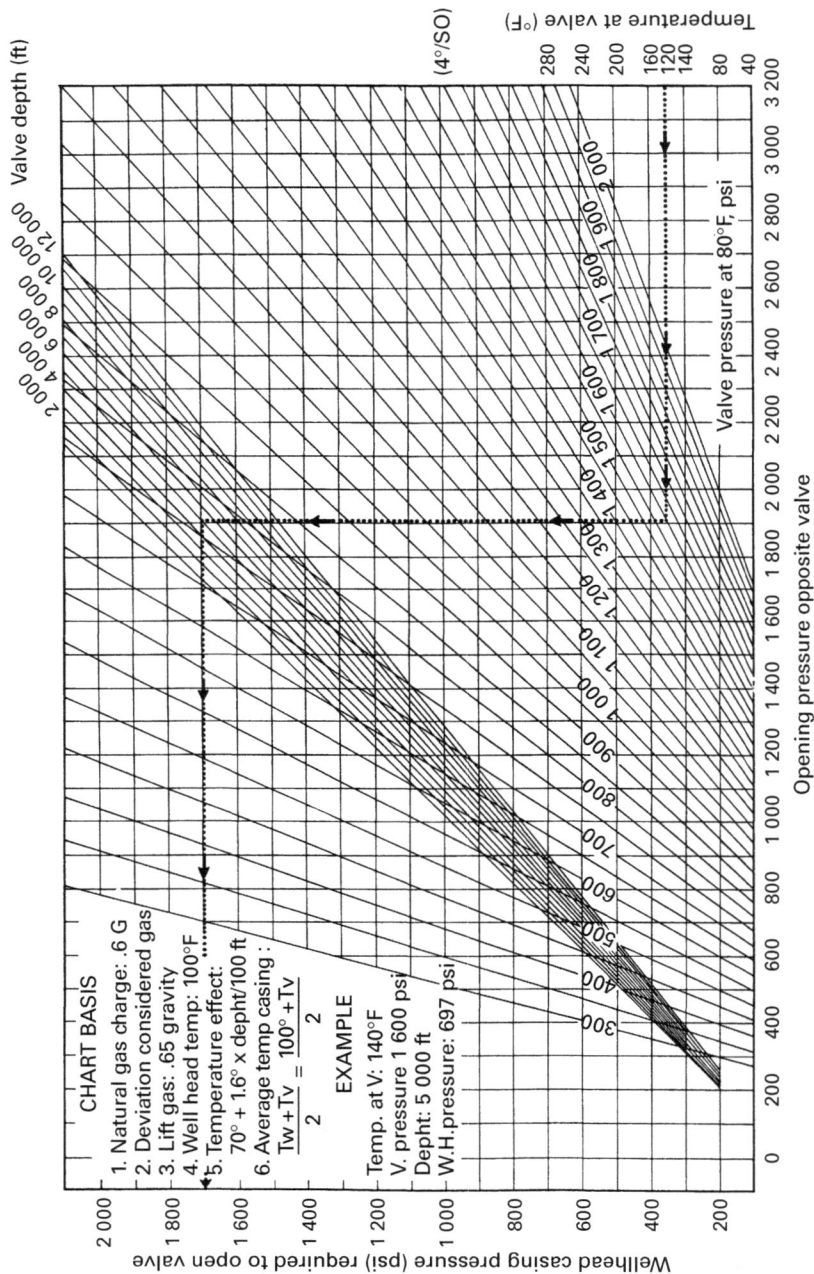

Figure P5b Determining operating pressures of balanced gas-lift valves [1].

CHART BASIS
1. Natural gas charge: .6 G
2. Deviation considered gas
3. Lift gas: .65 gravity
4. Well head temp: 100°F
5. Temperature effect: ......
   70° + 1.6° x depht/100 ft
6. Average temp casing :
   $\dfrac{Tw + Tv}{2} = \dfrac{100° + Tv}{2}$

EXAMPLE
Temp. at V: 140°F
V. pressure 1 600 psi
Depht: 5 000 ft
W.H.pressure: 697 psi

P3   DESIGN OF INTERMITTENT FLOW INSTALLATIONS [1]

3.1   Tubing size. Production rate

- In intermittent lift, a liquid slug is allowed to accumulate above the gas-lift valve. The valve then opens allowing sufficient gas to enter and propel the slug to the surface in piston shape.
- These installations must be designed for very low BHP wells in many cases. This may necessitate the use of chambers.
- The following production rates (see Table P6) are given as a guide as when to change from intermittent flow to continuous flow.

Table P6

Tubing size (in.)	Rate (bbl/D)	Rate (m³/d)
1.050	25-50	4-8
1.315	50-75	8-12
1	75-125	12-20
1.900	200	32
2 3/8	250	40
2 7/8	300	48
Chambers	400-600	64-96

3.2   Location of top valve depth

The location of the top valve can be extremely important and depends upon static BHP and if the well is to be loaded with "kill" fluid.

Of course, if the well is unloaded to the separator, then this pressure must be deducted and the first valve is placed at:

$$D_{v1} = \frac{p_{ko} - p_{wh}}{G_s} \tag{P6}$$

where

D_{v1} top first valve depth (ft)

p_{ko} available gas pressure for kick-off (psi)

p_{wh} surface back pressure on the tubing which must be loaded against (psi).

3.3 Analytical procedure for designing an intermittent installation for balanced valves

This problem may be worked out by the following analytical spacing procedure which neglects the gas column weight:

- **Step 1:**
$$D_{v1} = \frac{p_{ko} - p_{wh}}{G_s} \qquad (P7)$$

- **Step 2:**
$$D_{v2} = D_{v1} + \frac{\left[p_{so1} - (G_u \times D_{v1}) - p_{wh}\right]}{G_s} \qquad (P8)$$

- **Step 3:**
$$D_{v3} = D_{v2} + \frac{\left[p_{so2} - (G_u \times D_{v2}) - p_{wh}\right]}{G_s} \qquad (P9)$$

- **Step 4:**
$$D_{v4} = D_{v3} + \frac{\left[p_{so3} - (G_u \times D_{v3}) - p_{wh}\right]}{G_s} \qquad (P10)$$

and so on, where

D_{v1} depth from surface to valve 1 (ft)
p_{so1} surface operating pressure of valve 1 (psi)
p_{wh} tubing pressure on top of slug (psi)
G_s gradient of the fluid to be unloaded (psi/ft)
G_u unloading gradient (from Figs. P3 and P4) with the preselected design daily production rate and tubing size (psi/ft).

P

▼ Example

Assuming that the following informations are known:

p_{ko} kick-off pressure: 850 psi, (or 6 000 kPa)
p_{wh} 50 psi (or 0.35 MPa)
G_s 0.50 psi/ft (or 11.5 kPa/m)
G_u gradient of 0.04 psi/ft for 100 bbl/D design production rate and 2-in. ID tubing and 21/2-in. OD tubing.

Valve 1: $D_{v1} = \dfrac{850 - 50}{0.5} = 1\,600$ ft   (or 488 m)

Valve 2: $D_{v2} = 1\,600 + \dfrac{800 - (0.04 \times 160) - 50}{0.5} = 2\,972$ ft   (or 906 m)

and so on.

The following summary tabulation (Tables P7 and P8) would be the result, with:

p_t tubing pressure acting on valve seat in operation
p_d valve dome pressure (closing pressure) at valve operating temperature T_v
p_c casing pressure = 700 psi

$$p_d = p_t(1-R) - p_c R \qquad \text{(P11)}$$

where

$R = A_p/A_b$ valve port to effective bellows area ratio = 0.345
A_p area of valve port = 0.265 sq in.
A_b total effective bellows area = 0.77 sq in.

The following summary tabulation (Tables P7 and P8) would be the result:

- *In U.S. units*

Table P7

Valve No.	D_v (ft)	T_v (°F)	p_{so} (psi)	p_t (psi)	p_d (psi) (80°F - gas charge)
1	1 600	106	800	830	780
2	2 972	127	775	830	745
3	4 182	146	750	830	715
4	5 248	164	725	820	680
5	6 178	179	700	800	650
6	7 000	191	675	795	630

- *In metric units*

Table P8

Valve No.	D_v (m)	T_v (°C)	p_{so} (MPa)	p_t (MPa)	p_d (MPa) (27°C - gas charge)
1	488	41	5.5	5.7	5.4
2	906	53	5.3	5.7	5.1
3	1 275	63	5.2	57	4.9
4	1 600	73	5.0	5.6	4.7
5	1 883	82	4.8	5.5	4.5
6	2 133	88	4.6	5.5	4.3

▲

3.4 Designing a chamber gas-lift installation for intermittent flow

In low BHP, high PI wells, production may often be greatly increased and lift efficiency much improved by installation of a chamber type assembly. The formula for determining chamber length is:

$$L_c = \frac{(p_{vo} - p_w) - (p_{vo} - p_t)}{G_s(R_{ct} + 1)} \qquad \text{(P12)}$$

where

L_c chamber length (ft)

p_{vo} valve opening pressure at depth (psi)

p_w pressure at top of slug due to separator back pressure (psi)

p_t pressure of fluid head when transferred to the tubing at time of lift (psi)

G_s gradient of well fluid (psi/ft)

R_{ct} ratio of chamber housing (casing) volume to tubing volume.

▼ Example

Assuming that the following informations are known:
- Valve opening pressure at depth: 600 psi (or 4.15 MPa)
- Separator pressure: 50 psi (or 0.35 MPa)
- Fluid head to be lifted: 450 psi (or 3.1 MPa)
- Gradient of well fluid: 0.4 psi/ft (or 9 kPa/m)
- Ratio of chamber housing to tubing: 4.3

we have:

$$L_c = \frac{(600-50)-(600-450)}{0.4(4.3+1)} = 189 \text{ ft} \quad \text{(or 57.50 m)} \quad \blacktriangle$$

P4 DESIGN OF THE COMPRESSOR SYSTEM [1]

4.1 Factors to consider when designing a compressor system

- Number and location of wells, the battery location, location of all equipment, survey of terrain, etc.
- Individual gas lift valve design for each well and the type of lift anticiped (continuous or intermittent).
- Gas volume needed with an estimate of the peak demand at any time.
- Injection gas pressure needed at the wellhead, which determines the discharge pressure of the compressor.
- Separator pressure to be carried on each well and or lease which in turn will determine the suction pressure on the compressor.
- Gas distribution system: individual well lines, generally 2-in. ID (21/2-in. OD).
- Low-pressure gathering system.
- Availability of make-up gas: special precautions are necessary, if air is to be used.
- Availability of gas sales outlet
  Evaluation of system under freezing and hydrate conditions (Fig. P6).
- Sizing of the compressor.

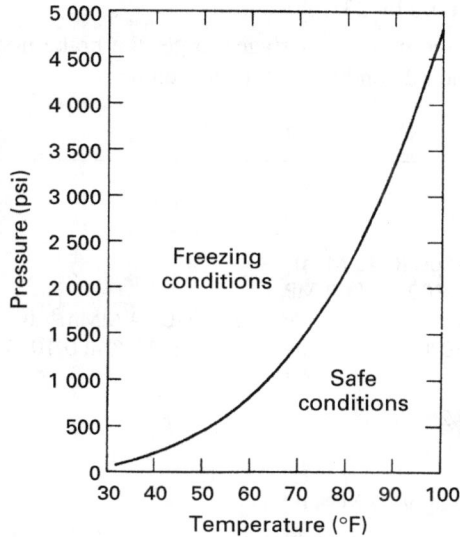

Figure P6 General correlation for hydrate formation [1].

4.2 Compressor selection

Basic relations

The basic relations are established between suction pressure, discharge pressure, compression ratio, and the number of compression stages:

$$C = \sqrt[n]{C'} = n\sqrt{\frac{P_d}{P_s}} \qquad (P13)$$

$$P_d = P_s C^n = P_s C' \qquad (P14)$$

where
   C absolute compression per stage (psi)
   C' over-all absolute compression ratio
   n number of stages
   P_d absolute discharge pressure (psi)
   P_s absolute suction pressure (psi).

Compressor input power

A simplified method for estimating compressor input power requirements for an integrally driven compressor is given, as follows:
1. Find the per stage compression ratio.
2. Multiply by 1.05 to correct for pressure drop and imperfect gas cooling.

3. Multiply the corrected ratio by 23.

4. Multiply the result by the number of stages to get the brake horsepower per million cubic feet of gas at standard conditions (bhp/MMcu ft).

▼ Example

Assuming that:
 Suction pressure P_s = 60 psi (0.42 MPa)
 Discharge pressure P_d = 700 psi (4.9 MPa)
 Volume of gas to be compressed = 1.5 MMcu ft/D (0.04 MMm³/d)
 Suction temperature = 80°F measured at 15.025 psi (27°C at 0.1035 MPa)

we obtain:
 1. $C = (700/60)^{1/2} = 3.42$
 2. $1.05 \times 3.42 = 3.60$
 3. $23 \times 3.60 = 82.8$
 4. $2 \times 82.8 \times 1.5 = 250$ bhp/1.5 MMcu ft/D

For belt-driven units, the input power should be increased by about 5%.   ▲

P5 ANNEX. BASICS OF PHYSICAL GAS APPLIED TO GAS LIFT [4]

P

5.1 Compressibility factor *(Z)*

Z is the ratio between the volume occupied by a gas at P and T conditions and the volume of an ideal gas *(PV = nRT)* in the same conditions.

At the atmospheric pressure P, gas behaves as an ideal gas.

Z is a function of the pseudo-reduced pressure P_{pr} and of the pseudo-reduced temperature T_{pr}:

$$P_{pr} = \frac{P}{P_{pc}} \tag{P15}$$

and

$$T_{pr} = \frac{T}{T_{pc}} \tag{P16}$$

where
 P and T are in absolute value (psi, °R)
 P_{pc} and T_{pc} pseudo-critical pression and temperature of gas, are determined on Fig. P7
     (*Chevron* chart) depending on specific gravity of gas.

Z is then determined using Fig. P8 (simplified *Camco* chart).

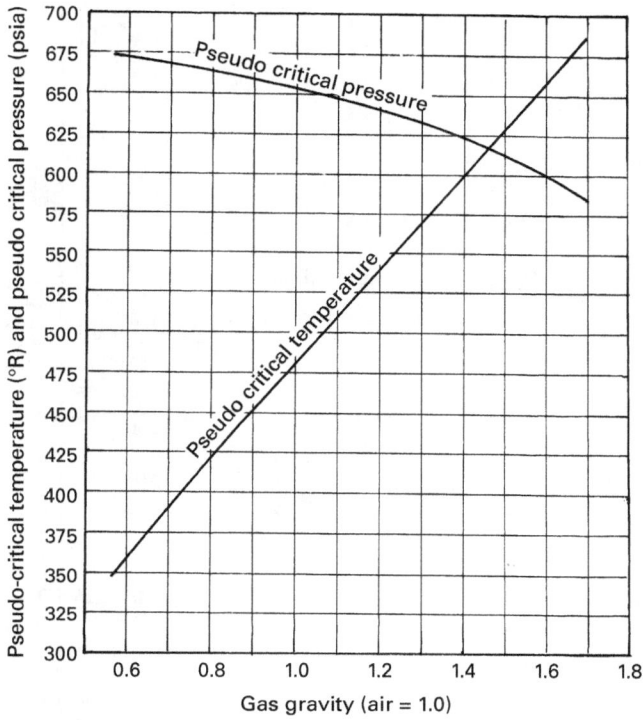

Figure P7 Pseudo-critical pressure and temperature vs. gas gravity [4].

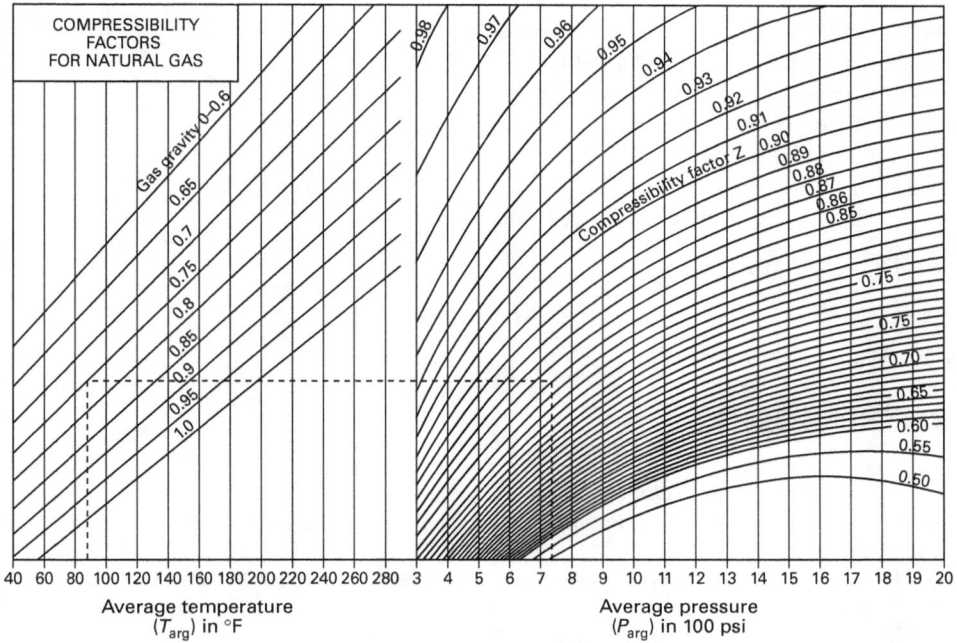

Figure P8 Compressibility factors for natural gas [4].

▼ Example

Specific gravity *S.G.* = 0.9
Temperature $T\,°F = 120°F$
Temperature $T\,°R = 120 + 460 = 580°R$
Gauge pressure $P_g = 1\,150$ psi
Atmospheric pressure $P_{atm} = 14.7$ psi (approximatively 15 psi)

$$P = P_g + P_{atm} = 1\,150 + 15 = 1\,165 \text{ psi}$$

Figure P6 gives: $\qquad P_{pc} = 660$ psi $\qquad\qquad T_{pc} = 450\,°R$

hence: $\qquad P_{pr} = \dfrac{1\,165}{660} = 1.77 \qquad\qquad T_{pr} = \dfrac{580}{450} = 1.29$

Camco chart (Fig. P8) gives: $\qquad\qquad Z = 0.69$ ▲

5.2 Calculation of the volume of compressed gas in a pipe

In gas lift, this calculation determines:
- Stocked HP and BP gas on the surface in a closed compressor gas lift system
- Required gas in a tubing to eject a plug of oil to the surface
- Volume of gas poured from the annular to the tubing, following a pressure drop in annular, with:

C internal volume of a pipe (cu ft)
Q volume of gas included in the pipe (st cu ft)
P_0 atmospheric pressure = 14.7 psia
P absolute pressure of gas in the pipe (st cu ft)
T_0 standard temperature = 60°F
T absolute temperature (°R)
Z gas compressibility factor at P and T conditions.

Hence:

$$Q = \frac{CPT_0}{ZP_0T} \qquad (P17)$$

REFERENCES

1 Brown KE et al. (1967) *Gas Lift, Theory and Practice*. Prentice-Hall, Inc.
2 Frick TC, Taylor RW (1962) *Petroleum Production Handbook,* Chapter 5, Kirkpatrick: Gas lift. SPE of AIME
3 Economides MJ, Hill AD, Ehlig-Economides C (1994) *Petroleum Production Systems*. PTR Prentice Hall
4 Bertain A (1987) *Séminaire Activation par Gas Lift*. ENSPM
5 Perrin D, Caron M, Gaillot G (1998) *Well Completion and Servicing*. Editions Technip, Paris.

Q

Electric Submersible Pump

Q

Q

Electric Submersible Pump

Q1 HYDRAULIC FUNDAMENTALS [1]

1.1 Pump Intake Pressure (PIP)

It is important to know the specific gravity or gradient of the liquid in the annulus.

$$PIP = \text{meter submergence} + \text{casing pressure (m)}$$

where equivalent meter of fluid per bar is observed.

There are two values for a pump intake to consider:
- **Required PIP:** This is the intake pressure necessary to properly feed the pump and prevent cavitation or gas locking.
- **Available PIP:** This pressure is a function of the system in which the pump operates. Available PIP is the operating submergence characteristic of each individual installation.

1.2 Fluid flow

Since most liquids are considered to be imcompressible, there is a definite relationship between the quantity of liquid flowing in a conduit and the velocity of flow:

$$Q = AV$$

where
- Q capacity
- A area of conduit
- V velocity of flow.

In a submersible pump installation, the heat generated in the motor is carried away by the well fluids moving at the exterior surface of the motor. Empirical data studies indicate that well bore fluids should pass the motor at a minimum rate of 30 cm per second in order to adequately dissipate the heat transferred through the motor housing.

1.3 Cavitation and gas locking

In a well where there is excessive free gas present, a certain intake pressure to control the amount of free gas handled by the pump must be maintained to prevent gas locking.

1.4 Typical characteristic curve

In a standard centrifugal pump, the characteristic curve for this type of pump can be changed either by keeping the speed constant and varying the impeller diameter or by keeping the speed constant, varying the impeller diameter and varying the speed.

A typical characteristic curve of an electrical oilfield submersible centrifugal pump is shown in Fig. Q1.

The mathematical relationship between these several variables are known as the **affinity laws** and can be expressed in Eqs. Q1 to Q9, as follows:

Diameter change only	Speed change only	Diameter and speed change
$Q_2 = Q_1 \dfrac{D_2}{D_1}$ (Q1)	$Q_2 = Q_1 \dfrac{N_2}{N_1}$ (Q2)	$Q_2 = Q_1 \dfrac{D_2 N_2}{D_1 N_1}$ (Q3)
$H_2 = H_1 \left(\dfrac{D_2}{D_1}\right)^2$ (Q4)	$H_2 = H_1 \left(\dfrac{N_2}{N_1}\right)^2$ (Q5)	$H_2 = H_1 \left(\dfrac{D_2 N_2}{D_1 N_1}\right)^2$ (Q6)
$\mathrm{bhp}_2 = \mathrm{bhp}_1 \left(\dfrac{D_2}{D_1}\right)^3$ (Q7)	$\mathrm{bhp}_2 = \mathrm{bhp}_1 \left(\dfrac{N_2}{N_1}\right)^3$ (Q8)	$\mathrm{bhp}_2 = \mathrm{bhp}_1 \left(\dfrac{D_2 N_2}{D_1 N_1}\right)^3$ (Q9)

where
$Q_1, H_1, \mathrm{bhp}_1, D_1$ initial capacity, head, brake power, diameter
$Q_2, H_2, \mathrm{bhp}_2, D_2$ new capacity, head, brake power, diameter.

1.5 Determining productivity of wells

The flow rate Q of a well depends on:
  (a) The difference between the available pressure which is the reservoir pressure P_g and the bottomhole back pressure P_f
  (b) The productivity index *IP*:

$$Q = IP\,(P_g - P_f) \tag{Q10}$$

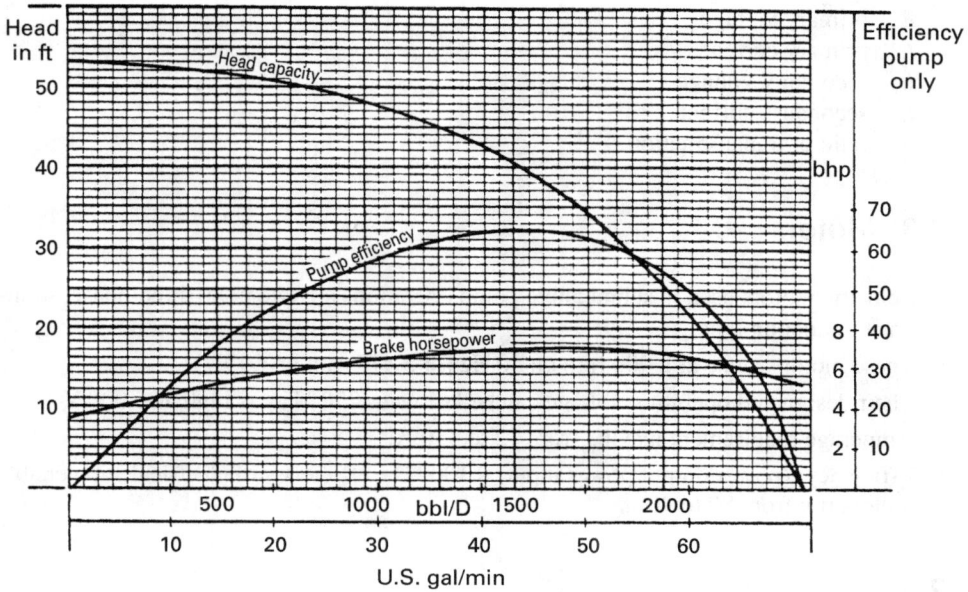

Figure Q1 Typical performance curve of a submersible pump at 3475 rpm.
Water test. Specific gravity: 1.0 [1].

Q2   ELECTRICAL FUNDAMENTALS [1]

2.1 Power factor

The power factor cos ϕ is the ratio of the real component to the actual current:

$$\text{actual power } (W) = 1.732 \times \text{volts} \times \text{amperes} \times \cos \phi \tag{Q11}$$

The power consumption is measured by the watt-hour meter. Large units require current and/or potential transformers so that the watt-hour meter reading must be multiplied by the ratio of these transformers to obtain the correct measurement of power consumption of the motor.

2.2 Transformers. Transformation ratios

Since transformer action is sustained under-load, the volt-ampere input must be at least equal to (and in an ideal transformer it is equal to) the volt-ampere output. Expressing these statements in equation form for the ideal transformer, we have:

$$E_p/E_s = I_s/I_p = a \tag{Q12}$$

323

where

E_p primary voltage
I_p primary current
E_s secondary voltage
I_s secondary current
a ratio of transformation.

2.3 Motors

The electric motors used in submersible pump operations are two pole, three-phase, squirrel cage, induction type.

- These motors rotate at 2 900 rpm on 50 Hz (or 3 475 rpm on 60 Hz).
- Their design and operating voltage can be as low as 230 V or as high as 5 000 V.
- Amperage requirement may be from 12 to 110 A.
- Efficiency: (power output/power input) × 100. The efficiency of electrical submersible motors run from 80 to 90%.

2.4 Frequency

The frequency can be changed by means of a variable speed unit often called "frequency converter".

The variation of the speed of the submersible pump, involves the following effects:

- Flow rate is a function of pump rpm
- Head is a function of (pump rpm)^2
- Brake horse power is a function of (pump rpm)^3.

2.5 Electric cable

Electric Submersible Pumps (ESP) electric cables are normally available from stock in conductor sizes number 1, 2, 4, 6, 8. These sizes are offered in both round and flat configuration.

2.5.1 Wire size

The wire sizes common to the submersible industry are for copper conductors listed in Table Q1.

The proper cable size is determined by combined factors of voltage drop, amperage and available space between tubing collars and casing.

The cables used in submersible applications are designed to maintain downhole voltage losses to a minimum. The voltage losses per 300 m (or 1 000 ft) for the cables size common to submersible applications are shown in Fig. Q2.

Table Q1

Number	Diameter (mm)	Area (mm^2)	Ohms/1 000 m	kg/1 000 m
1	7.34	42.3	0.489	377
2	6.53	33.5	0.617	299
4	5.18	21.1	0.978	188
6	4.11	13.3	1.555	118
8	3.25	8.3	2.425	74

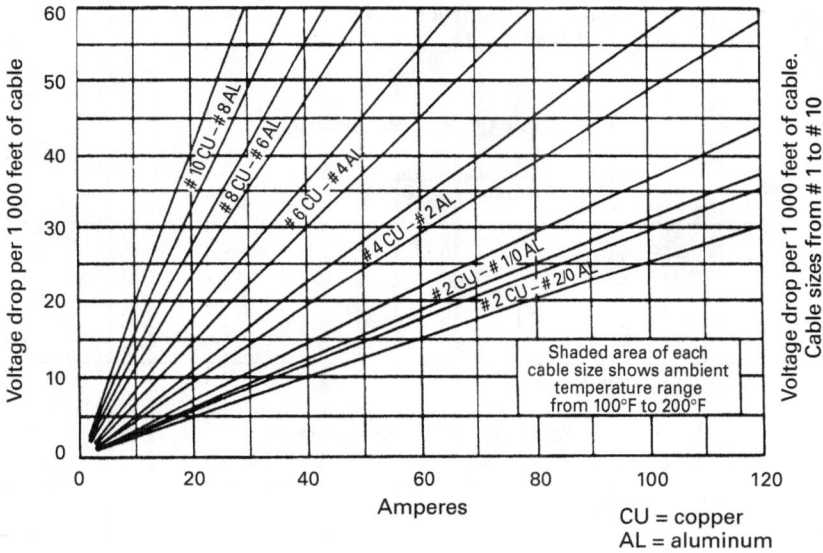

Figure Q2 Shaded area of each cable size shows ambient temperature from 100°F to 200°F (38°C to 93°C) [1].

2.5.2 Current carrying capacity

Current carrying capacity is usually the major factor in determining the conductor size. In Table Q2 are indicated the recommended maximum amperes for each cable size.

Table Q2

Copper conductor No.	Aluminum conductor No.	Maximum amperes
1	2/0	110
2	1/0	95
4	2	70
6	4	55

2.5.3 Cable type

Selection of cable type is primarily based on fluid conditions, bottomhole temperature and space limitations within the casing annulus.

- **Round cable:**
  - Polynitrile with copper or aluminum conductor, galvanized armor
  - Polyethylene with copper or aluminum conductor, unarmored.
- **Flat cable:**
  - Polynitrile with copper conductor, galvanized armor.
- **Physical and mechanical characteristics:**
  - Copper conductor, 3 000 V cable (round armored type) shown in Table Q3
  - Aluminum conductor, 3 000 V cable shown in Table Q4.

Table Q3 Copper conductor.

Size number	Weight (kg/m)	Diameter (mm)	Tensile strength (N)	Number of wires
1	2.60	34.11	17 000	7
2	2.46	32.13	13 600	7
4	1.74	28.95	8 650	7
6	1.31	24.64	4 500	solid

Table Q4 Aluminum conductor.

Size number	Weight (kg/m)	Diameter (mm)	Tensile strength (N)	Number of wires
2/0	2.07	36.20	9 350	7
1/0	1.73	33.91	7 450	7
2	1.29	30.35	4 700	7
4	1.10	27.43	2 950	7

2.5.4 Temperature limitations

The temperature limitations for wells containing hydrocarbons and gas are:
- Polyethylene jacketed cable: 60°C
- Polypropylene insulated with armor: 80°C
- EPR lead sheath: 120°C.

Some manufacturers has developed a new round, lead-sheathed power cable for electric submersible pumps. The new cable is rated for conductor temperature up to 230°C.

2.5.5 Cable length

The total cable length should be at least 30 m (100 ft) longer than the pump setting depth in order to make switchboard connections at a safe distance from the wellhead.

Q3 DESCRIPTION OF SUBMERSIBLE INSTALLATIONS [1]

3.1 General characteristics

The many sizes and types of submersible pumps are too numerous to list. Equipments are available to produce volumes in the range of 30 m³/d up to 4 000 m³/d.

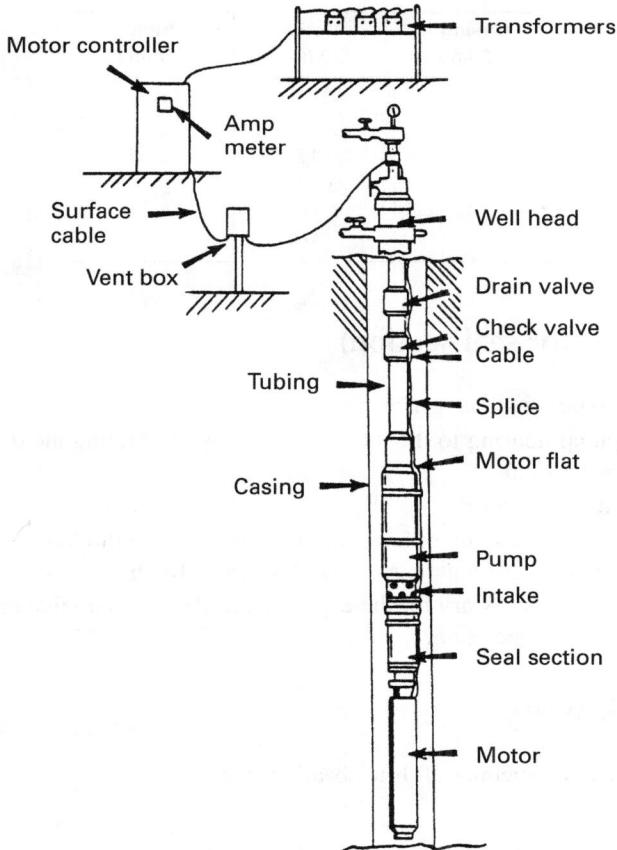

Figure Q3 Typical submersible installation [1].

The typical submersible subsurface pumping unit assembly (Fig. Q3) consists of:
- Three-phase electric motor
- Seal section
- Multi-staged centrifugal pump
- Motor flat cable
- Round power cable
- Drain valve and check valve.

3.2 Size

Table Q5 illustrates approximate outside diameters of equipment available from submersible pump suppliers for the different outside diameters (OD) of casings.

Table Q5

Casing OD (in.)	Motor OD (in.)	Pump OD (in.)
5 1/2	4 1/2	4
7	4 1/2 5 1/2	4 5 1/8
8 5/8 or greater	4 1/2 5 1/2 7 1/4	4 5 1/8 6 3/4

3.3 Protector (or seal section)

It performs four basic functions which are:
- Connect the pump housing to the motor housing by connecting the drive shaft of the motor to the pump shaft
- House the pump thrust bearing
- Seal the power end of the motor housing from the wellbore fluids while allowing pressure communication between the motor and wellbore fluids
- Provide the volume necessary for the expansion of the unit's oil due to heat generated when the motor is in operation.

3.4 Thrust bearings

It can be stated that the enemies of thrust bearings are:
- Heat-reduced viscosity
- Misalignment
- Foreign particles
- Vibration.

3.5 Electrical submersible motor

The electric motors used in submersible pump operations are two pole, three-phase, squir-rel cage, induction type. The motors are filled with a highly refined mineral oil with high dielectric strength. The design and operating voltage of these motors can be as low as 230 V or as high as 5 000 V. Amperage requirement may be from 12 to 110 A. The require-ment power is achieved by simply increasing the length of the motor section.

Figure Q4 shows the temperature rise for 100% load in water and crude oil for a fully loaded motor. It is obvious that fluid velocity is just as important as fluid ambient temperature.

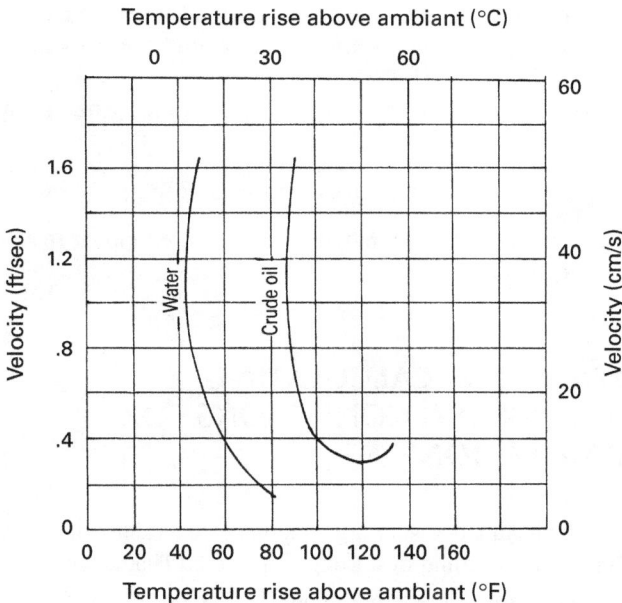

Figure Q4 Temperature rise above ambient vs. velocity [1].

Q4 PUMP SELECTION [1]

4.1 Pump

*   Refer to the manufacturer performance curve of the selected pump type and determine the number of stages required to produce the anticipated against the previously calcu-lated total head (Eq. Q13):

$$\text{total stages} = \text{total dynamic head/head per stage} \tag{Q13}$$

- Note that the characteristic curves are one-stage performance curves with water (specific gravity of 1) used as the testing fluid.

4.2 Gas separator

- For producing small to moderate volumes of gas with liquid, the built-in gas separator is recommended.

4.3 Motor

- All manufacturer motors can be safely operated when loaded substantially above nameplate ratings, the best practice is to restrict overload to ten per cent or less on initial selection.
- The brake power (bhp) required to drive a given pump is easily calculated by the following formula:

$$\text{bhp} = \text{total stages} \times \text{hp/stage} \times \text{sp gr} \qquad\qquad (Q14)$$

- The power shown on the water-pump performance curves applies to a liquid with a specific gravity (sp gr) of 1.

Q5   PROCEDURE FOR CALCULATING MOTOR STARTING CONDITIONS FOR ESP INSTALLATIONS [6]

Q

Since the motor can be installed at very high depth, considerable voltage drop can occur in the system both in steady state and in starting conditions. That is why it is important:
- To determine whether reliable starting of an ESP can be achieved
- To determine the effects of a weak distribution system on starting ESPs
- To determine whether reduced voltage starting is required
- To enable the coordination of electrical power capacity requirements of power systems.

5.1 Method for calculating motor starting conditions

A ten step calculation method can be implemented as follows:
1. Gather required equipment and well information.
2. Calculate transformer ohmic impedance.
3. Calculate cable ohmic impedance.
4. Calculate surface voltage and preliminary motor data.
5. Assume a motor voltage at starting (start with 70%).

6. Determine motor starting impedance (Fig. Q5).
7. Using simplified equivalent circuit model and motor impedance values determined in step 6, solve the circuit for motor terminal voltage.
8. Compare the calculated value (step 7) with the assumed value (step 5).
9. Iterate step 5 through step 8 until the assumed value and the calculated value agree.
10. Solve equivalent circuit for starting current, watt loss, power factor.

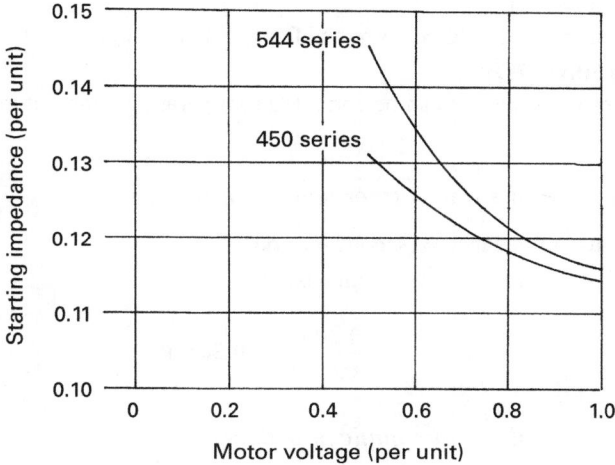

Figure Q5 Determination of motor starting impedance.

5.2 Example of calculation

In order to demonstrate the calculation method, a sample case is considered. The starting calculation can be made by following the ten step procedure.

Step 1. Gather required equipment and well information

A. Power distribution sag (*SA*)   5%

Estimate the percent of surface that the electrical distribution system is likely to sag during starting. Voltage sag is directly related to the "stiffness of the system". Factors affecting "stiffness" are primary conductor size, distance to substation, and the size of other loads on the primary. In absence of any feel for surface voltage sag, use 2%–5%.

B. Transformer data

– Transformer kVA (indicated on the nameplate)   250 kVA
– Transformer secondary voltage   1 300 V
– Transformer impedance (typically considered mostly inductive)   3%

C. **Cable data**
   – Cable size (measured in American Wire Gauge-AWG)   2
   – Cable length (*FT*)   8 500 ft
   – Power cable impedance (see Table Q6)

D. **Motor** (indicated on the nameplate)
   – Motor series   450
   – Motor horsepower   100 hp
   – Motor voltage (*V*)   1 130 V
   – Motor amperage (*A*)   55 A
   The motor resistance should be corrected for well temperature.

E. **Well temperature** (*TD*)   200°F
   Cable and motor resistances can be corrected for operating temperature
   effects.

Step 2. Calculation of the transformer ohmic impedance

- Transformer base impedance = (secondary volt)^2/VA.
- Ohmic impedance *(XT)* = % × base impedance:

$$XT = 3\% \times \frac{1\,300^2}{250\,000} = \textbf{0.203 ohm} \qquad (Q15)$$

Step 3. Calculation of the cable ohmic impedance

Using Table Q6, find the resistance and inductance for AWG 2 cable size:

$$R = 0.175 \text{ ohm/1 000 ft}$$

$$X = 0.034 \text{ ohm/1 000 ft.}$$

Table Q6 Typical power cable impedance [6].

AWG	Resistance R (ohm/1 000 ft and 77°F)	Inductance X (ohm/1 000 ft and 60 Hz)
1	0.139	0.032
2	0.175	0.034
4	0.271	0.036
6	0.431	0.038

In order to correct the cable resistance for temperature, calculate the average well temperature assuming 8°F/1 000 ft cooling from bottom to surface. This is done by using the following equation:

$$TA = TD - 0.5\frac{8 \times FT}{1\,000} \qquad (Q16)$$

where
- *TA*   average well temperature (°F)
- *TD*   estimated temperature at pump setting depth (°F)
- *FT*   feet of cable (ft)

$$TA = 200 - 0.5\frac{8 \times 8\,500}{1\,000} = \mathbf{166°F}$$

Correct *R* for temperature using the equation:

$$RN = R + \left[1 + 0.00214(TA - 77)\right] \tag{Q17}$$

where
- *RN*   corrected resistance (ohm/1 000 ft)

$$RN = 0.175 + \left[1 + 0.00214(166 - 77)\right] = \mathbf{0.208\ ohm\,/\,1\,000\ ft}$$

For the total cable length, the result then is:

$$RC = 8\,500 \times 0.208/1\,000 = \mathbf{1.77\ ohm}$$

$$XC = 8\,500 \times 0.034/1\,000 = \mathbf{0.289\ ohm}$$

where
- *RC*   cable resistance
- *XC*   cable inductance.

Step 4. Calculation of the surface voltage and preliminary motor data

The calculation is done by using the following equations:

$$PU = \frac{V}{\sqrt{3} \times A} \tag{Q18}$$

$$PU = 1\,130/(1.732 \times 55) = \mathbf{11.86\ ohm}$$

$$RU = 0.07 \times PU \tag{Q19}$$

$$RU = 0.07 \times 11.86 = \mathbf{0.830\ ohm}$$

$$RM = RU \times [1 + 0.00214 \times (TD - 77)] \tag{Q20}$$

$$RM = RU \times [1 + 0.00214 \times (200 - 77)] = \mathbf{1.048\ ohm}$$

$$RR = 0.82 \times PU \tag{Q21}$$

$$RR = 0.82 \times 11.86 = \mathbf{9.73\ ohm}$$

$$XR = 0.57 \times PU \tag{Q22}$$

$$XR = 0.57 \times 11.86 = \mathbf{6.76\ ohm}$$

$$ZR = \sqrt{(RC + RR)^2 + (XT + XC + XR)^2} \tag{Q23}$$

$$ZR = \sqrt{(1.77 + 9.73)^2 + (0.203 + 0.289 + 6.76)^2} = \mathbf{13.60\ ohm}$$

$$VS = ZR/PU \tag{Q24}$$
$$VS = 13.60/11.86 = \textbf{1.147 p.u.}$$

$$VT = VS \times V \tag{Q25}$$
$$VT = 1.147 \times 1130 = \textbf{1 296 V}$$

where
- PU motor base impedance
- RU motor starting resistance
- RM RU corrected for temperature
- RR motor running resistance
- XR motor running reactance
- ZR system running impedance
- VS surface voltage per unit
- VT surface voltage at steady state.

Step 5. Assumption of a motor terminal voltage at starting

To begin the first iteration, assume 70%, which equals 0.70 on a per unit notation.

Step 6. Determination of motor starting impedance

Using Fig. Q5 for a 450 series motor, at 0.70 p.u. the result is:

$$PZ = \textbf{0.121 p.u.}$$

where
- PZ motor starting impedance per unit.

Step 7. Calculation of the motor terminal voltage

$$ZM = PZ \times PU \tag{Q26}$$
$$ZM = 0.121 \times 11.86 = \textbf{1.435 ohm}$$

$$XM = \sqrt{ZM^2 - RM^2} \tag{Q27}$$
$$XM = \sqrt{1.435^2 - 1.048^2} = \textbf{0.98 ohm}$$

$$ZT = \sqrt{(RC + RM)^2 + (XT + XC + XM)^2} \tag{Q28}$$
$$ZT = \sqrt{(1.77 + 1.048)^2 + (0.203 + 0.289 + 0.98)^2} = \textbf{3.179 ohm}$$

$$TE = VS \times (1 - SA) \times ZM/ZT \tag{Q29}$$
$$TE = 1.147 \times (1 - 0.05) \times 1.435/3.179 = \textbf{0.492 p.u.}$$

where
- PZ motor starting impedance (per unit)
- ZM motor starting impedance (ohm)

XM   motor starting reactance
ZT   system starting impedance (ohm)
VS   surface voltage (per unit)
SA   estimated distribution sag
TE   calculated motor terminal (per unit).

Step 8. Comparison of the calculated value of motor terminal voltage with the assumed value

In step 5, a value of 0.7 p.u. has been assumed. In step 7, a value of 0.492 p.u. was just calculated. In order to properly solve the model, the assumed value must be reasonably close to the calculated value.

Therefore, the assumed value should be decremented and recalculated. For a second try, we will decrement to an assumed value of 0.52 p.u.

Step 9. Agreement of the assumed and recalculated values

From Fig. Q5, $PZ = 0.130$ p.u. Then:

$$ZM = PZ \times PU \tag{Q26}$$
$$ZM = 0.130 \times 11.86 = \mathbf{1.542\ ohm}$$

$$XM = \sqrt{ZM^2 - RM^2} \tag{Q27}$$

$$XM = \sqrt{1.542^2 - 1.048^2} = \mathbf{1.131\ ohm}$$

$$ZT = \sqrt{(RC + RM)^2 + (XT + XC + XM)^2} \tag{Q28}$$

$$ZT = \sqrt{(1.77 + 1.048)^2 + (0.203 + 0.289 + 1.131)^2} = \mathbf{3.252\ ohm}$$

$$TE = VS \times (1 - SA) \times ZM/ZT \tag{Q29}$$
$$TE = 1.147 \times (1 - 0.05) \times 1.542/3.252 = \mathbf{0.517\ p.u.}$$

Since the assumed value is 0.52 p.u. and the calculated value is 0.517 p.u., the solution is:

$$TE = \mathbf{0.52\ p.u.}$$

Step 10. Resolution of the equivalent circuit for starting current, power loss, and power factor

• Starting current *IS* can be calculated from the following equation:

$$IS = \frac{TE}{ZM} \times \frac{V}{\sqrt{3}} \tag{Q30}$$

$$IS = \frac{0.525}{1.542} \times \frac{1130}{1.732} = \mathbf{222.4\ A}$$

335

- Power loss KS in a three phase system is equal to $3\,R \times I^2$. For this model:

$$KS = 3 \times IS^2 \times (RC + RM)/1\,000 \tag{Q31}$$
$$KS = 3 \times (222.4)^2 \times (1.77 + 1.048)/1\,000 = \mathbf{418\ kW}$$

- Power factor PS can be defined as the ratio of resistance to impedance:

$$PS = (RM + RC)/ZT \tag{Q32}$$
$$PS = (1.048 + 1.77)/3.252 = \mathbf{0.866}$$

In conclusion, starting calculations are extremely helpful in verifying the proper application of electrical submersible pump systems. Where insufficient voltage is provided during unit starting, serious problems may occur which will result in equipment failure and costly pulling jobs.

Q6 NEW TECHNOLOGY

The thermal recovery presents some constraints for ESP artificial lift because of the high temperature applied on the motor, on electric parts, on power cable and on pump components as bearings and seals.

6.1 ESP Hot line production [7]

ESP manufacturers propose special designs for high temperatures such as "Hot line production" equipment with operating temperature rating up to 550°F (288°C) for power cable and motor.

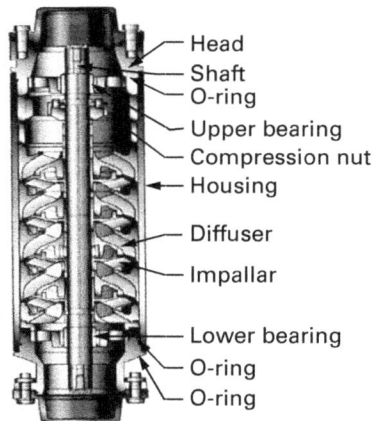

Figure Q6 Multistage centrifugal pump stage geometry radial flow or mixed flow [7].

6.2   ESP and coiled tubing deployed

ESP can be lowered down with a coiled tubing unit, the power cable can be located inside the coiled tubing or fixed along the coiled tubing from the motor to surface.

6.3   ESP High gas content handling

ESP are proposed with special impellers for handling high gas fraction, like the "Poseidon" axial flow design. Helicoaxial pumps have more efficient energy transfer resulting in trouble-free operation at high gas volume GVF. Such multiphase pump manages up to 75% free gas without gas locking.

REFERENCES

1   *Oilfield Centrilift-Hughes Submersible Pump Handbook*. Centrilift-Hugues, Inc.
2   API RP 11R (1986) *Electric Submersible Pump Installation*, Second Edition
3   API RP 11S (1986*) Operation, Maintenance and Troubleshooting of Electric Submersible Pump Installations,* Second Edition
4   API RP 11S1 (1987) *Electric Submersible Pump Teardown Report*, First Edition
5   API RP 11S2 (1990) *Electric Submersible Pump Testing,* First Edition
6   Vandevier J (1985) Procedure for calculating motor starting conditions for electric submersible pump installations. *Paper* presented at 1985 ESP Workshop, SPE Gulf Coast Local Section, Houston, TX.
7   Gulich JF (1999) Pumping Highly Viscous Fluids with Centrifugal Pumps. Parts 1 & 2, *World Pumps*, 395, pp. 9-11.

Q

Progressing Cavity Pumps

R

Progressing Cavity Pumps

R1   GENERAL DESCRIPTION [1]

1.1   Principle

A Progressing Cavity Pump (PCP) consists of a single helical rotor which rotates inside a double internal helical stator. The rotor is precisely machined from high strength steel; the stator is molded of resilient elastomer. Thus an interference fit can be obtained. When the rotor is inserted in the stator, two chains of lenticular, spiral cavities are formed. As the rotor turns within the stator, the sealed cavities spiral up the pump without changing size or shape and carry the pumped product.

For oil production, the stator is fixed to the tubing and the rotor is attached to a sucker rod string. The pump is driven by rotation of the sucker rod string. As an alternative, electrical submersible progressing cavity pumps can be offered, particularly on deviated wells.

1.2   Geometry

The geometry of pumps is defined by two numbers, the first being the number of lobes of the rotor, and the second being the number of lobes of the stator.

For example, the geometry of a pump with a single helical rotor and a double helical stator is described as a "1-2 pump".

A progressing cavity pump is defined by:

D	minor diameter of the rotor
E	eccentricity
$D + 2E$	major diameter
D and $D + 4E$	width of the double threaded helix in the stator
P_r	pitch length of the rotor
P_s	pitch length of the stator.

For a 1-2 pump:

$$P_s = 2P_r \qquad\qquad (R1)$$

Perspective

P_s

P_r

E

D

$D + 4E$

Stator centerline

Rotor centerline

D

Section

P_r

P_s

E

$D + 2E$

$D + 4E$

Pump assembly   Rotor   Stator

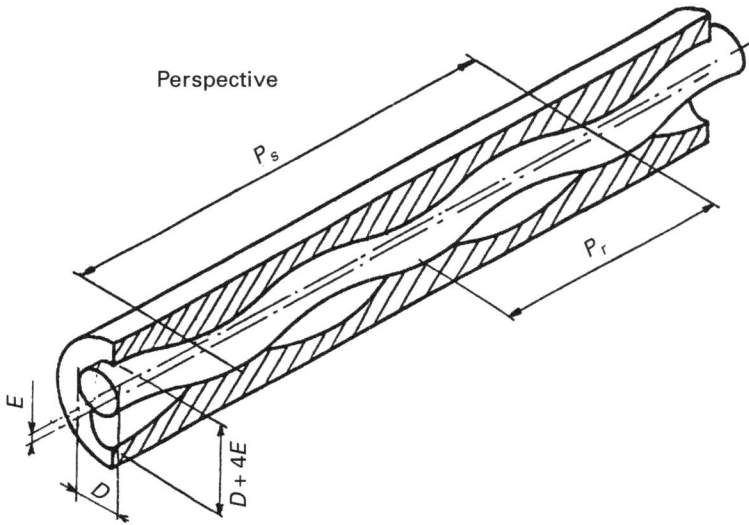

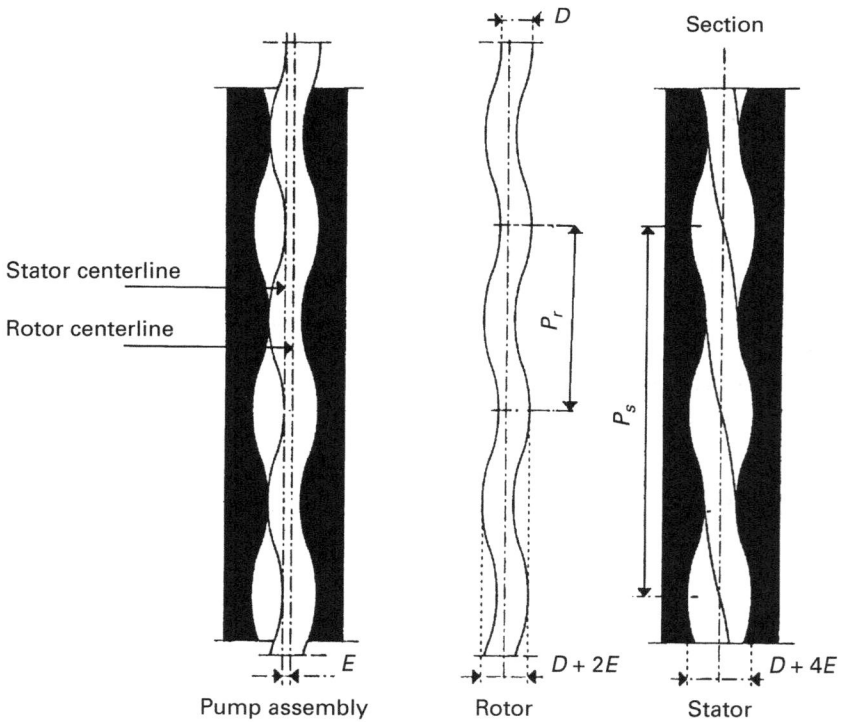

Figure R1 Rotor and stator geometry [1, 2].

For a multi-lobe pump (L_r: number of rotor lobes):

$$P_s = \frac{L_r + 1}{L_r} \times P_r \qquad \text{(R2)}$$

The number of cavities C is calculated as follows (H_s being the length of the stator):

$$C = L_r \left(\frac{H_s}{P_r} - 1 \right) \qquad \text{(R3)}$$

1.3 Rotor

For a 1-2 pump, the rotor of helical shape and of circular section is defined by:

D — diameter
E — rotor/stator eccentricity
$P_r = P_s/2$ — helix pitch.

1.4 Stator

For a 1-2 pump, the stator of helical inner shape is defined by:

D — minimum width of the section
$D + 4E$ — maximum width of the section
$P_s = 2P_r$.

1.5 Multi-lobe pumps

A multi-lobe pump is made of one rotor comprising L_r helices set one into the other, and inserted into a stator including $L_s = L_r + 1$ helices. The pump is called a "L_r-L_s pump".
    The pump kinematics ratio i is: $i = L_r/L_s = P_r/P_s$.

1.6 Fundamental parameters for maximum efficiency

The PCP is a very simple machine whose performance and operating time depend on some basic parameters:

* Elastomer behavior with:
  - Frictional wear (presence of abrasive products, sand, etc.)
  - Gaseous hydrocarbons which may diffuse into the elastomer and alter its mechanical properties
  - Temperature.
* Rotating speed being limited by:
  - The rotor imbalance which is equal to $M\varpi^2$
  - The friction speed of the rotor in the stator.

- Lubrication
  - The pumped liquid ensures lubrication between the stator and the rotor. If, for some reason, this lubrication is not achieved, the elastomer burns and the stator is damaged.

R2 PCP CHARACTERISTICS [1]

2.1 Pump displacement

- The displacement (V_0) is determined by the fluid volume produced in one revolution of the rotor:

$$V_0 = 4E \times D \times P_s \tag{R4}$$

- Calculated flow rate per minute (Q_c):

$$Q_c = 4E \times D \times P_s \times N \tag{R5}$$

- Actual pump flow rate (Q_a):

$$Q_a = Q_c - Q_s \tag{R6}$$

where Q_s is the leak rate.

The standard ISO 15136 codifies the daily flow rate of the pump at 500 rpm.

2.2 Head rating

The pump head rating ΔP is determined by:
- The number of cavities formed between the rotor and the stator
- The head rating developed into an elementary cavity δp:

$$\Delta P = \delta p \, (2n_p - 1) \tag{R7}$$

where n_p is the number of pitches P_s.

2.3 Mechanical resistant torque

It is the operational resistant torque (or running torque):

$$\Gamma_m = 1.63 \times 10^{-5} \frac{V \times \Delta P}{\rho} \tag{R8}$$

where
- Γ_m mechanical resistant torque (daN.m)
- V displacement (cm³)
- ΔP head rating (kPa)
- ρ efficiency (for an evaluation, $\rho = 0.7$).

2.4 Multi-lobe pump characteristics

The multi-lobe pump characteristics are:
- The theoretical rotating speed:
  - Rotating speed: $N_r = Q_0/V_0$ (rpm)
  - Nutation speed: $N_n = L_r \times N_r$ (rpm).
- The flow rate corresponding to one rotor rotation:

$$V_0 = S_0 \times P_s \times L_r \tag{R9}$$

- The theoretical torque:

$$\Gamma = \Gamma_0 \times \Delta P \times D \times E \times P_r \tag{R10}$$

2.5 Load on thrust bearing

Another characteristic, F_b, is the load on thrust bearing or the tensile strength on the drive string. Its value is:

$$F_b = \frac{\pi \times \Delta P \times (2E + D)^2}{4} \tag{R11}$$

2.6 Elastomers

2.6.1 Measurements characterizing physically an elastomer

- Utilization temperature limit
- Strength and behaviour in presence of sand
- Behaviour in presence of H_2S and CO_2
- Strength with aromatics
- Strength with water formation.

2.6.2 Measurements characterizing mechanically an elastomer

- Increase in weight (%)
- Shore hardness (°DIDC)
- Modulus 100% (MPa)
- Modulus 200% (MPa)
- Modulus 300% (MPa)
- Breaking strength (MPa)
- Ultimate elongation (%)
- Tearing strength (N/cm).

2.6.3 Elastomers recommended

- **Nitrile (NBR)**
  - Elasticity and flexibility at low temperature
  - Resistance to hydrocarbons and minor gas permeability.
- **Hydrogenated nitrile (HNBR)**
  Excellent resistance to:
  - 150°C water steam
  - Abrasion
  - Amine corrosion inhibitors
  - Acid gas H_2S and CO_2.
- **Fluorinated elastomers (Viton)**
  - Small swelling in the aliphatic and aromatic hydrocarbons
  - Excellent strength at temperature (up to 200°C)
  - Minimal permeability.

R3 SELECTION OF A PCP [1]

3.1 Well geometry

PCPs are adaptable to tubings with outside diameters ranging from 2 3/8 in. to 4 1/2 in.
- The stator must be able to go through the casing
- The rotor must be able to go through the tubing or any integrated part.

3.2 Reservoir characteristics

The flow rate Q of a well depends on:
- The difference between the available pressure which is the reservoir pressure P_g and the bottomhole back pressure P_f
- The productivity index IP:

$$Q = IP(P_g - P_f) \qquad (R12)$$

3.3 Characteristics of the produced effluents

3.3.1 Oil gravity and viscosity

- **Frictional pressure drop ΔP_f**

$$\Delta P_f = \frac{7.05 \times 10^{-4}}{(D+d)(D-d)^3} \times Q \times \mu_f \times L \times \frac{1}{\ln \frac{\mu_s}{\mu_f}} \left(\frac{\mu_s}{\mu_f} - 1 \right) \qquad (R13)$$

- **Resistant torque Γ_v generated by the viscosity**

$$\Gamma_v = 0.165 \times 10^{-8} \times \mu_f \times L \times N \times \frac{d^3}{D-d} \times \frac{1}{\ln \dfrac{\mu_s}{\mu_f}} \left(\frac{\mu_s}{\mu_f} - 1 \right) \qquad (R14)$$

where

Γ_v resistant torque generated by the viscosity (daN.m)
d drive string diameter (cm)
Q pumped flow rate (m^3/d)
L length of the tubing (m)
N rotating speed (rpm)
μ_f viscosity of the effluent at the inlet temperature (cp or mPa.s)
μ_s viscosity of the effluent at the surface (cp or mPa.s).

3.3.2 Fluid formation gas content

Refer to Chapter N, Paragraph 2.1.2.

3.3.3 Fluid temperature

- Relation with the fluid viscosity: the temperature evolution generates large variations of heavy oil viscosity, as shown in Fig. R2.

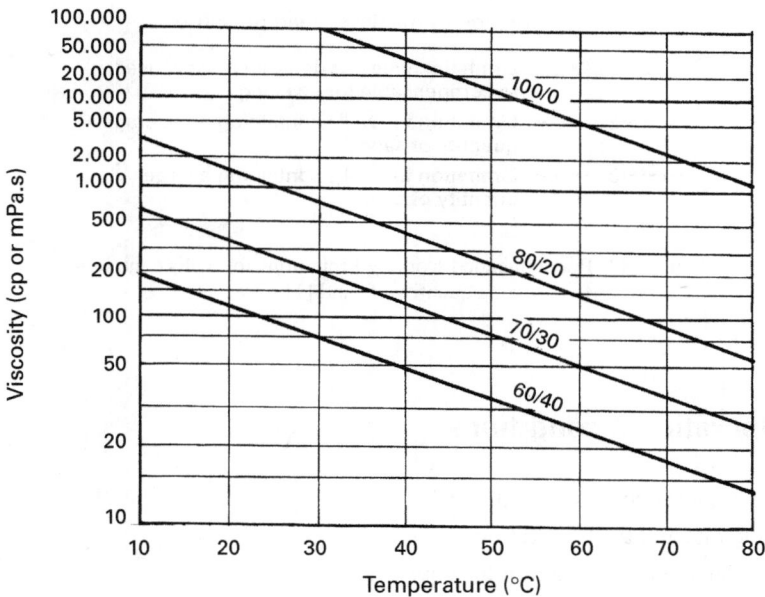

Figure R2 Heavy oil/kerosene dilution. Viscosity of heavy oil versus temperature and heavy oil/kerosene ratio [1].

- Critical temperature in relation with the elastomer
  - Nitrile elastomers: 110°C
  - Hydrogenated nitrile elastomers: 160°C.

3.3.4 Presence of sand

Figure R3 defines the application limits of pump models in presence of sand.

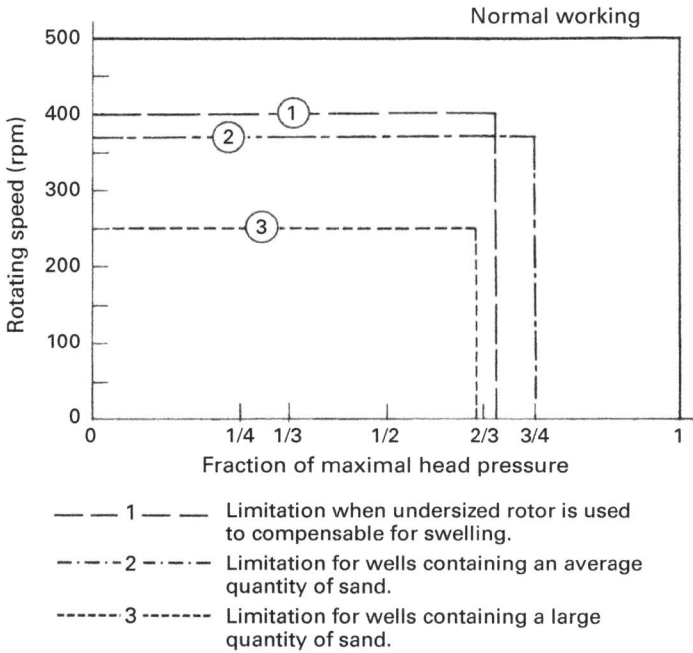

— — 1 — — Limitation when undersized rotor is used to compensable for swelling.

— · — · 2 — · — · Limitation for wells containing an average quantity of sand.

------ · 3 ------ Limitation for wells containing a large quantity of sand.

Figure R3 Recommended working limitations for wells containing a quantity of sand [1].

3.4 Operational conditions

The pump model is chosen according to:
- The flow rate and the head rating
- The diameter and the length of the equipment
- The drive systems.

Figure R4 defines the performance curves for a given pump.

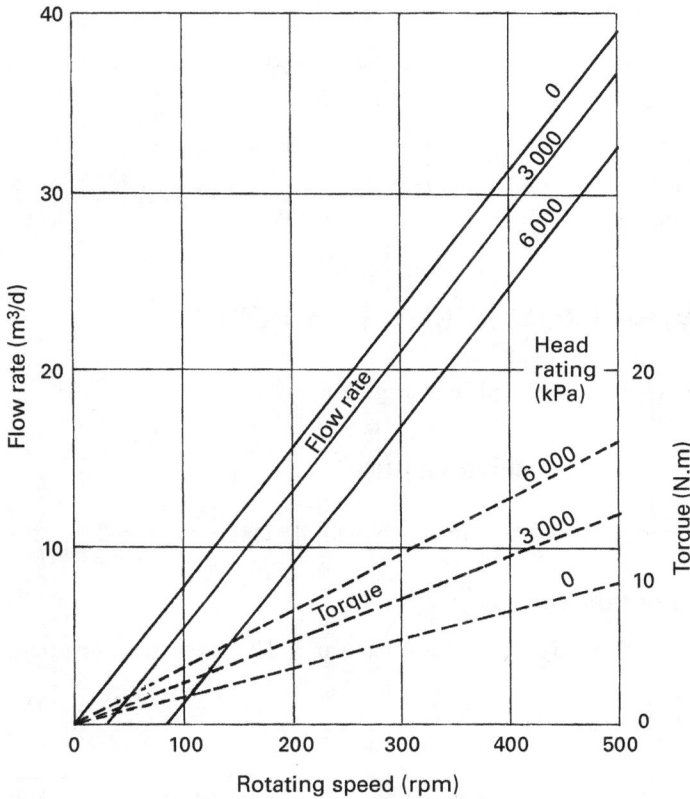

Figure R4 Example of PCP performance curves [1].

3.5 PCP identification

Progressing cavity pumps rotated by drive strings from the surface are identified in accordance with the established standard ISO 15136.

3.5.1 Stator code

This code is located no more than 0.8 m from the top of the stator, thus differentiating the top from the rotor stop.

vvv/hh/eee

where

    vvv   displacement in m³/d at 500 rpm
    hh   maximum head rating of the pump, in MPa
    eee   manufacturer's code for the elastomer.

3.5.2 Rotor code

The rotor head is identified with the following code:

$$vvv/hh$$

where
   vvv   displacement in m^3/d at 500 rpm
   hh   maximum head rating of the pump, in MPa.

R4   DRIVING FROM SURFACE OF PCP [1]

Figure R5 represents a PCP typical configuration.

4.1   Stresses in the drive strings

The stresses in the drive strings are due to the following parameters.

4.1.1 Weight of the rods

Sucker rods are currently used. They have the following characteristics, indicated in Table R1.

Table R1

Nominal diameter (in.)	Nominal diameter (mm)	Rods section (cm^2)	Weight in air (kg/m)
3/4	19.0	2.85	2.37
7/8	22.2	3.88	3.17
1	25.4	5.07	4.20
1 1/8	28.6	6.41	5.36

4.1.2 Thrust generated by the head rating of the pump

According to the pump model chosen, the load on the thrust bearing of the drive head is calculated with Eq. R11.

4.1.3 Mechanical resistant torque

The mechanical operational resistant torque Γ_m is calculated with Eq. R8.

4.1.4 Resistant torque generated by the viscosity of the effluent in the tubing

Resistant torque Γ_v is calculated with Eq. R14.

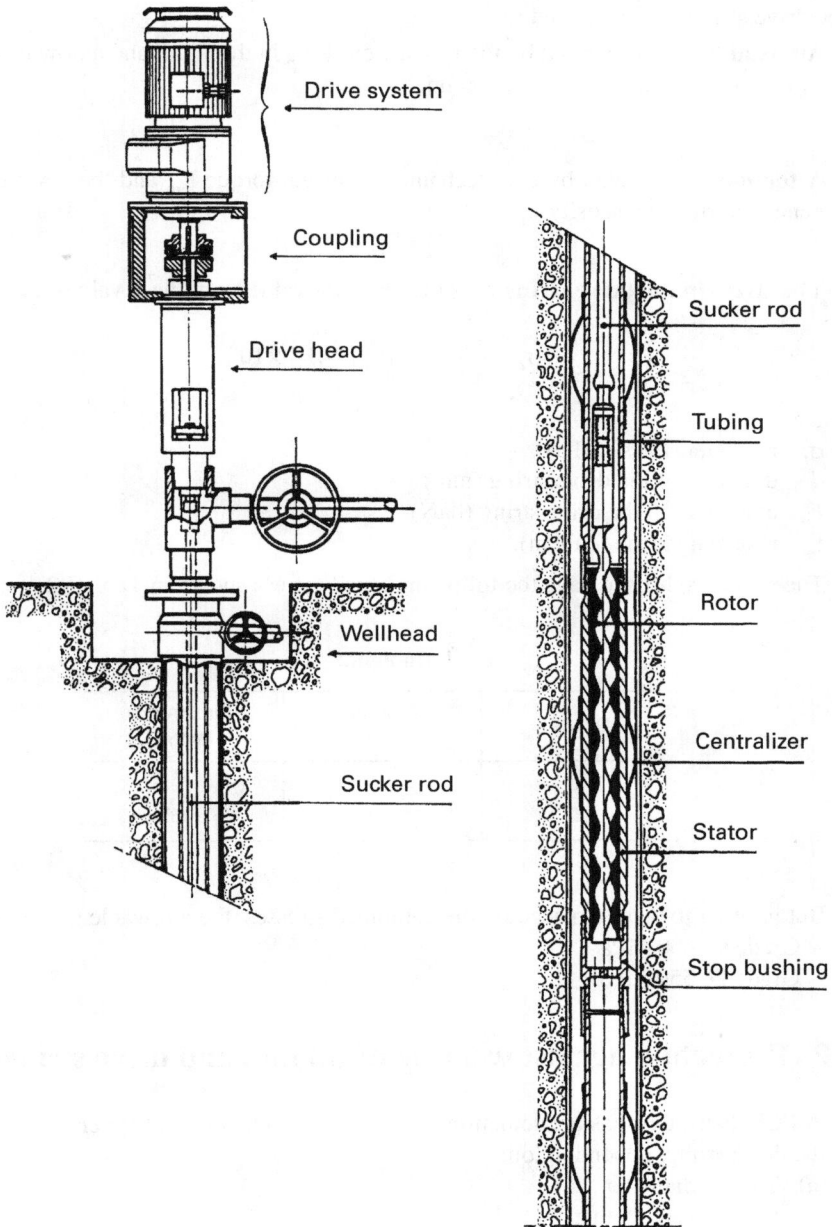

Figure R5 PCP typical configuration [1].

4.1.5 Total stresses in the drive strings

The drive strings are subjected to:

- **An axial load** F generated by their own weight F_p in the fluid, and a downward thrust F_b due to the head rating of the pump:

$$F = F_p + F_b$$

- **A torsion** Γ generated by the mechanical resistant torque Γ_m and the resistant torque generated by the viscosity Γ_v:

$$\Gamma = \Gamma_m + \Gamma_v$$

The stress in the drive string σ_t is the resultant value from these values, i.e.:

$$\sigma_t = \frac{0.4}{\pi d^3} \sqrt{F^2 d^2 + 64\Gamma^2} \times 10^6 \qquad \text{(R15)}$$

where

σ_t resultant stress (MPa)
d diameter of the drive string (mm)
F axial load in the drive string (daN)
Γ resistant torque (daN.m).

The sucker rods used have the following tensile stress shown in Table R2.

Table R2

Grade	Minimum (MPa)	Maximum (MPa)
C	620	793
D	793	965

R

But in order to take into account the combined stresses, the allowable grades are:
– Grade C: 400 MPa
– Grade D: 550 MPa.

4.2 Protection against wearing of tubings and drive strings

Each PCP distributor has a calculation program which can define the centralizer location on the drive string, depending on:
  (a) Tubing diameter
  (b) Drive string diameter
  (c) Tensile stress on the drive string
  (d) Well profile and the bending radius
  (e) Rotating speed
  (f) Physico-chemical characteristics of the pumped fluid.

4.3 Drive head

PCP distributors offer three types of drive heads:
- With a one-piece shaft: recommended in case of direct line coupling with a speed reducer.
- With a hollow shaft allowing a polished rod crossing screwed at the drive strings extremity. This type of drive head is recommended for drive systems designed with belts and pulleys fitted on a vertical rotation axis.
- Also with a hollow shaft, but with a return angle allowing the use of a drive system with horizontal rotation axis.

A drivehead brake can be installed to dissipate stored energy in order to limit and/or stop rotation of the drive string during shutdown events. Brakes may be of the following type: friction, hydraulic, electric motor, manual.

4.4 Drive system

All types of drive systems may be used:
- (a) Electric motor
- (b) Hydraulic drive
- (c) Internal-combustion engine, gas or diesel.

4.4.1 Determination of the hydraulic power of the pump

This power P depends on the total resistant torque (mechanical + viscosity), that is:

$$P = 1.05 \times 10^{-3} \times \Gamma \times N \qquad \text{(R16)}$$

where
- P hydraulic power (kW)
- Γ total torque (daN.m)
- N rotating speed (rpm).

4.4.2 Code

The Canadian Joint Industry PCP Steering Committee has proposed a code to identify drive heads according to their working characteristics:

$$TTTT/EEEE/LL/SSSS$$

where
- T   maximum torque of braking (N.m)
- E   capacity of energy dissipation (MJ)
- L   maximum axial load (tons), based on a 25 000 hours life duration
- S   maximum rotating speed (rpm), defined by the manufacturer.

The drive systems should also be defined by their maximum and minimum temperatures in use.

R5 THE INSERT PUMP [1]

The Insert PCP reduces downtime and rig expenses by allowing the pump to be installed and removed with the rod string, eliminating the need to pull tubing to replace the pump. The entire pump assembly can be installed or removed with a rod service rig or flush by, reducing pulling costs by 40-50%.

5.1 Description

Figure R6 represents a longitudinal section of an insert pump set at the production tubing end. As the tubing is lowered into the well, an anchor shoe is fitted in its end which enables the seating of the proper pump. The tubing may be equipped with centralizers or anchored into the casing.

The complete pump includes:

- A vertical anchor system
- A non-rotating mandrel completed by an element including seals
- The rotor and the joining polished rod fastened to the sucker rods
- The stator and the upper coupling on which the clamp of the joint rod rests during handling.

This assembly may be mounted in the workshop.

5.2 Evaluation of the tensile strength of the anchor system

In order to keep a well positioned pump into the anchor shoe which is located at the tubing end, the anchor system strength is:

- **The tensile strength** F_1 on the drive strings, generating a reaction on the stator under the head rating effect ΔP, generated in the pump:

$$F_1 = \frac{\pi \times \Delta P \times (2E + D)^2}{4} \tag{R17}$$

- **The force** F_2 on the joint section S_j:

$$F_2 = S_j \times \Delta P \tag{R18}$$

- **The screwing force** F_3 which tends to lift up the stator:

$$F_3 = \frac{\Gamma}{2E + D} \times \frac{P_s}{\sqrt{P_s^2 + 4\pi^2 (2E + D)^2}} \tag{R19}$$

So, the anchor system should withstand a force F:

$$F = F_1 - F_2 + F_3 \tag{R20}$$

The above relations are mentioned in bar, cm, daN.

Any anchor system should be bench tested in the workshop before running in well.

Output
production

Centralizer

Clamp part

Joining
polished rod

Tubing

Rotor

Stator

Casing

Seals

The back
seal mandrel

Anchor shoe

Input

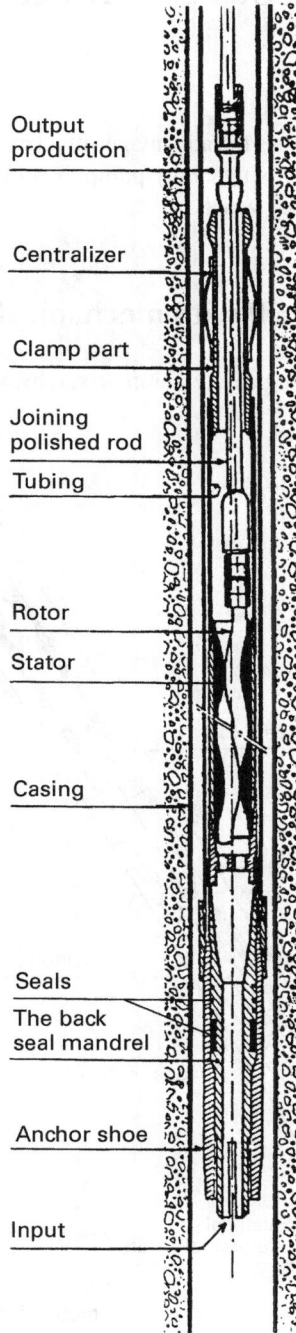

Figure R6 Insert progressing cavity pump [1].

R6   THE ELECTRICAL SUBMERSIBLE PCP [1]

6.1   Introduction

A method of artificial lift combines the advantages of an electric submersible pumping system with the benefits of a progressing cavity pump. It is mainly used:
- (a)   In horizontal and deviated wells
- (b)   In high sand content wells.

6.2   Description of a PCP with mechanical speed reducer

Figure R7 represents a progessing cavity pump driven by an electrical submersible motor.

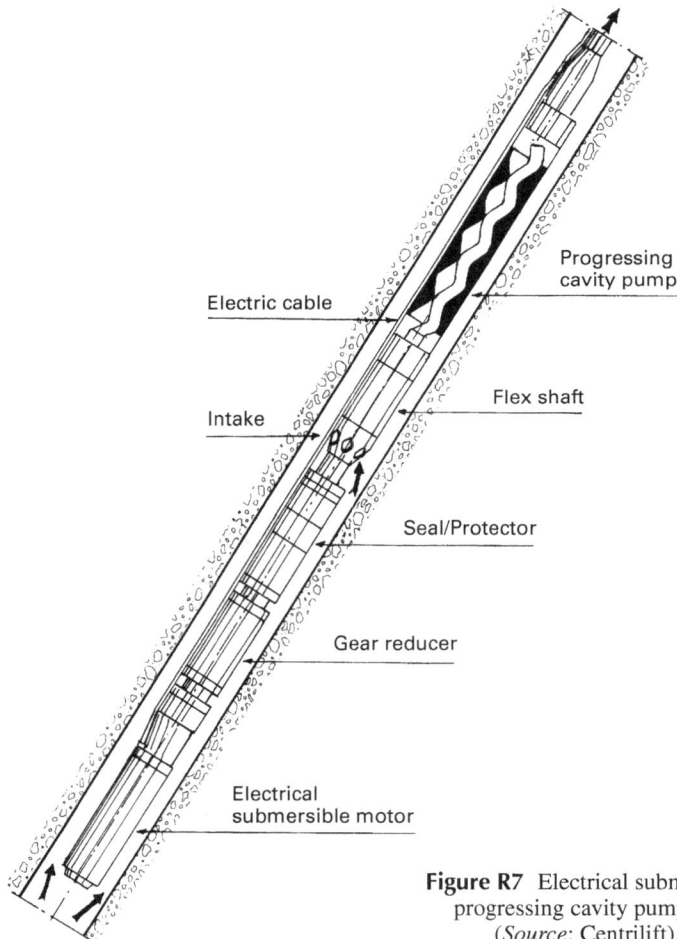

Figure R7   Electrical submersible progressing cavity pump [1]. (*Source*: Centrilift).

It includes:

- **Progressing cavity pump**. Supplied by any of the PCP manufacturers/distributors.
- **Flex shaft and intake**. Designed to absorb the eccentric rotation of the rotor and allow for fluid flow into the pump.
- **Seal**. Isolates the motor and gear reducer oil from the well fluids and houses the thrust bearing that handles the thrust load from the pump.
- **Gear reducer**. A planetary gear assembly that reduces the speed of rotation to speeds acceptable to the PCP.
- **Motor**. Standard ESP motor.
- **Cable**.
- **Surface drives**. Variable frequency controllers or switchboards may be used, usually in conjunction with a transformer.

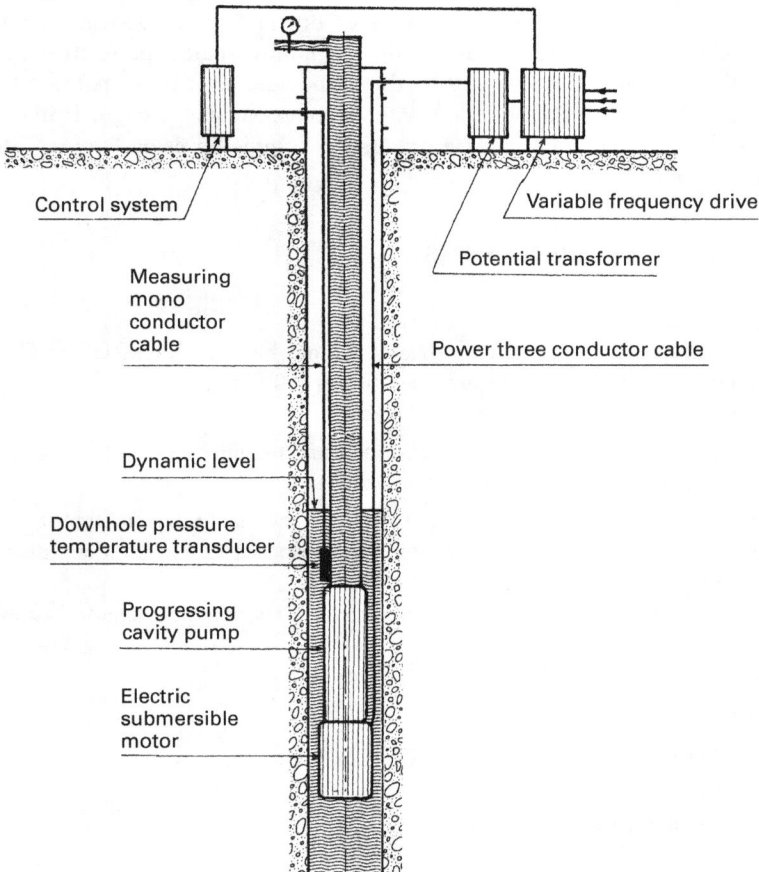

Figure R8 Well equipped with an electrical submersible progressing cavity pump [1].

357

6.3 Description of a PCP with electronic speed reducer

This equipment, shown in Fig. R8, is able to produce from a well at optimum conditions, which means keeping a constant submergence level whatever the reservoir productivity. This characteristic is achieved by a continuous measurement of the dynamic pressure and in adapting the pump flow rate to any variation of the submergence level by subtraction or addition.

R7 NEW PCP TECHNOLOGY

7.1 Coiled tubing drive system [4]

More commonly used for well stimulations, sand cleanouts and drilling operations, coiled tubing is also ideal as a driverod to a PC pump. Coiled tubing can be used as a drive string on a producing PC pump because the continuous hollow shaft is perfectly suited to transmitting torque. Torque is transmitted by the outside diameter of a shaft, not the interior, and therefore, the interior can be hollow without compromising strength, in high torque rod driven wells it is possible to increase production volume by using the inner diameter of drive rod (Fig. R9).

7.2 Thermal recovery [5, 7, 8]

The PCM Vulcain ™ pump
To access untapped reserves of extra heavy oil and bitumen, PCM Oil & Gas develops solutions adapted to high temperature opérations (350°C/660°F).

To overcome the limitations:

- Chemical compatibity between the elastomer and the pumped fluid (typically with aromatics, resulting in swelling and/or hardening)
- Maximum allowable temperature, above which the elastomer-chain structure changes irreparably (for nitrile or fluorocarbonated compounds, this temperature is approximately 160°C).

Consequently, the elastomer PCP stator is replaced with a metal stator formed by use of a hydroforming process;

Characteristics of the PCM Vulcain ™ pump are:

- Metallic stator, metallic rotor
- Max. operating temperature: 350°C (660°F)
- Max. head: 150 bar – 2100 psi
- Very efficient at low submergence.

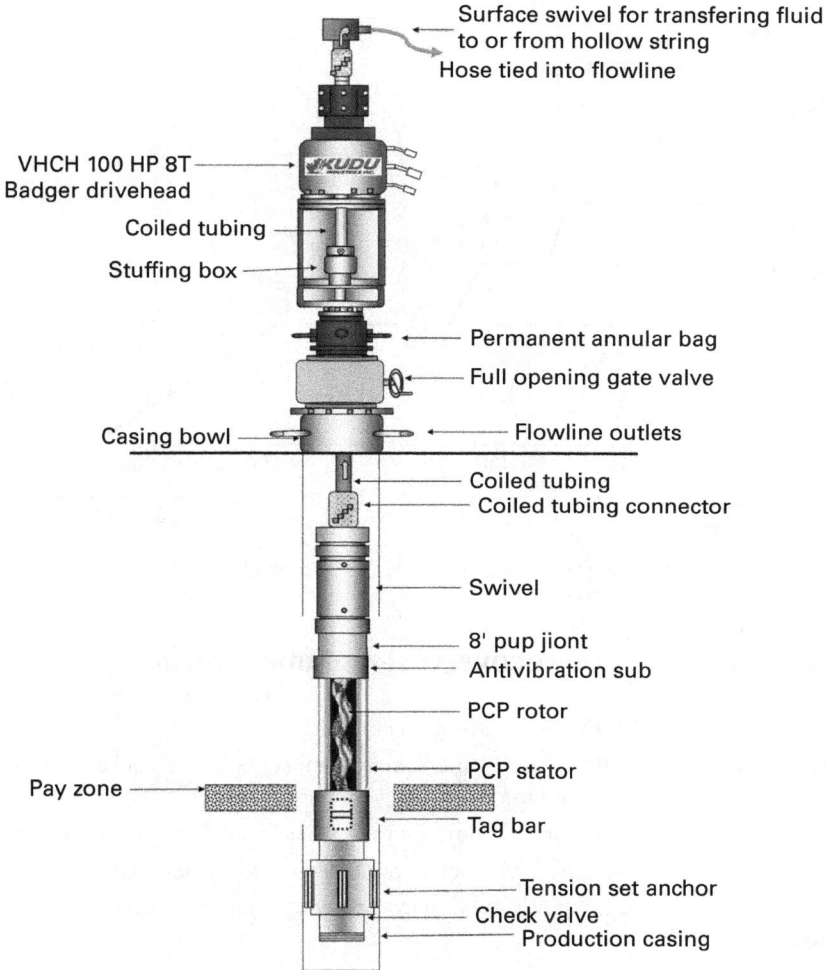

Figure R9 Scheme of a coiled tubing drive system [4] (Kudu Industries Inc.).

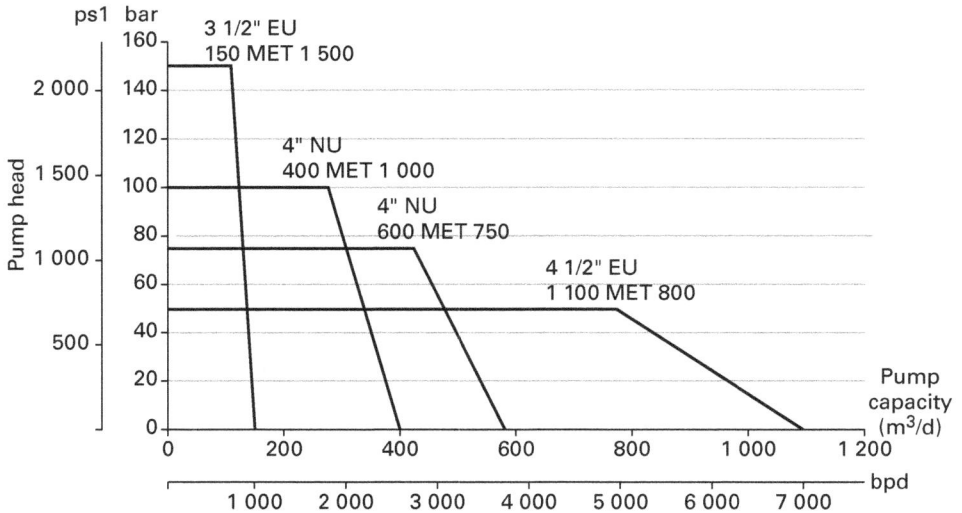

Figure R10 Range of PCM VulcainTM pumps [5, 7].

7.3 PCP for heavy oil recovery: steam injection [6]

The Kaechele Even Wall PC Pump system (Fig. R11)
- This new technology allows use of the same components of the sucker-rod pumping system for both, pumping and injecting.
- Steam is injected into the reservoir through hollow sucker-rods and a hollow rotor.
- The injected steam builds up a steam chamber that rises to to the top of the reservoir.
- This system allows handling of slow flowing heavy oil containing sand at temperatures up to 300°C (572°F).

7.4 New Progressing Cavity Pump (NPCP) for multiphase and viscous liquid production [9]

The NPCP is composed of a Progressing Cavity Pump (PCP) and a system of Hydraulic Regulators (HR) installed inside the pump in between the cavities. The HR are self-regulated devices that recirculate the fluid between the cavities in order to control the pump thermo-hydraulic response and to avoid excessive built up of heat, which might result in premature failure of the pump's stator.

The benefits are multiple:
- It uniformizes the pressure across the pump length, which stabilizes the temperature
- It compensates the compasse gas volume of progressing cavities
- It protects the stator and therefore, improves the pump's performances.

360

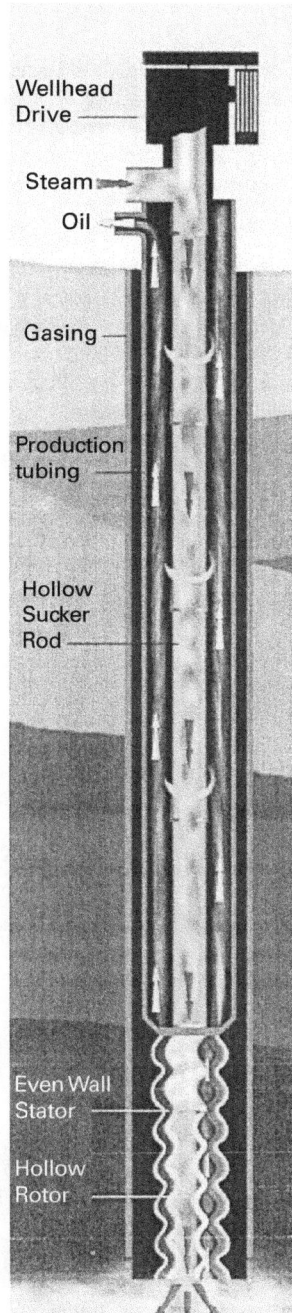

Figure R11 The Kaechele Even Wall PC Pump [6]

REFERENCES

1   Cholet H (1998) *Progressing Cavity Pumps*. Editions Technip, Paris
2   ISO 15136 (2001 à 2006) *Standard for Progressing Cavity Pump Systems for Artificial Lift in the petroleum industry. Part 1: Pumps* (2001) – *Part 2: Drive heads* (2006)
3   Revard JM (1995) *The Progressing Cavity Pump Handbook*. PennWell Books
4   www.kudupump.com/products-and-services/new-technology-ct-pcp.html
5   www.pcm-pump.com/oil-gas/vulcain.html
6   www.w-kaechele.com/25.html
7   Beauquin JL, Boireau C, Lemay L, Seince L (2005) Development status of a metal progressing cavity pump for heavy oil and hot production wells *Paper SPE 97796* of the 2005 SPE International Thermal Operations and Heavy Oil Symposium, Calgary, 1-3 November. And Denney D (Ed.), *Journal of Petroleum Technolology*, May 2006
8   Chalier G (2008) *Metal PCP Pushes Up Pumping Window for Heavy Oil Hot Production ; Joslyn field case*. World Heavy Oil Congres, Edmonton, Alberta, Canada, March 2008.
9   Bratu C, Seince L (2005) New Progressing Cavity Pump (NPCP) for Multiphase and Viscous Liquid Production, *Paper SPE/PS-CIM/CHOA – PS2005-97833*.

R

S

Hydraulic Pumping

S

Hydraulic Pumping

S1 GENERAL INTRODUCTION

Hydraulic pumping applies the Pascal's principle to activiting wells by transmitting pressure generated on the surface to the bottom of a well by a working fluid in order to actuate:

- An engine with a reciprocating piston driven by a power fluid connected by a short shaft to a piston in the pump end.
- A jet pump equipped with a nozzle that leads into a venturi, in order to carry the fluid from the pay zone by means of the working fluid.
- A turbine pump where a turbine drives a centrifugal pump.

S2 PISTON-TYPE HYDRAULIC PUMPING [1, 2]

2.1 Principle

A subsurface hydraulic piston pump is a closely coupled reciprocating engine and pump. The unit is installed below the working fluid level in a well, as shown in Fig. S1. High pressure power fluid is directed to the engine through one conduit and spent power fluid and well production are directed to the surface through another conduit. The high pressure power fluid causes the engine to reciprocate much like a steam engine except that the power fluid is oil or water instead of steam. The pump, driven by the engine, pumps the fluid from the wellbore.

S

2.2 Advantages

Hydraulic pumping:

- Represents one of the deepest methods of lift (5 500 m)
- Can handle deviated wells
- Is easily adapted to automation
- Facilitates the addition of inhibitors
- Is suitable for pumping heavy crudes

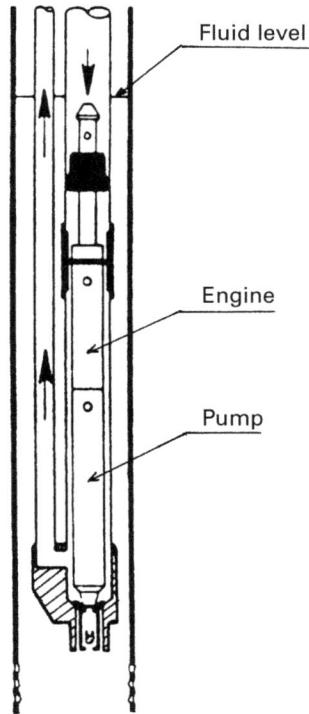

Figure S1　Subsurface hydraulic pump-piston type [1].

- One well or multiple well units are available
- Simple wellheads accommodate closely spaced wells, covered or cellered wellheads and wells in visually sensitive areas.

2.3 Power fluid systems

There are two basic types of power fluid systems:

- The closed power fluid (CPF) system in which the surface and subsurface power fluids stay in a closed conduit and do not mix with the produced fluid (Fig. S2a).
- The open power fluid (OPF) system in which the power fluid mixes with the production fluid downhole and returns to the surface as commingled power fluid and production (Fig. S2b).

The power fluid is either water or oil.

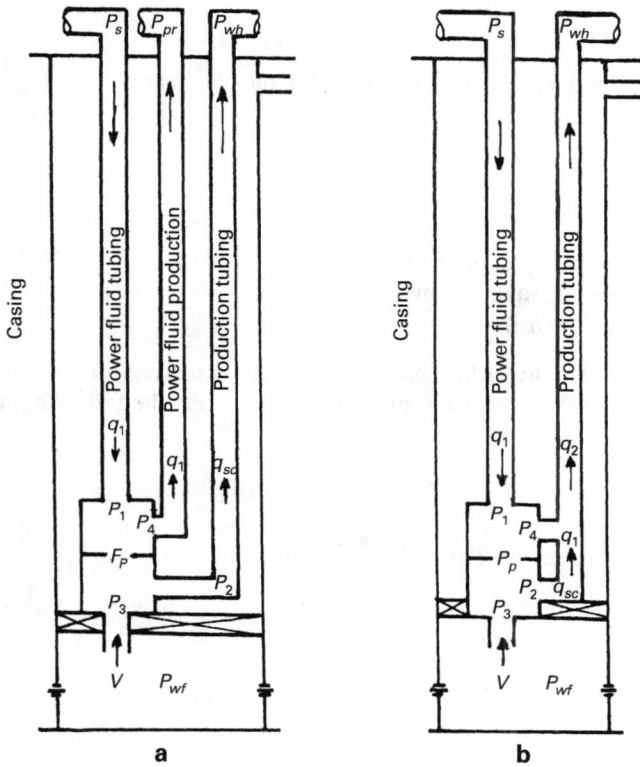

P_1	Power fluid pressure	P_{pr}	Power return back pressure
P_2	Pump discharge pressure	P_{wh}	Wellhead pressure
$P_3 = P_{wf}$	Intake pressure	q_1	Power fluid rate
P_4	Engine discharge pressure	q_{sc}	Production rate
P_s	Surface operating pressure	V	Intake volume

Figure S2 Pressures affecting a hydraulic pump [2].
a. CPF. **b.** OPF.

2.4 Tubing arrangements

They are two types of tubing arrangements:

- **Fixed type of pump**: the pump is attached to the power fluid tubing and lowered into the well by this tubing.
- **Free type pump**: the pump fits inside the power fluid tubing and is free to be circulated to the bottom and back out to the surface again.

Either type can be a CPF or an OPF system, but the main difference is that the free pump size is limited by the tubing size, while any fixed pump size is adaptable to the tubing, provided that the pump fits inside the casing.

2.5 P/E ratio

It is the ratio of the net pump piston area to the net engine piston area:

$$P/E = \frac{A_p - A_r}{A_e - A_r} \qquad (S1)$$

where
A_p area of pump piston (sq in. or cm^2)
A_e area of engine piston (sq in. or cm^2)
A_r area of rod (sq in. or cm^2).

The *P/E* ratio is related to the surface pressure required for a given lift. To limit the surface pressure to the generally acceptable maximum of 30 MPa, the following maximum value is recommended:

- *In U.S. units*

$$\text{maximum } P/E = \frac{10\,000}{\text{net lift}} \quad \text{(net lift, in ft)} \qquad (S2)$$

- *In metric units*

$$\text{maximum } P/E = \frac{3\,050}{\text{net lift}} \quad \text{(net lift, in m)} \qquad (S3)$$

The net lift *(NL)* (ft or m) given by:

$$NL = D_p - (P_3/G_f) \qquad (S4)$$

where
D_p pump setting depth (ft or m)
P_3 pump intake pressure (psi or MPa)
G_f flowing gradient of the fluid in the production conduit (psi/ft or MPa/m).

In the special case in which the pump is set at the bottom of the well:

$$NL = D - (P_{wf}/G_f) \qquad (S5)$$

where
D well depth (ft or m)
P_{wf} flowing bottomhole pressure (psi or MPa).

Generally, for deep wells with low bottomhole pressures, P_{wf}/G_f is compared to D and, therefore, can be neglected.

Usually, when more than one pump size can be used, the one with the greatest lift capability (lowest *P/E* ratio) is chosen. This reduces the surface operating pressure, thereby reducing slippage in the bottomhole pump.

S   *Hydraulic Pumping*

2.6 Pump displacement

The production rate is given by:

$$q_3'' = q_3' N \qquad (S6)$$

where
q_3' pump displacement (bbl/D/spm or m^3/d/spm)
N pump speed (spm).

Normally, q_3'' is referred to as the theoretical production rate. It is equal to the actual production rate only if the pump operates at 100% efficiency. Good design practice is to use 85% pump efficiency and to select a pump that will operate below 85% of its rated speed. Hence:

$$V = q_3'' \eta_p \qquad (S7)$$

or

$$V = q_3' N \eta_p \qquad (S8)$$

where
η_p pump end efficiency
V volume of the produced fluid rate (liquid + gas) at the intake pressure.

2.7 Engine displacement

The engine being coupled to the pump, the engine piston moves at the same speed as the pump piston. The theoretical power fluid rate is given by:

$$q_1'' = q_1' N \qquad (S9)$$

where q_1' is the engine displacement (bbl/D/spm or m^3/d/spm).
The engine-end efficiency is the ratio of the theoretical rate to the actual rate, or

$$\eta_e = q_1''/q_1 \qquad (S10)$$

or

$$q_1 = q_1' N/\eta_e \qquad (S11)$$

where
q_1 actual power fluid rate required to produce an actual fluid rate V
η_e engine-end efficiency, estimated at about 90%.

2.8 Pressure calculations

The various pressures involved in a CPF and in an OPF system are shown in Fig. S2. The pressure available to drive the engine is P_1, while the engine must discharge against P_4. The pump end must discharge against P_2 while being filled with P_3.

The pump friction F_p depending on the pump type, the percentage of rated speed and the viscosity of the power fluid, must be subtracted.

369

This force balance is shown as follows:

$$\left(P_1 - P_4\right) - \left(P_2 - P_3\right)\left(\frac{A_p - A_r}{A_e - A_r}\right) - F_p = 0 \tag{S12}$$

or
$$(P_1 - P_4) - (P_2 - P_3)\,(P/E) - F_p = 0 \tag{S13}$$

This equation is equally valid for an OPF system and a CPF system. In an OPF system, however, P_4 is identical to P_2. Thus, the equation can be written as:

$$(P_1 - P_2) - (P_2 - P_3)\,(P/E) - F_p = 0 \tag{S14}$$

2.9 Input power

The input power requirement is estimated from the following equation:

- *In U.S. units*
$$HP = 1.7 \times 10^{-5}\, q_1\, P_s \tag{S15}$$
where
  HP horsepower (hp)
  P_s surface operating pressure (psi).

- *In metric units*
$$WP = 1.13 \times 10^{-5}\, q_1\, P_s \tag{S16}$$
where
  WP power (kW)
  P_s surface operating pressure (MPa).

2.10 Subsurface trouble-shooting guide [1]

The following list will help as a guide for analyzing and trouble-shooting the subsurface pumping unit.

Indication	Cause	Remedy
1. Sudden increase in operating pressure–pump stroking.	(a) Lowered fluid level which causes more net lift. (b) Paraffin build-up or obstruction in power oil line, flow line or valve. (c) Pumping heavy material, such as salt water or mud. (d) Pump beginning to fail.	(a) If necessary, slow pump down. (b) Run soluble plug, hot oil or remove obstruction. (c) Keep pump stroking–Do not shut down. (d) Retrieve pump and repair.
2. Gradual increase in operating pressure–pump stroking.	(a) Gradually lowering fluid level. Standing valve or formation plugging up. (b) Slow build-up of paraffin. (c) Increasing water production.	(a) Surface pump and check. Retrieve standing valve. (b) Run soluble plug or hot oil. (c) Raise pump SPM and watch pressure.

Indication	Cause	Remedy
3. Sudden increase in operating pressure– pump not stroking.	(a) Pump stuck or stalled. (b) Sudden change in well conditions requiring operating pressure in excess of triplex relief valve setting. (c) Sudden change in power oil-emulsion, etc. (d) Closed valve or obstruction in production line.	(a) Alternately increase and decrease pressure. If necessary, unseat and reseat pump. If this fails to start pump, surface and repair. (b) Raise setting on relief valve. (c) Check power oil supply. (d) Locate and correct.
4. Sudden decrease in operating pressure (Speed could be increased or reduced).	(a) Rising fluid level–Pump efficiency up. (b) Failure of pump so that part of power oil is bypassed. (c) Gas passing through pump. (d) Tubular failure, downhole or in surface power oil line. Speed reduced. (e) Broken plunger rod. Increased speed. (f) Seal sleeve in bottomhole assembly washed or failed. Speed reduced.	 (b) Surface pump and repair. (d) Check tubulars. (e) Surface pump and repair. (f) Pull tubing and repair bottomhole assembly.
5. Sudden decrease in operating pressure– pump not stroking.	(a) Pump not on seat. (b) Failure of production unit or external seal. (c) Bad leak in power oil tubing string. (d) Bad leak in surface power oil line. (e) Not enough power oil supply at manifold.	(a) Circulate pump back on seat. (b) Surface pump and repair. (c) Check tubing and pull and repair if leaking. (d) Locate and repair. (e) Check volume of fluid discharged from triplex. Valve failure, plugged supply line, low power oil supply, excess bypassing, etc., all of which could reduce available volume.
6. Drop in production– pump speed constant.	(a) Failure of pump end of production unit. (b) Leak in gas vent tubing string. (c) Well pumped off–Pump speeded up. (d) Leak in production return line. (e) Change in well conditions. (f) Pump or standing valve plugging. (g) Pump handling free gas.	(a) Surface pump and repair. (b) Check gas vent system. (c) Decrease pump speed. (d) Locate and repair. (f) Surface pump and check. Retrieve standing valve. (g) Test to determine best operating speed.
7. Gradual or sudden increase in power oil required to maintain pump speed. Low engine efficiency.	(a) Engine wear. (b) Leak in tubulars–Power oil tubing, bottomhole assembly, seals or power oil line.	(a) Surface pump and repair. (b) Locate and repair.

Indication	Cause	Remedy
8. Erratic stroking at widely varying pressures.	(a) Caused by failure or plugging of engine.	(a) Surface pump and repair.
9. Stroke "down-kicking" instead of "up-kicking".	(a) Well pumped off–Pump speeded up. (b) Pump intake or downhole equipment plugged. (c) Pump failure (balls and seats). (d) Pump handling free gas.	(a) Decrease pump speed. Consider changing to smaller pump end. (b) Surface pump and clean up. If in downhole equipment, pull standing valve and back flush well. (c) Surface pump and repair.
10. Apparent loss of, or unable to account for, system fluid.	(a) System not full of oil when pump was started due to water in annulus U-tubing after circulating, well flowing or standing valve leaking. (b) Inaccurate meters or measurement. (c) Leaking valve, power oil or production line or packer. (d) Affect of gas on production metering. (e) Pump not deep enough.	(a) Continue pumping to fill up system. Pull standing valve if pump surfacing is slow and cups look good. (b) Recheck meters. Repair if necessary. (c) Locate and repair. (d) Improve gas separation. (e) Lower pump.
11. Well not producing: (a) Pressure increase, stroking (b) Pressure loss, stroking	(a) Engine plugging, flow line plugged, broken engine rod, suction plugged. (b) Standing valve leaking. Tubular leak.	(a) Surface unit and repair. Locate restriction in flow line. Pull standing valve. (b) Pull standing valve. Check tubulars.

S3   JET PUMPS [3]

3.1   Description of equipment

A typical example of a subsurface jet pump is shown in Fig. S3 with details in Fig. S4. The power fluid enters the top of the pump from the tubing and passes through the nozzle, where virtually all of the total pressure of the power fluid is converted to a velocity head[1]. The jet from the nozzle discharges into the production inlet chamber, which is connected to the pump intake for formation fluids. The production fluid is entrained by the power fluid, and the combined fluids enter the throat of the pump.

In the confines of the throat, which is always of larger diameter than the nozzle, complete mixing of the power fluid and the production fluid takes place. During this process, the power fluid loses momentum and energy. The resultant mixed fluid exiting the throat

---

1. To prevent cavitation, the minimum cross-section of the ejector-diffuser throat annulus must be determined; this depends on the specified flow rate and the submergence depth of the pump. Jet pumps can exhaust in two-phase pumping.

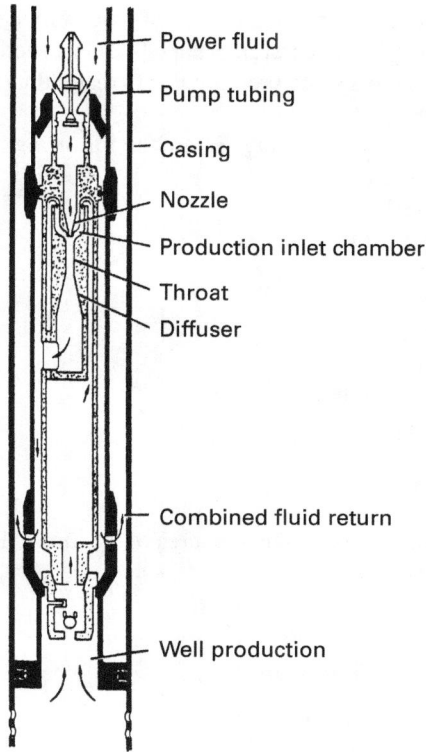

Figure S3 Jet pump [2].
Production by casing/tubing annular.

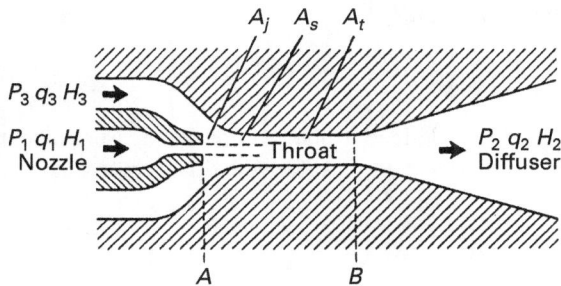

Figure S4 Jet pump nomenclature [2].

has sufficient total head to flow against the production return column gradient. Much of this total head, however, is still in the form of a velocity head. The final working section of the jet pump is, therefore, a carefully shaped diffuser section of expanding area that converts the velocity head to a static pressure head greater than the static column head, allowing flow to the surface.

The parameter represented on Fig. S4 are the following:

P_1 power fluid pressure
P_2 discharge pressure
$P_3 = P_{wf}$ intake pressure
A_j nozzle area
A_s net throat area
A_t total throat area
q_1 power fluid rate
q_2 total liquid rate in return column
V intake volume.

3.1.1 Dimensionless area

The ratio R of the nozzle area to the total area of the throat (Fig. S4) is called the area ratio, or:

$$R = A_j/A_t \tag{S17}$$

3.1.2 Dimensionless flow rate

The dimensionless flow rate M is defined by:

$$M = V/q_1 \tag{S18}$$

where

V volume of the produced fluid rate (liquid + gas)
q_1 power fluid rate.

When pumping slightly compressible fluids such as liquids, V can be considered constant and equal to the surface rate.

3.1.3 Dimensionless head

The dimensionless head H is defined as the ratio of the pressure increase experienced by the production fluid to the pressure loss suffered by the power fluid (refer to Figs. S4 and S5):

$$H = \frac{P_2 - P_3}{P_1 - P_2} \tag{S19}$$

where

P_1 power fluid pressure
P_2 discharge pressure
P_3 intake pressure.

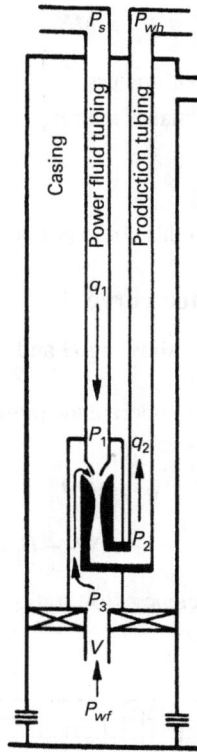

Figure S5 Jet pump schematic [2].

The parameters represented on Fig. S5 are:

P_1 power fluid pressure
P_2 discharge pressure
$P_3 = P_{wf}$ intake pressure
P_s surface operating pressure
P_{wh} wellhead pressure
q_1 power fluid rate
q_2 total liquid rate in production tubing
V intake volume.

3.1.4 Efficiency

In pumping liquid, the efficiency η_p is:

$$\eta_p = \frac{V}{q_1} \times \frac{P_2 - P_3}{P_1 - P_2} \qquad (S20)$$

or:

$$\eta_p = MH \tag{S21}$$

Therefore, it depends of various parameters:

- Geometry (shapes defined by the manufacturers)
- Type of fluid: power and pumped
- Flowrate, pressures.

Each manufacturer proposes formulas corresponding to their equipments.

3.1.5 Dimensionless performance curve

An example of these equation results showing H and η_p versus M for several values of R is represented in Fig. S6.

It is good field practice to attempt to operate the pump at its peak efficiency. In that case, the M and H ratios will be fixed; hence:

$$q_1 = V/M_p \tag{S22}$$

and:

$$P_3 = (1 + H_p)P_2 - H_pP_1 \tag{S23}$$

where M_p and H_p are the peak efficiency flow ratio and the peak efficiency head ratio, respectively.

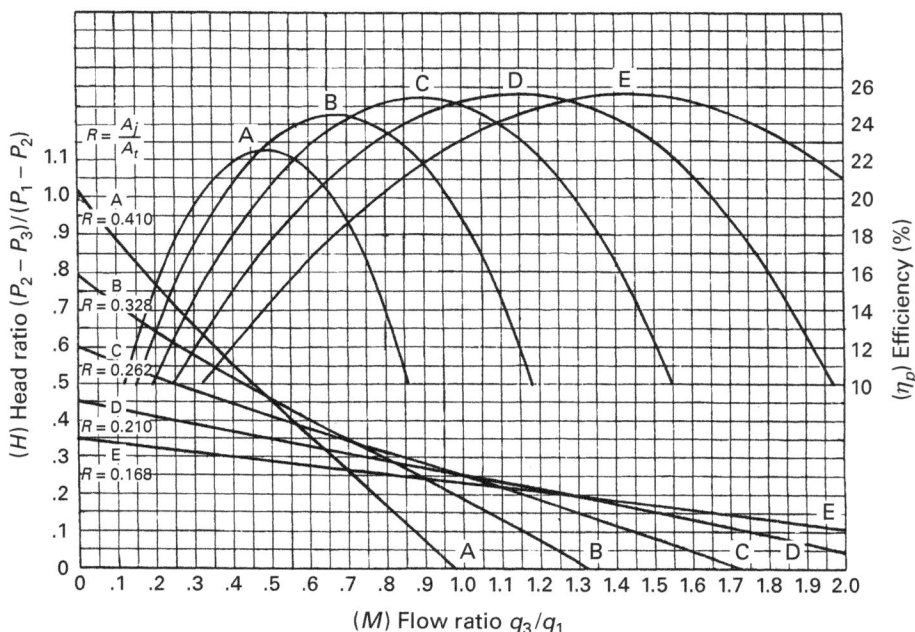

Figure S6 Example of jet pump characteristics: H vs. M and efficiency [2].

3.2 Power fluid and pressure

Similar to hydraulic pumps, jet pumps utilize either water or oil as a power fluid. The actual power fluid rate is a function of the pressures P_1 and P_2, of the flow area of the nozzle A_j, and of the specific gravity of the power fluid γ_1. When everything is measured in common oil-field units, the power fluid rate can be estimated from the following equation:

- In U.S. units

$$q_1 = 1214.5 A_j \sqrt{\frac{P_1 - P_3}{\gamma_1}} \tag{S24}$$

- In metric units

$$q_1 = 354.1 A_j \sqrt{\frac{P_1 - P_3}{\gamma_1}} \tag{S25}$$

where
q_1 fluid rate (bbl/D or m^3/d)
P_1 and P_3 pressures (psi or MPa)
A_j nozzle area (sq in. or cm^2).

In normal operations, the surface operating pressure should not exceed 4 000 psi or 28 MPa.

3.3 Input power

The input power requirement is estimated from the following equations:

- In U.S. units
$$HP = 1.7 \times 10^{-5} q_1 P_s \tag{S26}$$
where
HP horsepower (hp)
P_s surface operating pressure (psi)

- In metric units
$$WP = 1.13 \times 10^{-5} q_1 P_s \tag{S27}$$
where
WP power (kW)
P_s surface operating pressure (MPa).

REFERENCES

1   Brown KE, Wilson P (1980) *The Technology of Artificial Lift Methods*, Vol. 2b, Ch. 5, Hydraulic pumping. Piston type. PennWell Books

2 Brown KE et al. (1984) *The Technology of Artificial Lift Methods*, Vol. 4, Ch. 5, Production optimization of oil and gas wells by nodal systems analysis. PennWell Books

3 Brown KE, Petrie H (1980) *The Technology of Artificial Lift Methods*, Vol. 2b, Ch. 6, Jet pumping. PennWell Books

4 Coberly CJ, Brown FB (1962) *Petroleum Production Handbook*, Vol. 1, Ch. 6, Hydraulic pumps. SPE of AIME

5 Corteville J, Hoffmann F, Valentin E (1986) Activation des puits: critères de sélection des procédés. *Revue de l'Institut Français du Pétrole*, Vol. 41, No. 6. Editions Technip, Paris

6 Perrin D, Caron M, Gaillot G (1998) *Well Completion and Servicing*. Editions Technip, Paris

7 Corteville JC, Ferschneider G, Hoffmann FC, Valentin EP (1987) Research on jet pumps for single and multiphase pumping of crudes. *Paper SPE 16923* presented at the 62nd Annual Technical Conference of SPE, Dallas, TX, September 1987.

S

T

Multiphase Pumping and Metering

T

Multiphase Pumping
and Metering

T1 MULTIPHASE PUMPING

When the production of a marginal field or a group of remote wells is considered with an existing central gathering system the traditional options for field development are:

- Natural flow
- Artificial lift
- In-field separation with crude oil transfer pumps, gas to flare, or gas compression systems.

With the recent field deployment of numerous multiphase pumps, new approaches to field development and production have been demonstrated.

Many different pumping technologies are emerging:

- The Helico-Axial Multiphase Pump
- The Rotary Screw Pump
- The Progressing Cavity Pump.

They are presented in Paragraphs 1.4 to 1.6.

1.1 Multiphase pumping vs separation [4]

A comparison between multiphase pumping and separation is presented in Table T1 where it is reported the required equipment for each option.

1.2 Typical field characteristics for multiphase pumping

Multiphase pumps have been successfully deployed now in a broad range of field conditions, environments and climates. They are used in extreme cold weather conditions such as Siberia, and Canada as well as hot condition in Middle East, North Africa, and South East Asia, South America. They are in operation onshore, offshore on platforms and, since 1997, subsea. The deepest subsea pump (water depth: 1 700 m) was installed in 2007 in

Table T1 Equipment: separation vs multiphase pumping [4].

Equipment	Separation	Multiphase pumping
Main equipment	Separator Compression module Pump motor set Gas pig launcher Oil pig launcher	Multiphase pump package Multiphase pig launcher
Bulk	Piping Instrumentation Electrical equipment	Piping Instrumentation Electrical equipment
Pipelines	Gas Liquid	Multiphase pipeline

Gulf of Mexico. The largest pumps in capacity are operating in Siberia (3600 m^3/h per pump).

Multiphase pumps are used with a broad range of process conditions such as high GVF (98%), high liquid viscosity (heavy oil production), high temperature (steam assisted production). Figure T1 gives an overview of the performance range (differential pressure versus total volumetric flow) of the various pump types. Limits can be slightly different for each manufacturer, due to their size standardization. The pumps can be also used in parallel or in series to provide higher capacities and pressure rises.

A number of application cases are described in reference [12].

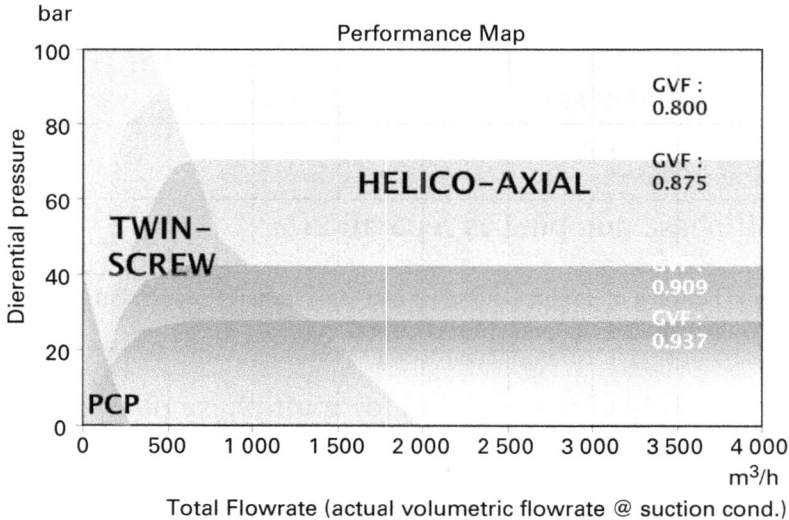

Figure T1 Performance range of the various pump types [12].

1.3 Field parameters and pump selection [5]

In order to size a multiphase pump for a particular application, the following data is required as a minimum:
  (a) Oil flow rate
  (b) Standard Gas Oil Ratio (GOR at standard conditions)
  (c) Water Cut (WC)
  (d) Pump suction pressure
  (e) Required pump discharge pressure.

The pump size is determined by the total volumetric flow rate at suction condition. To estimate this total flow rate one needs to determine the Gas Volume Fraction (GVF) or the Gas Liquid Ratio (GLR) at suction conditions. The following relationship can be used:

$$GVF = \frac{GLR}{1 + GLR}$$

1.4 The helico-axial multiphase pump [4–6, 11]

1.4.1 Introduction

The helico-axial multiphase pump is a rotodynamic system. The compression is not obtained with a volumetric compression device, but by a transfer of energy.

The helico-axial multiphase pump is an inline multistage barrel pump (Fig. T2). Each stage or compression cell comprises a rotating helico-axial impeller and a stationary diffuser. The pressure rise is a function of the number of stages, and flow rate a function of the diameter of the compression cell.

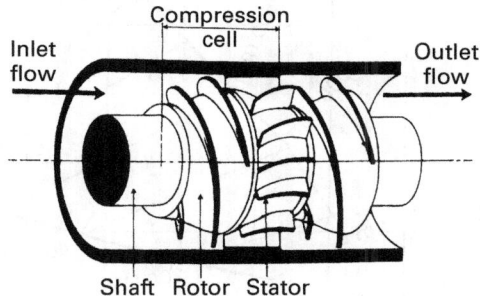

Figure T2 Helico-axial compression cell [4–6].

1.4.2 Characteristics

The main advantages of the helico-axial multiphase pump are:
• Ability to handle any GVF ranging from 0 (100% liquid) to 1.0 (100% gas) on a continuous basis)

383

- Mechanical simplicity and reliability (one single shaft, rotodynamic principle)
- Compactness
- Self-adaptation to flow changes
- Great tolerance to solid particles
- Possibility to use various driver types.

1.4.3 Industrial pump P 302

The following design specifications:
- Suction pressure: 0.5 to 1.5 MPa (70 to 220 psi)
- Required discharge pressure: to 3.5 MPa (to 500 psi)
- GVF (at suction conditions): 0.66 to 0.91
- Total flow at suction: 2400 to 8750 m³/d (15 000 to 55 000 bbl/D)
- Speed: 3000 to 6800 rpm
- Hydraulic power: 100 to 500 kW.

The performance of a helico-axial pump for given suction conditions (GVF, pressure level) is shown on figure T3.

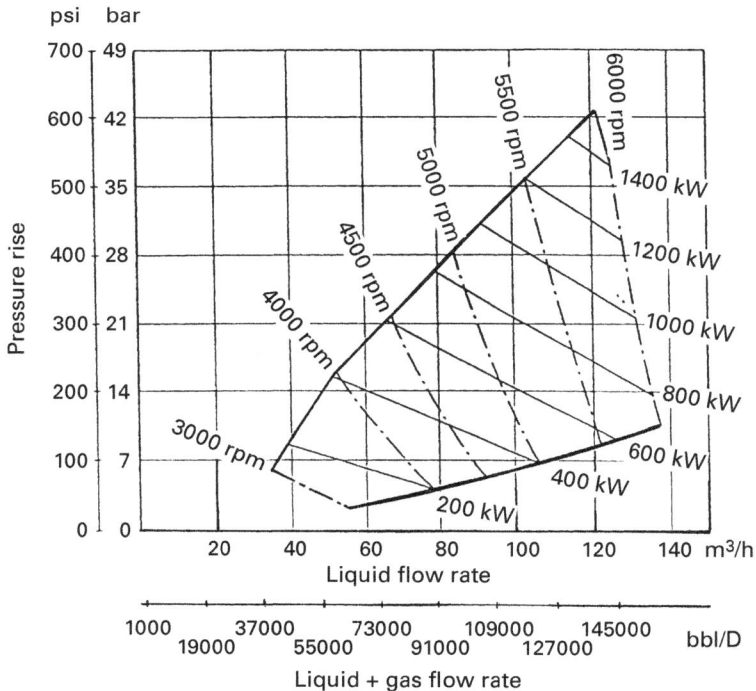

Figure T3 Multiphase pump (P 302) performance curves [4].

1.4.4 Poseidon pump

The progressive accumulation of field experience secured the oil operators to apply the technology on a large scale. Two very high capacity Poseidon pumps were installed in 1999 on the Dunbar platform (North Sea) operated by Total. These two pumps, running in parallel, boost the pressure of the low-pressure wells in the Alwyn South area to reach the pressure of the pipeline which exports the production from Dunbar to the Alwyn North Platform. The design characteristics of each pump are [13] :

- Total flowrate: 1200 m^3/h
  - Inlet GVF: 30 to 90%
  - Inlet pressure: 50 to 70 bar g
  - Discharge pressure: 125 bar g
  - Drive power: 4500 kW.
- The special stage design allows fluid and gas axial flow which significantly reduces the possibility to gas lock.
  - Does not replace the traditional pump or the gas separator
  - Primes the main production pump and pushes the gas/liquid mixture the production pump stages
  - Increases the mixture to reduce the gas volume
  - Does not induce separation of gas and liquid.

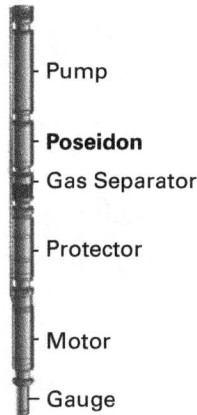

Pump

Poseidon

Gas Separator

Protector

Motor

Gauge

Figure T4 Typical Downhole ESP System with Poseidon module for high GVF application [19].

Poseidon Applications

- To produce Gassy wells
- Whenever there is production limitations due to the amount of gas pumps are capable to handle
- To solve gas locking problems with existing gas handling technology
- The Poseidon offers unique solution when gas venting is not possible:
  - Sub Sea Applications

   – Reda Coil Applications
   – Some government regulations do not allow gas venting in some conditions
   – Straddle packers.
* Offers the best chance to pass a gas slug through the pump
* Gas well dewatering.

The initial downhole Poseidon pump trials highlighted the need for shaft radial bearings with high lubricity in order to handle gas slugs. As a result the commercialised design includes silicon carbide bearings impregnated with graphite. [19]

1.5　Rotary screw pumps [7, 8]

Different types of rotary screw pumps exist; they can be adapted to various applications.

1.5.1　Single screw pumps, also known as progressing cavity pumps use an elastomeric stator and flexible joint eccentric rotation metallic screw. This equipment is presented in the following paragraph 1.6.

1.5.2　Two screw or twin screw pumps are specially suited to very low available inlet pressure applications and more so if the required flow rates are high. They produce a smooth, pulse free flow over a wide range of viscosities, temperatures, and pressures at high volumetric efficiency.

The vast majority of twin screw pumps are of the double suction design presented in Figure T5.

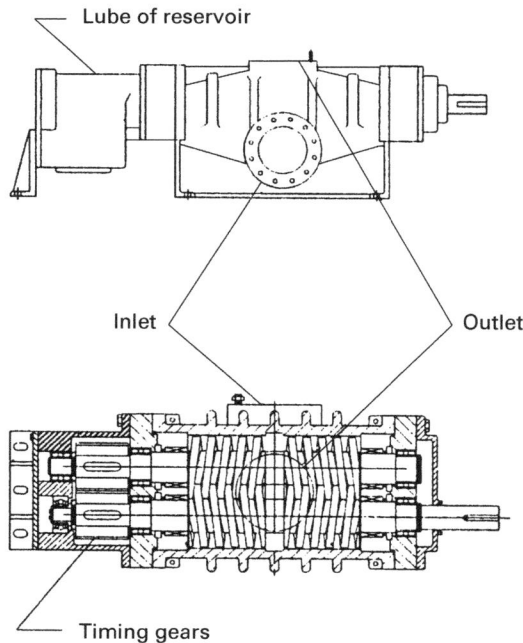

Figure T5　Twin screw pump construction [7].

The opposed thread arrangement provides inherent axial hydraulic balance due to its symetry. The more common and better design keeps the timing gears and bearings external to the liquid pumped. They need not rely upon the lubricating qualities of the pumped liquid nor its cleanliness. Four mechanical shaft seals keep these bearing and operating in a controlled environment.

1.5.3 Three screw pumps are the largest class of multiple screw pumps in service today.

They are manufactured in two basic styles, single suction and double suction (Fig. T6).

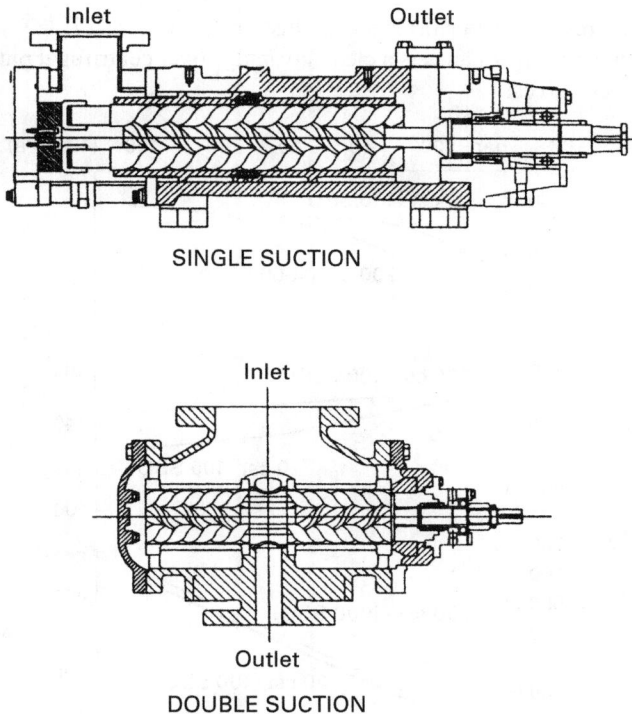

Figure T6 Conventional three screw pump designs [7].

Design

- **The single suction design** is used for low to medium flow rates and low to very high pressure.
- **The double suction design** is really two pumps in parallel in one casing. They are used for medium to high flow rates at low to medium pressure.

Characteristics

- **The theoretical flow rate** is a function of speed, of the screw set diameter and of the lead angle of the threads. Basically, **flow rate** is a function of the cube of the center screw diameter.
- **Slip flow,** the volumetric inefficiency due to clearances, differential pressure and viscosity, is a function of the square of the power rotor diameter.
- **Speed** is ultimately limited by the application capability to deliver flow to the pump inlet at a sufficient pressure to avoid cavitation.

Performances

Typical multiple screw pump performance is illustrated in Figure T7.

The high efficiency performance is a clear advantage over centrifugal pumps where liquid viscosity exceeds 20 cst.

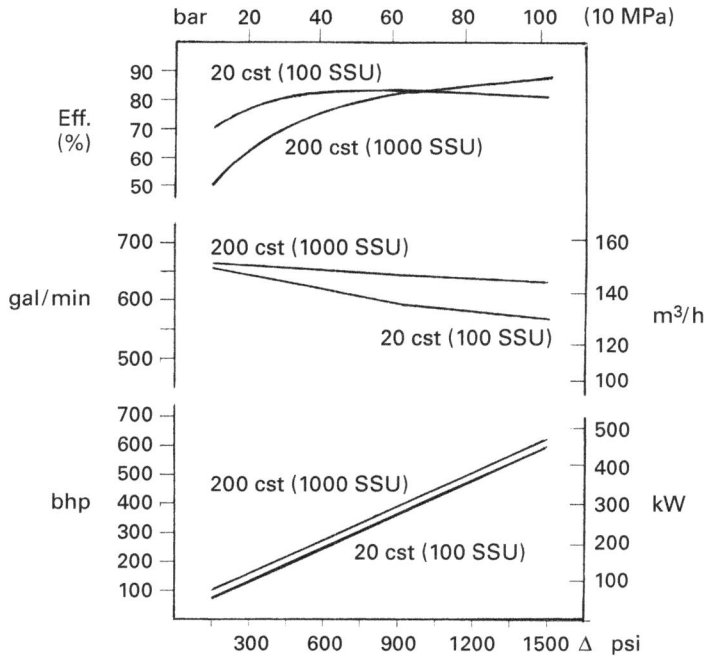

Figure T7 Typical multiple screw pump performance [7].

1.6 Progressing cavity pumps [10]

The principle of the progressing cavity pump (PCP) is presented in Chapter R.

Due to their unique design and principle of operation, progressing cavity pumping systems provide certain additional benefits not achieved with other types of multiphase pumps.

1.6.1 Benefits

Key benefits for these systems are as follows:
- Multiphase fluids with sand cuts of 8% can be handled without filtering or pre-separation.
- Multiphase fluids with oils of low-to-high API gravity can be handled without significant system changes.
- Their high suction lift capabilities allow PCP to be used in vapor recovery installations.
- The ability to handle variable flow regimes, including slug flow, without hesitation results in lower corrosive damage to the system suction piping.
- For high gas fractions (>90%), there is no need to separate and recirculate some produced fluids.

1.6.2 Applications

The largest commercially available progressing cavity multiphase pump is capable of generating flows up to 4 600 m^3/d (29 000 bbl/D). However, for higher flow rates some arrangements are proposed.

Parallel progressing cavity pump arrangement

This arrangement increases the pump inlet volume flow, virtually eliminates system downtime and provides increased flexibility to adjust to changing well conditions.

Otherwise, if one of the pumps requires routine maintenance, it may be brought off-line for repair while the remaining pumps continue to operate, thus preventing a total loss of production.

Cascaded progressing cavity pump arrangement

As a solution to the disproportionate pressure distribution phenomena, a cascaded concept provides for the arrangement of a least two PCPs connected in series. The flow volume rate of the second pump is less than the flow volume of the first pump. It is modified by altering the cavity sizes of the rotor/stator pumping elements or by adjusting the pump speed. It can be modified by a combination of the above adjustements.

As the multiphase fluid is introduced into the system (Fig. T8), the first pump acts a compressor to reduce the fluid volume of the gas. Since the liquid present is incompressible, it now becomes a larger percentage of the total fluid volume. The heat generated by the gas compression does not remain within the pump chamber long enough to adversely affect its elastomer stator. The interconnecting piping between the two pumps contains heat transfer mechanisms, such as cooling fins.

Due to the higher percentage of liquid present in the compressed flow, the second pump, which may have a greater number of stages than the first pump, acts as a booster to create the high differential pressures necessary to transfer the fluid downstream for futher processing.

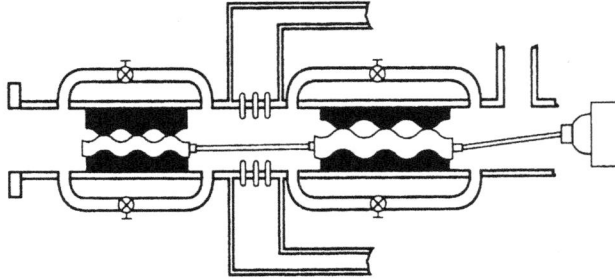

Figure T8 Cascaded progressing cavity arrangement [10].

T2 FLOW METERING GENERAL EQUATIONS [1]

Several documents served as references for the edition of the Standard concerning the surface flow metering. In particular:

- AGA: American Gas Association. Report No. 3
- API: American Petroleum Institute. API 14.3
- GPA: Gas Processors Association. GPA 8185-90
- AFNOR: French standardization. AFNOR X 10-101.

2.1 Field of application [1]

2.1.1 Applicable fluids

AGA standard applies to steady-state mass flow conditions for fluids that, for all practical purposes, are considered to be clean, single phase, homogeneous and Newtonian, and have pipe Reynolds numbers of 4 000 or greater. All gases, most liquids, and most dense phase fluids associated with the petroleum, petrochemical, and natural gas industries are usually considered Newtonian fluids.

2.1.2 Types of meters

AGA standard provides design, construction and installation specifications for flange-tapped, concentric, square-edged orifice meters of nominal 2-in. schedule 160 and larger pipe diameters.

An orifice meter is a fluid flow measuring device that produces a differential pressure to infer flow rate. The meter consists of the following elements indicated on Figure T9.

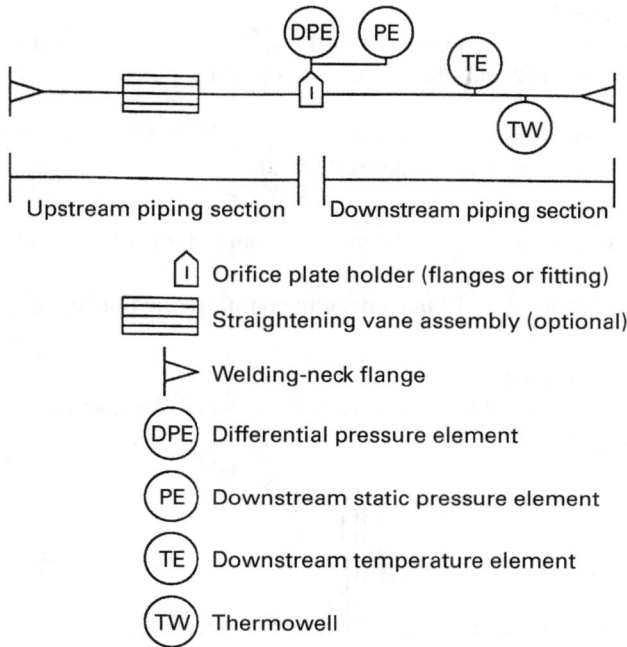

Figure T9 Orifice meter [1].

2.2 Flow rate determination [1]

2.2.1 Definitions

- **Orifice flow rate** q_m, q_v, Q_v
  It is the mass or volume of flow through an orifice meter per unit of time.

- **Orifice plate coefficient of discharge** C_d
  It is the ratio of the true flow to the theoretical flow and is applied to the theoretical flow equation to obtain the actual (true) flow. It is derived from experimental data.

- **Velocity of approach** E_v
  It is a mathematical expression that relates the velocity of the flowing fluid in the orifice meter approach section (upstream meter tube) to the fluid velocity in the orifice plate bore.

$$E_v = \frac{1}{\sqrt{1 - \beta^4}} \tag{T1}$$

and

$$\beta = d/D \tag{T2}$$

where
  d orifice plate bore diameter calculated at flowing temperature T_f
  D meter tube internal diameter calculated at flowing temperature T_f.

391

- **Expansion factor** Y

  It is an empirical expression used to correct the flow rate for the reduction in the fluid density that a compressible fluid experiences when it passes through the orifice plate bore.

  For incompressible fluids, such as water at 60°F (15.56°C) and atmospheric pressure, the empirical expansion factor is defined as 1.00.

- **Density** $\rho_{t,p}$, ρ_b

  The flowing fluid density $\rho_{t,p}$ is the mass per unit volume of the fluid being measured at flowing conditions T_f, P_f.

  The base fluid density ρ_b is the mass per unit volume of the fluid being measured at base conditions T_b, P_b.

- **Differential pressure** ΔP

  It is the static pressure difference measured between the upstream and downstream flange taps (Fig. T10).

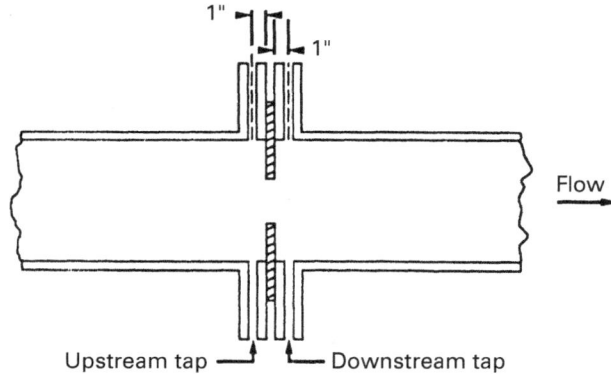

Figure T10 Flange-tapped orifice meter. Orifice tapping location [1].

2.2.2 Orifice flow equation

The practical orifice meter mass flow equation q_m is a simplified form that combines the numerical constants and unit conversion constants in a unit conversion factor N_1:

$$q_m = N_1 \times C_d \times E_v \times Y \times d^2 \times \sqrt{\rho_{t,p} \times \Delta P} \tag{T3}$$

where

Symbol	Represented quantity (see paragr. 2.2.1)	U.S. units	Metric units
q_m	Mass rate	lbm/sec	kg/sec
N_1	Conversion factor	5.25021 E–01	3.51241 E–05
C_d	Coefficient of discharge	dimensionless	dimensionless
E_v	Velocity of approach	dimensionless	dimensionless
Y	Expansion factor	dimensionless	dimensionless
d	Orifice diameter	in.	mm
$\rho_{t,p}$	Density	lbm/sec	kg/m^3
ΔP	Differential pressure	lbf/in.^{2}	kPa

The volumetric flow rate at flowing (actual) conditions can be calculated using the following equation:

$$q_v = q_m/\rho_{t,\,p} \tag{T4}$$

The volumetric flow rate at base (standard) conditions can be calculated using the following equation:

$$Q_v = q_m/\rho_b \tag{T5}$$

The mass flow rate q_m can be converted to a volumetric flow rate at base (standard) conditions Q_v if the fluid density at base conditions ρ_b can be determined or is specified.

T3   FLOW METERING PRACTICAL DATA

3.1   Gas flow rate

The gas flow rate Q_G is expressed in st cu ft/h (60°F and 14.73 psi):

$$Q_G = Fb \times Y \times Ft_f \times F\gamma \times Fp_v \times \sqrt{hw \times \left(P_f + P_a\right)} \tag{T6}$$

3.1.1   Input

D	internal pipe diameter (in.)
d	orifice diameter (in.)
P_f	gas static pressure (psi)
P_a	atmospheric pressure
T_f	gas temperature (°F)
hw	gas differential pressure (inches of water)
fCO_2	CO_2 fraction (dimensionless)
fH_2S	H_2S fraction (dimensionless)
γ_G	gas specific gravity (air = 1)
fnak	Pitot tube constant.

3.1.2 Reduced gas specific gravity γ_{red}

If $fCO_2 \leq 0.3$, then:

$$\gamma_{\text{red}} = \gamma_G - fCO_2 \times (0.6 - 0.7\,fH_2S) - fH_2S \times (0.25 - fH_2S) \tag{T7}$$

else: $$\gamma_{\text{red}} = \gamma_G - 0.1 - fCO_2 \times (0.3 - 0.7\,fH_2S) - fH_2S \times (0.25 - fH_2S) \tag{T8}$$

3.1.3 Specific gravity factor $F\gamma$

$$F\gamma = \sqrt{\frac{1}{\gamma_G}} \tag{T9}$$

3.1.4 Flowing temperature factor Ft_f

Using the Fahrenheit to Rankine conversion:

$$T = T_f + 459.688$$

then: $$Ft_f = \sqrt{\frac{520}{T}} \tag{T10}$$

3.1.5 Supercompressibility factor Fp_v

The basic formula is:

$$P = P_f \times (1 - fCO_2 \times 0.4) \times (1 - fH_2S \times 0.5) \tag{T11}$$

- If $\gamma_{red} < 0.7$, then:

$$K_1 = 344\,400 \qquad K_2 = 1.785$$

  else: $$K_1 = 916\,000 \qquad K_2 = 1.188$$

$$Fp_v = \sqrt{\frac{1 + \left(P \times K_1 \times 10^{K_2 \times \gamma_{\text{red}}}\right)}{T^{3.825}}} \tag{T12}$$

with
  $K = 1.29$ average ratio of specific heats
  $P_a = 14.7$ psi.

- If Pitot tube, then:

$$B = \frac{4 \times d}{\pi D} \tag{T13}$$

$$Y = 1 + \left[(1 - B)^2 \times 0.011332 - 0.00342\right] \times \frac{hw}{P_f \times K} \tag{T14}$$

$$Q_G = \text{fnak} \times D^2 \times Y \times Ft_f \times F\gamma \times Fp_v \times \sqrt{hw \times (P_f + P_a)} \tag{T15}$$

3.1.6 Basic orifice flow factor *Fb*

- For flange taps:

$$B = \frac{530}{\sqrt{D}}$$ (T16)

- Orifice diameter ratio:

$$\beta = \frac{d}{D}$$ (T17)

$$E = d \times \left(830 - 5\ 000\beta + 9\ 000\beta^2 - 4\ 200\beta^3 + B\right)$$ (T18)

- Coefficient of discharge: K_O.

  Evaluation of K_e (flange taps):

$$FK_4 = \left(0.364 + \frac{0.076}{\sqrt{D}}\right) \times \beta^4$$ (T19)

If $\beta < 0.07 + \dfrac{0.5}{D}$, then:

$$FK_3 = 0.4 \times \left(1.6 - \frac{1}{D}\right)^5 \times \sqrt{\left(0.07 + \frac{0.5}{D} - \beta\right)^5}$$ (T20)

else: $FK_3 = 0$

If $\beta < 0.5$, then:

$$FK_2 = \left(\frac{0.034}{D} + 0.009\right) \times \sqrt{(0.5 - \beta)^3}$$ (T21)

else: $FK_2 = 0$

If $\beta > 0.7$, then:

$$FK_1 = \left(\frac{65}{D^2} + 3\right) \times \sqrt{(\beta - 0.7)^5}$$ (T22)

else: $FK_1 = 0$

Also:

$$K_e = 0.5993 + \frac{0.007}{D} + FK_4 + FK_3 - FK_2 + FK_1$$ (T23)

and:

$$K_O = \frac{K_e \times 10^{-6}}{(15E + 1) \times d}$$ (T24)

- Basic orifice flow factor is:

$$Fb = 338.17 \times d^2 \times K_O$$ (T25)

3.1.7 Downstream expansion factor *Y*

Considering $X = \dfrac{hw}{27.68 \times P_f}$ (P_f in psi):

$$Y = \sqrt{1 + X} - \left(0.41 + 0.35\beta^4\right) \times \frac{X}{K \times \sqrt{1 + X}}$$ (T26)

3.2 Corrected oil flow rate V_o (bbl)

3.2.1 Input

γ_o standard oil gravity at T_{sg} (dimensionless)
T_{sg} oil gravity measurement temperature (°F)
F_{shr} shrinkage factor (dimensionless)
T_{shr} shrinkage temperature (°F)
V_{tk} tank volume (bbl)
T_{tk} tank temperature (°F)
BSW   basic sediment and water (dimensionless)
F_{omc} oil meter correction factor (dimensionless)
F_{vc} volume correction factor (initialized at 0) (dimensionless)
V_s oil measured at meter (floco or rotron) (bbl)
P_{sep} separator pressure (psi)
T_o oil temperature (°F).

3.2.2 Calculation

$$\gamma_{60} = \gamma_o + \left(0.00069 - 0.000372 \times \gamma_o\right) \times \left(T_{sg} - 60\right) \tag{T27}$$

$$°API = \frac{141.5}{\gamma_{60}} - 131.5 \tag{T28}$$

where
°API   API degree of oil
γ_{60} specific gravity at 60°F.

- If **Case No. 1** (oil meter/no gauge tank/no shrinkage tester), then:

$$R_{S1} = \left(\frac{1.797}{\gamma_o - 1.838}\right) \times P_F \tag{T29}$$

For convergence test CT = 1 try loop, do:

$$R = °API \times \left(0.02 - R_{S1} \times 3.57 \times 10^{-6}\right) + 0.25 \tag{T30}$$

$$N = \frac{°API}{80} - T_{tk} \times 0.00091 \tag{T31}$$

$$A = 10^N \tag{T32}$$

$$B = \left|\frac{P_F}{19.6} \times A\right| \tag{T33}$$

$$R_{S2} = R \times B^{1.205} \tag{T34}$$

If $\left| R_{S2} - R_{S1} \right| > 0.1$, convergence test on 10 try loops are executed:

$$F = \left| R_{S2} \times \sqrt{\frac{R}{\gamma_o}} + 1.25 T_{tk} \right| \tag{T35}$$

$$F_{shr} = 1 - \frac{1}{0.9759 + 0.00012 \times F^{1.2}} \tag{T36}$$

$$F_{vc} = 1$$

$$V_o = V_s \times \left(1 - F_{shr}\right) \times \left(1 - BSW\right) \times F_{omc} \times F_{vc} \tag{T37}$$

- If **Case No. 2** (oil meter/gauge tank/no shrinkage tester):

$$F_{vc} = 1 - 10^{-4} \times \left(0.066 \times {}^{\circ}API + 2.75\right) \times \left(T_{tk} - 60\right) \tag{T38}$$

$$V_o = V_s \times \left(1 - F_{shr}\right) \times \left(1 - BSW\right) \times F_{omc} \times F_{vc} \tag{T39}$$

- If **Case No. 3** (oil meter/no gauge tank/shrinkage tester):

$$F_{vc} = 1 - 10^{-4} \times \left(0.066 \times {}^{\circ}API + 2.75\right) \times \left(T_{shr} - 60\right) \tag{T40}$$

$$V_o = V_s \times \left(1 - F_{shr}\right) \times \left(1 - BSW\right) \times F_{omc} \times F_{vc} \tag{T41}$$

- If **Case No. 4** (no oil meter/gauge tank/no shrinkage tester):

$$F_{shr} = 0$$

$$F_{vc} = 1 - 10^{-4} \times \left(0.066 \times {}^{\circ}API + 2.75\right) \times \left(T_{tk} - 60\right) \tag{T42}$$

$$V_o = V_{tk} \times \left(1 - F_{shr}\right) \times \left(1 - BSW\right) \times F_{vc} \tag{T43}$$

3.3   Corrected water flow rate V_w (bbl)

3.3.1 Input

V_{H_2O}	water measured at meter (floco) (bbl)
F_{wmc}	water meter correction factor (dimensionless)
V_s	oil measured at meter (floco or rotron) (bbl)
BSW	basic sediments and water (dimensionless)
BSW_{H_2O}	water portion of BSW (dimensionless)
F_{shr}	shrinkage factor (dimensionless)
F_{omc}	oil meter correction factor (dimensionless)
F_{vc}	volume correction factor (initialized at 0) (dimensionless).

3.3.2 Calculation

$$V_w = V_{H_2O} \times F_{wmc} + V_s \times (1 - F_{shr}) \times F_{omc} \times F_{vc} \times BSW \times BSW_{H_2O} \qquad \text{(T44)}$$

3.4 Remarks

3.4.1 Shrinkage

The shrinkage factor F_{shr} allows to subtract the volume of dissolved gas at process conditions from oil measured at meter V_s.

3.4.2 Volume correction factor F_{vc}

The volume correction factor F_{vc} allows to report the volume at measuring conditions (P/T at measuring end) to the standard conditions:

$$F_{vc} = 1 - \frac{(0.066 \times °API + 2.75) \times (T_{ref} - T_{std})}{10\ 000} \qquad \text{(T45)}$$

where
  T_{ref} reference measuring temperature after shrinkage correction
  T_{std} standard temperature (60°F or 15°C)
  $°API$ see Eq. T28.

3.4.3 Four cases of well test

Flow measured with an oil meter and a shrinkage tester

- The shrinkage is measured on the shrinkage tester in the final degassing at ambient pressure and at temperature T_{shr}.
- The oil meter correction factor F_{omc} is previously estimated with water.
- The volume correction factor F_{vc} reports the volume of $T_{ref} = T_{shr}$ at T_{std}.

Flow measured with an oil meter and a tank

- The shrinkage is included in the F_{omc} value.
- F_{omc} is estimated during the test by simultaneous measurings at the meter and in the tank (degassed oil) at ambient pressure and at temperature T_{tk}.
- The volume correction factor F_{vc} reports the volume of $T_{ref} = T_{tk}$ at T_{std}.

Flow only measured with an oil meter

- The shrinkage is calculated.
- The oil meter correction factor F_{omc} is previously estimated with water.
- The volume correction factor is included in the F_{shr} value.

Flow only measured with a tank

- The shrinkage is not reported, because a degassed oil volume is measured at ambient pressure and at temperature T_{tk}.
- The volume correction factor F_{vc} reports the volume of $T_{ref} = T_{tk}$ at T_{std}.

T4   FLOW REGIMES [2]

4.1 In vertical wells

Transitions between flow regimes in the vertical tubing of an oil well are illustrated in Fig. T9, which shows the different hydrodynamic flow regimes which may occur in vertical liquid-gas multiphase flows.

It should be noted that Figure T11 is a schematic illustration which is intended to show the transitions between the flow regimes as the superficial gas velocity increases from the bottom of the well up to the wellhead. In real production tubing it is rare that more than two or three flow regimes are present at the same time.

Figure T12 represents the multiphase flow map for a vertical flow.

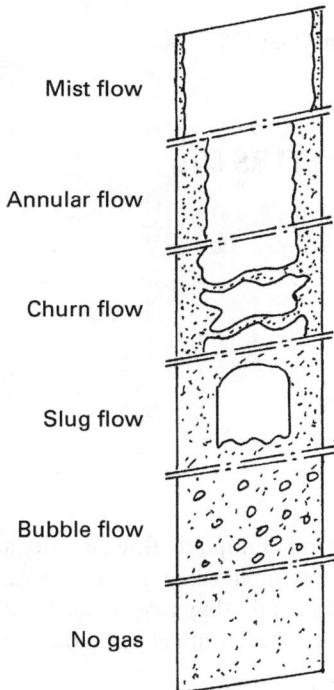

Mist flow

Annular flow

Churn flow

Slug flow

Bubble flow

No gas

Figure T11
Schematic transitions between flow regimes in oil wells [2].

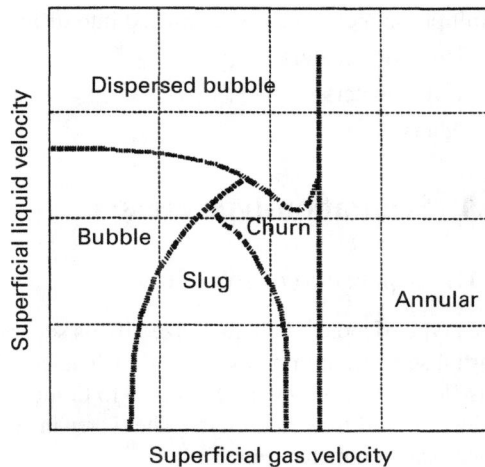

Dispersed bubble

Bubble

Churn

Slug

Annular

Superficial liquid velocity

Superficial gas velocity

Figure T12 Multiphase flow map, vertical flow [2].

4.2 In horizontal wells

Figure T13 is a qualitative illustration of how flow regime transitions are dependent on superficial gas and liquid velocities in horizontal multiphase flow.

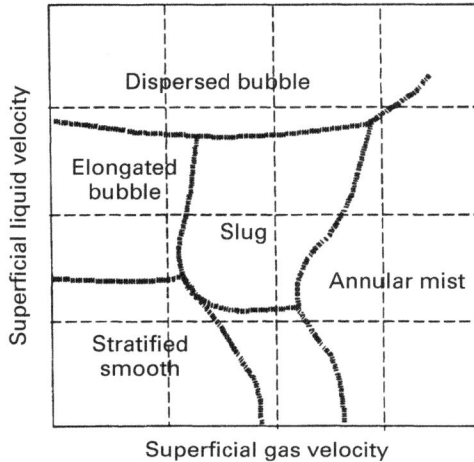

Figure T13 Multiphase flow map, horizontal flow [2].

T5 CLASSIFICATION OF MULTIPHASE METERS [2]

Multiphase meters can be classified into different categories:
- Separation meters
- In-line meters
- Others.

5.1 Separation-type meters

5.1.1 Separation of total flow

This type of meter is characterized by its separation of the total multiphase flow, usually a partial separation into gas and liquid. The gas flow is then measured using a single-phase gas-flow meter with good tolerance to liquid carry-over. The liquid flow rate is measured using a liquid flow rate meter. The water-in-liquid ratio may be determined by an on-line water fraction meter (Fig. T14).

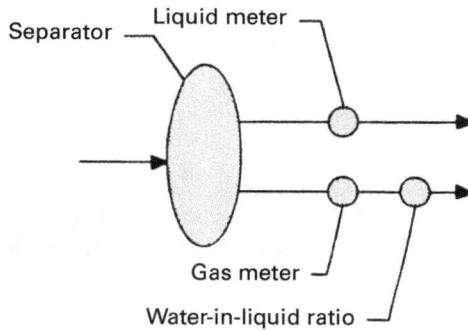

Figure T14   Principle design of a full separation type meter [2].

5.1.2 Separation in sample line

This type of meter is characterized by the fact that separation is not performed on the total multiphase flow, but on a bypassed sample flow. The sample flow is typically separated into gas and liquid, whereafter the water-in-liquid ratio in the sample stream can be determined using an on-line water fraction meter. Total multiphase flow rate and gas liquid must be measured in the main flow line (Fig. T15).

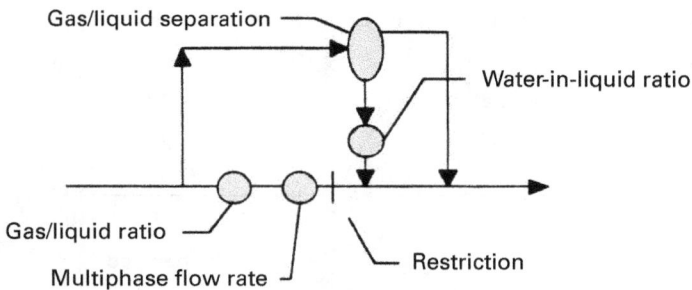

Figure T15   Principle of a multiphase meter with separation in sample line [2].

- Gas/liquid ratio (GLR):
  - gamma absorption
  - vibrating tube
  - neutron interrogation
  - weighing.
- Multiphase flow rate:
  - cross-correlation using radioactive, acoustic or electrical signals
  - differential pressure using venturi, V-cone or Dall tube
  - mechanical, e.g. positive displacement or turbine.

- Water-in-liquid ratio (WLR):
  - electrical impedance
  - vibrating tube.

5.2 In-line meters

The characteristics of in-line multiphase meters are such that **the complete measurement** of phase fractions and phase flow rates **is performed directly in the multiphase flow line**, without any separation of the flow.

The volume flow rate of each phase is represented by area fraction multiplied by the velocity of each phase. This means that a minimum of six parameters have to be measured or estimated. Some multiphase meters assume that two, or all three, phases travel at the same velocity, thus reducing the required number of measurements. In this case either a mixer must be employed or a set of calibration factors established (Fig. T16). In-line multiphase meters commonly employ a combination of two or more of the following measurement techniques:

(a) Microwave technology
(b) Capacitance
(c) Gamma absorption
(d) Neutron interrogation
(e) Cross-correlation using radioactive, acoustic or electrical signals
(f) Differential pressure using venturi, V-cone or other restriction
(g) Positive displacement/turbine meter.

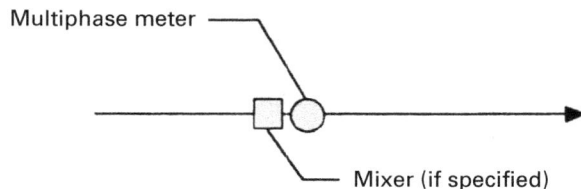

Figure T16 Principle design of in-line multiphase meters [2].

5.3 Other categories of multiphase meters

Other categories of multiphase meters include advanced signal processing systems, estimating phase fractions and flow rates from analysis of the time-variant signals from sensors in the multiphase flow-line. Such sensors may be acoustic, pressure or other types. The signal processing may be a neural network or other pattern-recognition or statistical signal-processing system, for example.

There are also multiphase metering systems which have been developed on the basis of process simulation programs combined with techniques for parameter estimation. Instead of predicting the state of the flow in a pipeline at the point of arrival, its pressure and temperature can be measured at the arrival point and put into the simulation program. The pressure and temperature of an upstream or downstream location also have to be measured. When the pipeline configuration is known along with properties of the fluids, it is possible to make estimates of phase fractions and flow rates.

T6   Performance specification [2, 14, 15, 16, 17, 18]

A fill-in form for summarizing the performance specification of a multiphase meter is proposed in Tables T2a and b. The following items should also be included.

6.1 Sketch showing important details of installation requirements

- Horizontal/vertical upwards/vertical downwards flow.
- Mixer/not mixer.
- Straight upstream/downstream length.

6.2 Rated conditions of use

- The list in the form is not exhaustive, other parameters important for the particular meter should be included.
- Flow regimes that the meter is designed to handle should be listed.
- The interval in which the influence parameters are allowed to vary, still maintaining the uncertainty specifications, should be specified.

6.3 Influence quantities

- The list includes the same parameters as listed in "Rated conditions of use"; all influence parameters important for the particular meter should be included.

6.4 Operating range

- Superficial velocity axes should be linear, from zero to any required upper velocity range.
- Secondary flow rate axes may be used to denote or select a suitable meter size.
- Operating range should be marked on the graph, and may be divided into as many sub-areas as required.

Table T2a

Multiphase meter performance specification

Manufacturer: _____
Meter type: _____
Date: _____
Reference: _____

Required installation configuration schematic:

Rated conditions of use:

Pressure:		Temperature:	
Oil density:		Oil viscosity:	
Gas density:		Gas viscosity:	
Water density:		Water salinity:	
Flow regimes:			

Influence quantities:

Quantity	Influencing	Effect
Oil density:		
Gas density:		
Water density:		

Flow regime:		

Table T2b

Operating range:

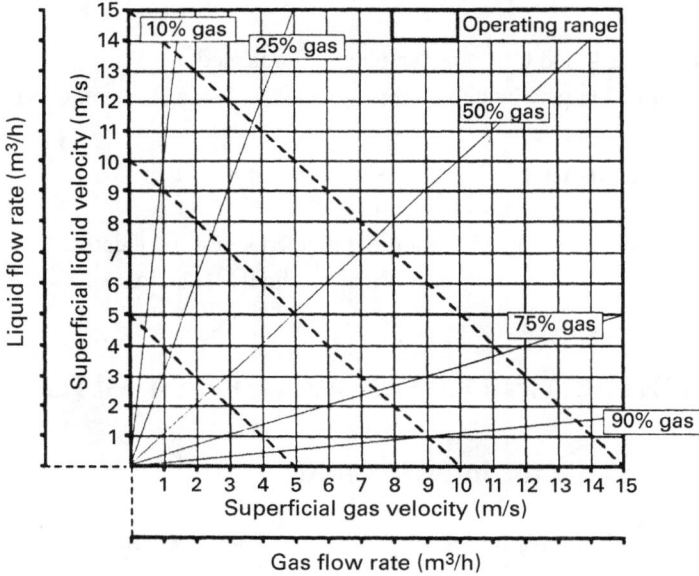

Uncertainty specification:

Sub - Range	WLR Range					
	[%]	Oil	Water	Gas	Liquid	WLR
A	0-x					
B	0-x					
C	0-x					
D	0-x					
E	0-x					
A	x-100					
B	x-100					
C	x-100					
D	x-100					
E	x-100					
Calibration requirements:						
Reference:						

6.5 Uncertainty specification

- Uncertainties should be given for each sub-range of the operating range.
- Uncertainties in phase flowrates should preferably be given relative to actual phase flowrates.
- Absolute deviations in WLR and GVF may be given as indicated.
- The uncertainty specification may be quoted for as many WLR-ranges as required.

6.6 Additional information

- Method of meter calibration/special calibration requirements must be identified.
- Reference to more comprehensive product information should be indicated.

T7 MEASUREMENT TECHNIQUES OF MULTIPHASE METERS [3]

7.1 Basic measurements for inferring flow

For a single-phase liquid or gas travelling through a pipe of cross-sectional area A at an average velocity V, the volumetric flow rate Q can be calculated by $Q = AV$. When an oil, water and gas mixture is flowing through the same pipe the calculation of the volumetric flow rates is complicated by the distribution and the velocity of each phase. A simple way to estimate the volumetric flow rates is to assume that each phase occupies a fraction of the total cross sectional area at any instant determined by the following relationships:

$$f_o = A_o/A \qquad f_w = A_w/A \qquad f_g = A_g/A$$
$$f_o + f_w + f_g = 1$$

where f_o, f_w and f_g are the volume fraction of oil, water and gas in the mixture. The volumetric flow rate of each phase and the total (mixture) flow rate is then determined by:

$$Q_o = Af_oV_o \qquad Q_w = Af_wV_w \qquad Q_g = Af_gV_g$$
$$Q_t = Q_o + Q_w + Q_g$$

where V_o, V_w and V_g are the superficial velocity of oil, water and gas phases in the mixture.

7.2 Measurement techniques

The main methods used are the following.

7.2.1 Dielectric volume fraction measurements

In-situ fraction measurements use either dielectric measurements or gamma ray density measurements. Because there is a large difference between the dielectric constants of oil and water, the dielectric constant of an oil/water mixture can determine the oil and water fraction.

The dielectric constant of an oil-water mixture can be established by measuring its attenuation in an electromagnetic field. The frequency of the electromagnetic radiation can range from megahertz to gigahertz. Depending on the frequency used, the dielectric devices will be referred to as a capacitance or a microwave device.

7.2.2 Microwave-based volume fraction measurements

Phase and amplitude changes in the microwave signal are used to establish water fraction.

7.2.3 Gammay ray attenuation volume fraction measurements

Gamma rays are produced by chemical nuclear sources that decay with time. When the gamma rays pass through an oil, water and gas mixture, they are attenuated by the electrons and nuclei of molecules in the mixture in accordance with the gamma ray attenuation relationships in the equations nearby.

Measurements can be taken at a single gamma ray energy level or at two different energy levels. With the second option, measurements provide a second relationship between phase fractions if the attenuations are different enough in oil oil and water at the two levels. This allows to determine the phase fractions without usage of an additional measurement technique.

7.2.4 Measuring component velocities

With volume fractions determined, the other parameters (e.g., component velocities) must be measured to determine flow rates. Venturi meters and cross-correlation meters are the most commonly used tools for component velocity measurements.

T

7.3 Metering techniques used in major commercial multiphase meters [3, 9, 12]

Table T3 summarizes the principal measurement techniques used in most currently installed multiphase meters. There are, of course, other types of meters not shown in this table that are under development and have not yet reached commercial installation.

Table T3 Main commercial multiphase meters [12].

Manufacturers	Commercial Meter Names	Phase Fraction Measurement	Flow or Velocity Measurement	Comments
Agar Corp.	MPFM 300 MPFM 400	indirect from mass and volume flow-rates; water cut from microwave	positive displace-ment (oval gears) 2 Venturis	high GVF with gas bypass loop
FlowSys AS FMC	TopFlow WellSense	capacitance/con-ductivity and indi-rect from mass flowrate	cross-correlation	available sizes from 1 in. to 8 in. subsea version
Haimo	MFML 2000 L	dual gamma densi-tometer	cross-correlation	
Jiskoot	Mixmeter	dual gamma densi-tometer (^{137}Ce, ^{241}Am)	differential pressure through a flow homogeniser	available sizes from 2 in. to 6 in.
Kvaerner Cisro	Duet	simple gamma densitometer plus a dual gamma den-sitometer	cross-correlation	
PSL	Esmer	capacitance mea-surements	absolute and differ-ential pressure (Venturi)	special signal anal-ysis available sizes from 2 in. to 6 in.
Roxar	MPFM 1900 VI	electr. impedance and simple gamma densitometer (^{137}Ce)	Venturi and cross-correlation	available sizes from 2 in. to 12 in. subsea version
TEA Sistemi Spa	Lyra	gamma densitom-eter impedance meter	Venturi or orifice	Vega version for wet gas
3-Phase Measure-ment AS (Framo Engineer-ing/Schlumberger)	PhaseWatcher Vx Phase Tester	High frequency dual gamma densi-tometer	Venturi	high GVF with gas extraction by phase splitter subsea version

REFERENCES

1   Orifice metering of natual gas and other related hydrocarbon fluids. Part 1: General equations and uncertainty guidelines. Part 2: Specification and installation requirements. 3: Natural gas applications. Part 4: Background, development, implementation, procedure, and subroutine documentation for empirical flange-tapped discharge coefficient equation. Compressibility factors of natural gas and other related hydrocarbon gases. *AGA - API - GPA. From AGA catalog No XQ 9212,* Novembre 1992.

2   Dahl E, Corneliussen S, Couput JP, Dykesteen E, FrØysa KE, Malde E, Moestue H, Moksnes PO, Scheers L, Tunheim H (2005) *Handbook of Multiphase Metering,* Norwegian Society for Oil and Gas Measurement, Norwegian Society of Technical and Scientific Professionals, Olso

3   Parviz Mehdizadeh (1998) Multiphase Meters: Delivering improved production measurements and well testing today. *Hart's Petroleum Engineer International,* May 1998

4   de Salis J, Oudin JC, Falcimaigne J (1994) Helico-axial multiphase pump. Product development. Selection tools and applications. *Paper OSEA 94046,* 10th Offshore South East Asia Conference held in Singapore, 6-9 December 1994

5   de Salis J, de Marolles C, Falcimaigne J, Durando P (1996) Multiphase pumping. Operation & control. *Paper SPE 3659* presented at the 1996 SPE Annual Technical Conference and Exhibition held in Denver, Colorado

6   Gié P, Buvat P, Bratu C, Durando P (1992) Poseidon multiphase pump: Field tests results. *Paper OTC 7037* presented at the 24th Annual OTC in Houston, Texas

7   Brennan JR (1996) *High Performance Rotary Screw Pumps.* Document of Imo Industries Inc., Monroe, NC

8   Worthington, Sier-Bath Two Screw Pump. *Document 2166-S3*

9   Hammer EA, Johansen GA (1997) *Basic Principle in Multiphase Metering. Advantages and Disadvantages, Properties and Potentials.* BHR Group 1997, Multiphase '97, 601–602

10 Mirza (1999) *Progressing Cavity Multiphase Pumping Systems: Expanding the Possibilities.* BHR Group 1999, Multiphase '99, 77–84

11 de Salis J, Heintzé E, Charron Y (1999) *Dynamic Simulation of Multiphase Pumps.* BHR Group 1999 Multiphase '99, 11–43

12 Falcimaigne J, Decarre S (2008) *Multiphase Production: Pipeline Transport, Pumping and Metering,* Editions Technip, Paris

13 Leporcher E, Delaytermoz A, Renault JF, Gerbier A, Burger O (2001) Deployment of Multiphase Pumps on a North Sea Field, *Paper SPE 71536* presented at the 2001 SPE Annual Technical Conference and Exhibition held in New Orleans, Louisiana

14 API (2002) *Use of Subsea Wet-gas Flowmeters in Allocation Measurement Systems.* API Recommended Practice 85, Dallas

15 API (2004) *State of the Art – Multiphase Metering.* API Publication 2566, Dallas

16 API (2005) *Recommended Practice for Measurement of Multiphase Flow.* API Recommended Practice 86, Dallas

17 ISO (2002) *Allocation of Gas and Condensate in the Upstream Area.* Draft version of report by TC 193 – SC3 – WG 1.

18 UK Department of Trade and Industry (2003) *Guideline Notes for Petroleum Measurements under the Petroleum (Production) Regulations.* Oil and Gas Division, Issue 7, December 2003, London.

19 Camilleri L (2006) *Multiphase Pumping – A Technology Snapshot.* Journées annuelles du Pétrole, Paris, 2006

20 Falcimaigne J, Brac J, Charron Y, Pagnier P, Vilagines (2002) Multiphase Pumping: Achievements and Perspectives, *Oil & Gas Science and Technology* – Revue IFP, Vol. 57, Editions Technip, Paris

U

Deposit Treatment

U

U

Deposit Treatment

U1 ASPHALTENE DEPOSITION

Asphaltene deposition causes serious problems in production operations, from the reservoir, through production tubing, and in surface facilities.

In principle, a more efficient solution to asphaltene deposition problem could be achieved by means of squeeze treatments with asphaltene inhibitors, in analogy to the treatments currently performed for controlling inorganic scale.

1.1 Definition [1]

Asphaltenes are defined as highly condensed polyaromatic structures or molecules, containing heteroatoms (i.e., S, O, N) and metals (e.g., V, Ni), that exist in petroleum in an aggregated state in the form of suspension and are surrounded and stabilized by resins (i.e., natural peptizing agents).

1.2 Where are found asphaltene deposits? [2–6]

1. In surface facilities (pipelines and separators). Asphaltene deposition affects all the flow lines and it occurs regardless of the temperature conditions. There is a natural sedimentation of the asphaltenes in the vessels of a gas/oil separation process, which can fill up completely in only a few weeks if care is not taken. As the asphaltenes adhere particularly strongly to the metal walls, these must therefore be protected.
2. In production tubing. Deposits were subsequently found in the tubing in which deposits form at depth corresponding to the bubble pressure of produced oil.
3. Afterwards the asphaltene deposit zone can migrate to bottomhole and well neighboring formation as reservoir depletion proceeds.

1.3 Causes of asphaltene deposition [2, 4, 7, 8]

Asphaltene deposition deep in the reservoir occurs only in reservoirs where asphaltene flocculation is possible by depressuring the oil. Asphaltene flocculation is described as a

413

thermodynamic transition inducing the formation of a new liquid phase with a high asphaltenic content; this phase being the asphaltenic deposit.

- Asphaltene deposition occurs when gaseous saturated hydrocarbons are used to displace oil in EOR.
- Mixing of crude with light oils or gases used in miscible flooding could lead to asphaltene precipitation.
- Acid stimulation can cause organic deposition.
- Asphaltene flocculation is caused by temperature, pressure, and composition changes.

Deposited asphaltenes can reduce effective hydrocarbon mobility by:

1. Blocking the pore throats
2. Adsorbing onto the rock, thereby altering the formation wettability from water-wet to oil-wet
3. Increasing hydrocarbon viscosity by nucleating water-in-oil emulsions.

When asphaltene flocculation occurs in the rock matrix, some asphaltenes may drop out in the pores because of their large size; others may be carried by the flowing fluid until they arrive simultaneously at the pore throats to bridge and reduce effective permeability (Fig. U1).

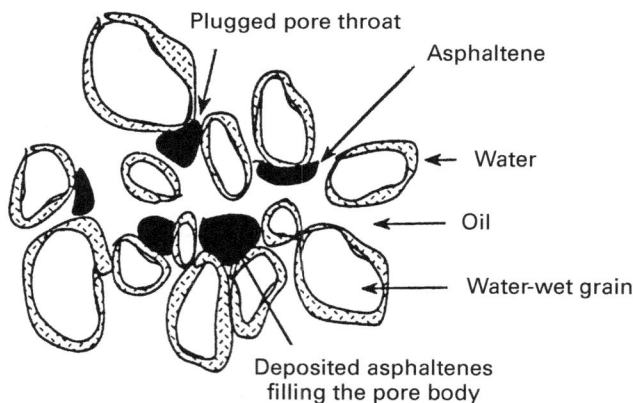

Figure U1 In situ asphaltene deposition causing physical blockage [4].

Some flocculated asphaltenes, especially the most polar and charged particles, attach to negatively charged, water-wet sands and alter their intrinsic wettability toward more oil-wet tendancy (Fig. U2).

Once plugging has occurred, it is necessary to obtain a sample of the deposit. The nature of the solids is fundamental in the selection of the treatment.

Solubility tests in organic solvents, including low cost and readily available petroleum cuts are carried out in order to decide the most cost effective treatment.

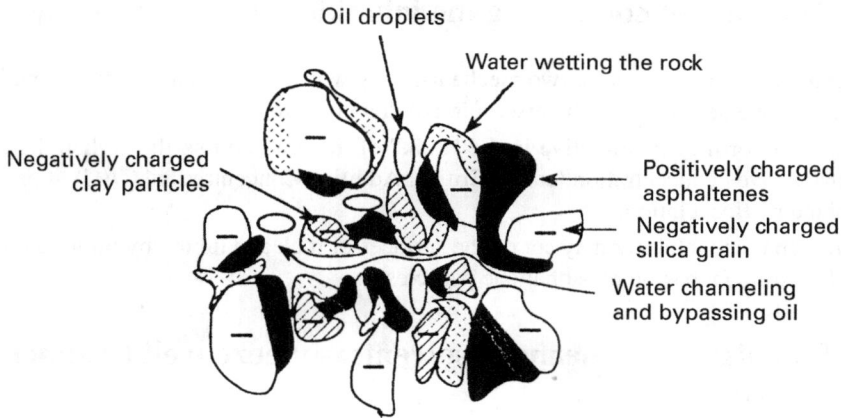

Figure U2 Asphaltenes adsorbed on the rock, causing wettability changes [4].

1.4 Mechanical removal of asphaltene deposits [5]

This technique consists in disposing of scales by making use of tools such as high-pressure lances, cutting heads, expanding brushes, expanding scrapers, etc. It is undoubtedly to be preferred in all the cases in which access to the site does not present particular problems and the removal time is not particularly long. However, the use of this technique involves some drawbacks. For instance, whenever the access of operator to pieces of equipment such as separators, desalters, stabilization columns, etc., is required, prior draining of these elements is implied. As to other facilities, the disassembly and the extraction of some of their components (e.g. tube bundle, demister, etc.), as well as their subsequent reassembly, must be envisaged and the removal time is consequently longer.

1.5 Chemical removal in the near wellbore and oil treating plants [5]

Aromatic solvents available on the market and generally used in the removal of asphaltenes are:

- Toluene, xylene or light petroleum distillates.
- OLG (gas oil from coal tar distillation) is an AGIP-patented industrial product. This product has a high power of asphaltene dissolution (up to 95% of their weight compared to 40% for toluene).
- Cosolvents, i.e. mixtures of several selected solvents or chemical additives (see Paragr. 1.7).

In many cases, the chemical cleaning of the equipment is undoubtedly to be preferred to the mechanical methods since it is more cost-effective.

1.6 Assumption concerning the inhibition mechanisms [14]

Conceptually, there are at least two mechanisms by which a chemical inhibitor could prevent asphaltene deposition in the reservoir and downhole tools:

1. The inhibitor may be effective in the "bulk" of the crude oil so that, when dissolved above a given concentration (called Critical Additive Concentration, CAC), it prevents asphaltene flocculation.
2. The asphaltene inhibitor may act on the surfaces of rock and tubing by limiting the rate of deposition/adhesion of asphaltene particles.

1.7 Stimulation by means of solvents, squeeze well treatments
[3, 9–11]

A series of chemical additives have been evaluated for their ability to improve the natural solvent characteristics of the aromatic well-stimulation solvent (see Table U1).

Table U1

Solvent	Concentration	Remarks
n-Butylamine	0.5% (volume) minimum	Economic
Alkyl phenol		
Xylene	100%	
HAS (High Aromatic Solvent)		40% more economic than xylene

1.8 Bottomhole injection of chemicals

Special polymers or surfactants are generally used [12, 13].

1. Specific surfactants (developed by Anticor Chimie and Elf Aquitaine) can be chosen for their electrochemical behavior and their chemical structure ; they form a complex system of bonds with asphaltenes more stable than those existing between asphaltenes and resins.
2. One inhibitor developed by Shell Chemicals, an oil-soluble polymeric dispersant in a mineral oil (Asphaltene Inhibitor B) having the following characteristics also be used:
   – Recommended concentration for field applications $\cong$ 500 ppm
   – With the reblended, newly developed Asphaltene Inhibitor B, the amount of inhibitor is reduced by 45%, and costs are reduced by 15%.

416

1.9 Screening solvents or chemicals [3]

Aromatic solvents are to be preferred for asphaltene deposit dissolution. But, in the case of especially hard asphalt deposits, their efficiency can be increased by addition of 1% to 5% polar chemicals such as amines or alcohols.

The principle is to adjust the polarity of aromatic solvent to be used to the physico-chemical properties of the asphaltene fraction contained in the deposit considered. For such an optimization, a laboratory study is recommended as described in reference [3] which details the guidelines used for screening the chemicals available.

1.10 Economic balance [14]

The economy of injection of chemicals can be evaluated, assuming that:
1. During asphaltene deposition, well production decreases linearly over time
2. Solvent washes completely recover well production decline due to asphaltene damage
3. In the presence of an inhibitor above the Critical Additive Concentration (CAC), well production remains constant
4. When the inhibitor concentration falls below the CAC, well production decreases linearly over time with the same slope as without any inhibitor.

The economic balance is calculated as follows:
- **Balance term for wash treatments**

$$[\text{maximum production}(\tau) - \text{loss}(\tau) - \text{single wash cost}]$$
$$\times \text{number of treatments per year}$$

where
(τ) wash treatment life (months).

- **Balance term for squeeze treatments**

$$[\text{maximum production} (\alpha + \beta) - \text{loss} (\beta) - \text{single squeeze cost}]$$
$$\times \text{number of treatments per year}$$

where
(α) squeeze treatment life (months)
(β) time up to next squeeze treatment (months).

- **Cost ratio**

$$\text{cost ratio} = \frac{\text{squeeze cost per year}}{\text{solvent-wash cost per year}}$$

U

U2 HYDRATES

2.1 Basics of hydrate formation [15]

Hydrates can only form when three main conditions are met:

1. Water must be present. Hydrates are 80–90 wt% water formed into a lattice structure similar to that of ice.
2. Hydrocarbons must be present. The hydrate structure is stabilized at relatively high temperature compared to ice by the presence of small molecules trapped in the lattice. Molecules such as methane, ethane, propane and butane in addition to nitrogen and carbon dioxide stabilize the structure.
3. Hydrates form at temperatures of around 5–25°C depending on the pressure. Unlike ice, when the pressure is increasing the hydrate formation temperature increases.

2.2 Hydrate prevention [15]

2.2.1 Design philosophy for hydrate prevention

1. Identify the hydrate formation conditions.
2. Determine which areas are likely to have hydrate problems in the current system design.
3. Investigate design options which will prevent/reduce problems.
4. Investigate operational options which will prevent/reduce problems.
5. Evaluate most favourable options which may be a combination of design and operational changes and implement these in a revised design.
6. Determine which areas of the revised design could still be subject to a hydrate blockage in the event of some other system failure.
7. Establish whether the system is flexible enough to recover from a hydrate blockage in any of these locations.

2.2.2 Design options for hydrate prevention

The main design strategies for preventing or reducing hydrate problems are:

1. To keep the fluid temperature above the hydrate temperature.
2. To add chemicals to the water in order to change the hydrate formation temperature (methanol and glycol are often used).
3. To add chemicals to the water in order to slow down hydrate formation (methanol depresses the hydrate formation more than glycol).
4. To add chemicals to the water in order to change the hydrate crystal formation and to prevent hydrates agglomerating and forming a blockage: threshold hydrate inhibitors (THI).

5. To remove water from the system. It is particularly important to deshydrate gas entering the gas lift because these tend to be operated at fairly high pressure and are uninsulated.
6. To keep the system pressure below the hydrate formation pressure.

2.2.3 Operational options for hydrate prevention

When restarting a pipeline which has not been completely inhibited there are a number of guidelines which apply:

1. The pipeline should be depressurized if this has not already been done.
2. If a number of wells feed into the same pipeline then these should be brought on-stream in sequence starting with the well which has the leanest gas (highest hydrate formation temperature) and lowest water content. Starting with a low flowrate keeps the pressure low and therefore reduces the hydrate formation temperature. Once the pipeline has warmed up the flow rate and pressure can be increased.
3. A one-off slug of methanol should be injected at the wellhead before flowing any gas or liquid from the well. This will help to prevent hydrate formation in the tree and manifold valves.

2.3   Thermodynamic inhibitors [16]

Figure U3 shows the methane hydrates equilibrium (dissociation) temperature decrease with methanol (CH_3OH) content in the water phase. Figure U4 indicates how monoethyleneglycol (MEG) depresses this same temperature. Figure U5 illustrates the effect of salt (either pure sodium chloride or mixed with calcium chloride).

With 20 weight percent of inhibitor, the reduction of hydrates equilibrium temperature is given in Table U2.

Table U2

Inhibitor (20%)	Reduce hydrates equilibrium temperature by:	Remarks
Methanol (CH_3OH)	10°C	Prohibition for environmental reasons
Monoethyleneglycol (MEG)	6°C	
Salt	10°C	Corrosion problems

2.4   Kinetic inhibitors [16–18]

Kinetic inhibitors are water soluble chemicals (mainly polymers) which can act by different mechanims:

  (a) By delaying hydrate nucleation
  (b) By slowing down crystal growth
  (c) By preventing hydrates agglomeration.

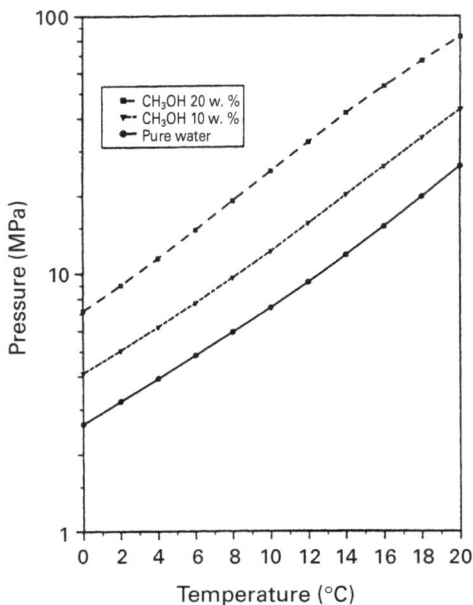

Figure U3 Methanol influence on methane hydrates equilibrium temperature (calculated) [16].

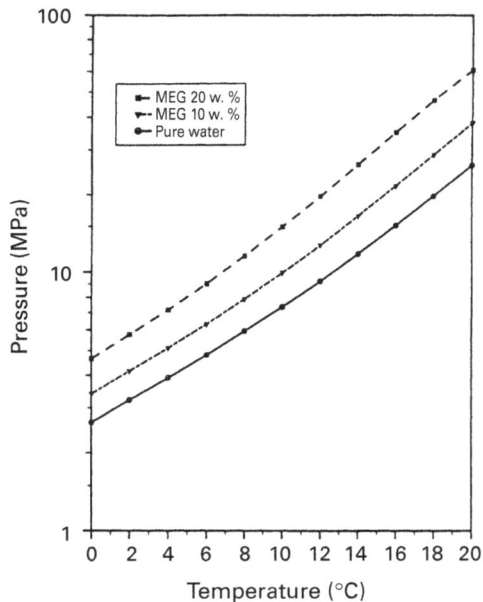

Figure U4 MEG influence on methane hydrates equilibrium temperature (calculated) [16].

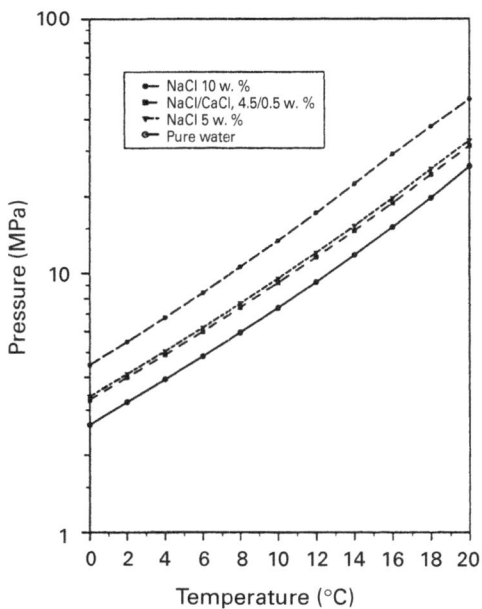

Figure U5 Salt influence on methane hydrates equilibrium temperature (calculated) [16].

U

420

Kinetic inhibitors can:

- Either increase crystals in suction time up to values which may surpass the residence time of multiphase fluids in flow and pipe lines
- Or decrease hydrates growth in order to delay exploitation and transport facilities plugging.

TR Oil Services have developed a new hydrate inhibitor, HYTREAT 530. It has the following advantages over methanol and MEG:

- It is 5% more cost effective than methanol and gives 74% cost saving over MEG on a once through treatment basis.
- It allows to reduce offshore chemical delivery costs considering the large reduction in chemical injection volume.
- HYTREAT 530 storage tank is significantly smaller than the one required for methanol storage.
- It does not contaminate the downstream processes.

2.5 Guidelines for use of kinetic or thermodynamic inhibitors [26]

This guidelines is a stepwise protocol to determine whether the use of inhibitors might be suitable.

1. If the field is mature, record the current hydrate prevention strategy. Record the existing or planned procedures for dealing with an unplanned shutdown. Provide a generic description of the chemistry of the scale and corrosion inhibitors used.
2. Obtain an accurate gas, condensate, and water analysis during a field drill test. Estimate how these compositions will change over the life of the field. Estimate the production rates of gas, oil, and water phases over the life of the field.
3. Generate the hydrate pressure-temperature equilibrium line with several prediction methods. If the operating conditions are close to the hydrate line, confirm the prediction with experiment(s).
4. Determine the water production profile over field life.
5. Consider the pipeline topography along the ocean floor to determine where water accumulations will occur at dips, resulting in points of hydrate formation.
6. Simulate the pipeline pressure-temperature profile using a simulator to perform hydraulic and heat transfer calculations in the well, flow lines, and separator over the life of the field.
7. Determine the water residence times in all parts of the system, especially in low points of the pipeline.
8. Estimate the subcooling ΔT (at the lowest temperature and highest pressure) relative to the equilibrium line over all parts of the system, including fluid separators and water handling facilities. List the parts of the system which require protection.

9. If $\Delta T < 8°C$ (14°F), consider the use of kinetic inhibitors. If $\Delta T > 8°C$ (14°F), consider the use of standard thermodynamic inhibitors or anti-agglomerants.
10. Perform economic calculations (capital and operating expenses) for four options (a) drying, (b) methanol, (c) monoethylene glycol, and (d) kinetic inhibitors.
11. Determine if inhibitor recovery is economical.
12. Design the hardware system to measure: (a) temperature and pressure at pipe inlet and outlet, (b) water monitor for rates at receiving facility, and (c) the chemical check list below.
    (a) Has the inhibitor been tested with systems at the pipeline temperature and pressure?
    (b) Consider the environmental, safety, and health impact of the chemical.
    (c) Determine physical properties such as:
        – flash point (which should be <60°C or 135°F)
        – viscosity (which should be <200 cp or mPa.s at lowest *T*)
        – density
        – pour point (which should be > −10°C or 15°F).
    (d) Determine the minimum, maximum, and average dosage of inhibitor.
    (e) Determine the storage and injection deployment methods.
    (f) Determine the material compatibility with gaskets, seals, etc.
    (g) Determine compatibility with other production chemicals.
    (h) Determine the compatibility with the process downstream including cloud point, foaming, and emulsification tendencies.

At an early stage in the inhibitor design process, it will be worthwhile to consider obtaining laboratory data and involving a service company to provide field support of process hydrate inhibition.

2.6 Dispersant additives [16, 17]

Contrary to thermodynamic inhibitors or kinetic inhibitors, dispersant additives do not prevent the formation of hydrate crystals but enable their transport in suspension. Dispersant additives are oil soluble components and act according to the following steps:
    (a) Formation of a water-in-oil emulsion by decreasing the water/oil interfacial tension
    (b) Prevention of water droplets coalescence
    (c) Prevention of aggregation between hydrate particles.

By decreasing interfacial tension and preventing coalescence between water droplets, additives act as emulsifiers. When hydrate formation occurs, water droplets are transformed into hydrate particles which are dispersed in the hydrocarbon liquid phase. Additives act as dispersant additives by preventing the formation of large aggregates.

Institut Français du Pétrole (IFP) has developed a process which aims to transport hydrate crystals as a suspension in well streams. This process is based on the use of surfactants which are likely to create a water emulsion in liquid hydrocarbons (condensate or crude). Thus, hydrates are formed at the water droplets/liquid hydrocarbons interface, and remain dispersed in this latter phase if the emulsion is sufficiently stable and there is no agglomeration of the solid crystals.

Dispersant additives are sometimes called "emulsifiers" or "anti-agglomerators". Table U3 gives some characteristics of dispersant additives.

Table U3

Dispersant additives	Adding weight%	Licensee
EMULFIP 102 B	1%	FINA
IPE 201	0.5%	IFP

2.7 Limitations of low-dosage additives [17]

2.7.1 Main limitations

Additives present different limitations depending on their mechanism of action (see Table U4).

Table U4 Main limitations of kinetic inhibitors and dispersant additives.

Kinetic inhibitors	Dispersant additives
Sub-cooling temperature (temperature gap between the equilibrium and the actual hydrate formation temperature). At present time, kinetic inhibitors are efficient for sub-cooling up to 10°C.	**Water-cut.** The maximum water cut is expected to be around 50%. This limitation is caused by the rheological properties of suspensions with high solid fraction and may depend on flow regime conditions.
Residence time. The residence time of fluids in the pipeline should not be too high. Therefore, the acceptable length of pipeline should typically be less than 50 km. For shut-down of flow in pipeline, injection of methanol may be necessary.	

2.7.2 Field of application

As shown in Fig. U6, the application of kinetic inhibitors and dispersant additives is limited in terms of water cut and sub-cooling, respectively.

For both sub-cooling and high water cut, traditional methods such as injection of methanol (or glycol), heating or insulation remain useful.

U

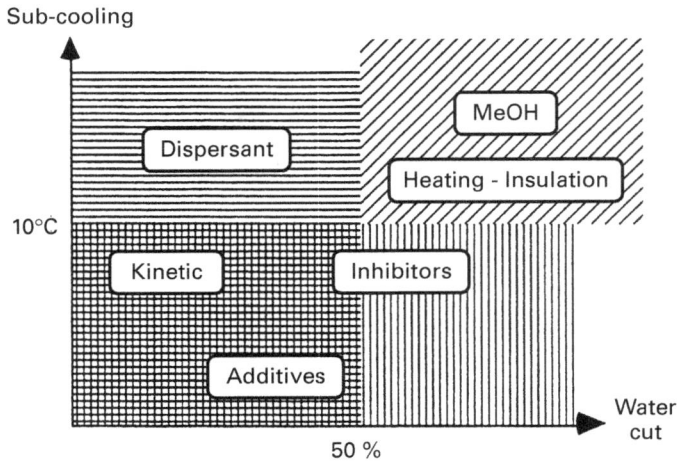

Figure U6 Application range of low-dosage aditives [17].

2.8 Economic comparison between different inhibition processes [16]

1. The use of methanol is characterized by considerable losses of methanol and a high operating cost.
2. The injection of glycol (MEG) involves a high investment cost because of the inhibitor regeneration while the operating cost is low.
3. The use of dispersant additives corresponds to a lower investment cost than that of the methanol process as well as to lower operating costs.

U3 PARAFFINS (WAXES)

3.1 Definition [1]

Paraffins are primarily aliphatic hydrocarbons (both straight and branched chain) that change state from liquid to solid during conventional oil production and processing operations. In addition to aliphatics, field wax deposits usually contain aromatic, naphthenic, resin, and asphaltenic molecules as well. The combined mass is called "wax". Paraffin waxes usually melt between 45 and 70°C (110 and 160°F). Field waxes contain molecules that can have melting points in excess of 95°C (200°F).

Consequently, paraffins are deposited when the average temperature of crude oil decreases below approximately 50°C.

It is the Wax Appearance Temperature (WAT).

3.2 How to predict a wax problem [19]

The following steps are generally observed:
- Downhole and/or surface sampling.
- Specific analyses on the liquids recovered at surface conditions.
- Use of a thermodynamic model to determine the WAT and the crystallized fraction as a function of temperature, on live oils or condensates. The model is adjusted beforehand to reproduce the surface conditions.
- When necessary, use of a kinetic model to evaluate the amount of wax deposited in tubings and pipes.

3.3 Sampling [19]

The quality of the prediction is highly dependent on the quality of the sample.

Because of the risk of deposition during the course of the test, downhole sampling is generally preferred to surface sampling. One major difficulty is to avoid contamination by drilling fluids, especially where oil-based mud is used.

3.4 Analysis [19]

In addition to standard PVT analysis, the evaluation of potential wax problems requires additional information:
- "Carburane" type analysis by Gas Chromatography (GC) is used, giving a semi-detailed composition up to C20, about 300 components are identified.
- N-paraffin distribution from C21 to approximatively C50/60.
- Saturate, aromatic, naphthenic, aliphatic (SARA) distribution in heavy fraction C_{15+}.
- WAT and crystallized fraction as determined by differential scanning calorimetry.

3.5 Modelling [19, 20]

Three models are used to simulate the waxy behaviour of crude oils:
- A thermodynamic model: CRYSPAR. It calculates wax appearance temperature (WAT) and the fraction of solid precipitated when waxy crude oils are cooled below the WAT.
- A dynamic deposition model: CIRE.
- A combined model: PARASYM.

U

3.6 How to prevent wax deposition [19]

The prevention of wax deposition is dependent on the field where the treatment will be used.

In existing fields:
- Chemical injection to reduce pigging and scraping frequency.
- Average concentration rate optimized to around 70 ppm.

In new fields:
- Thermal solutions:
  - Induction heating of tubing
  - Thermal insulation of tubing and pipes
  - (Magnets).

U4 SCALES [21, 24]

Deposits of scale are severe problems in many areas. Scale deposition can be attributed to such factors as pressure drop, temperature change, and mixing of incompatible waters, particularly in injection wells for secondary recovery or disposal purposes.

4.1 Scales usually encountered

Tables U5a, b and c give the chemical formulas and mineral names of the oil well scale deposits usually encountered.

Table U5a

Water soluble scale	Chemical formula	Mineral name
Sodium chloride	NaCl	Halite

Table U5b

Acid soluble scales	Chemical formula	Mineral name
Calcium carbonate	$CaCO_3$	Calcite
Iron carbonate	$FeCO_3$	Siderite
Iron sulfide	FeS	Trolite
Iron oxide	Fe_3O_4	Magnetite
Iron oxide	Fe_2O_3	Hematite
Magnesium	$Mg(OH)_2$	Brucite

Table U5c

Acid insoluble scales	Chemical formula	Mineral name
Calcium sulfate	$CaSO_4.2H_2O$	Gypsum
Calcium sulfate	$CaSO_4$	Anhydrite
Barium sulfate	$BaSO_4$	Barite
Strontium sulfate	$SrSO_4$	Celestite
Barium strontium sulfate	$BaSr(SO_4)_2$	....

4.2 Scale removal

Usually a scale, that is an acid soluble deposit, will be removed by using a hydrochloric acid solution. Sometimes it has been found very effective to use a hot acid treatment in which the acid is reacted with magnesium or aluminum in the wellbore to heat up the entire system.

Particular cases:

*   Iron is perhaps the most common problem encountered in acid removal of scales. Additives are available which are designed to control the precipitation of the iron when the acid becomes spent.
*   The chemical treatment for removal of calcium sulfate is slow in comparison with the removal of acid soluble scales. A first step is a treatment with a chemical solution which converts the acid insoluble calcium sulfate into an adhering acid soluble scale.

4.3 Identification of scale

4.3.1 Laboratory or field test for scale

Table U6 is a summary of easily run laboratory or field tests to determine the type of scale.

Table U6 [24].

Scale	Additive for test	Results
Acid soluble scale		
Calcium carbonate	HCl	Bubbles vigorously
Iron carbonate	HCl	Bubbles–Solution yellow
Iron sulfide	HCl	Bubbles–H_2S gas
Iron oxide	HCl	Dissolves–Solution yellow
Magnesium hydroxide	HCl	Dissolves–Solution clear
Acid insoluble scale		
Calcium sulfate	1. Soak overnight in $NaHCO_3$	
	2. Treat with HCl	Bubbles vigorously
Barium sulfate	–	Not soluble
Strontium sulfate	–	Not soluble
Sodium chloride	Fresh water	Dissolves

4.3.2 Guide to field identification of oilfield scales

Physical appearance and chemical analysis may be used. Table U7 is a practical guide for field identification of oilfield scales.

Table U7 [21].

Physical appearance	Acid solubility	Indicated composition and origin
1. White or light-colored		
1.1 Hard, compact, fine granular	Insoluble.	$BaSO_4$, $SrSO_4$, $CaSO_4$ Incompatible waters.
1.2 Compact, with long, pearly crystals	Powder dissolves slowly with no gas bubbles. Solution gives SO_4 test with $BaCl_2$.	Gypsum, $CaSO_4.2H_2O$. Incompatible waters supersaturation.
1.3 Compact, fine grain, or crystals which break into rhombohedrons	Easily soluble in HCl with gas bubbles.	$CaCO_3$ or mixture of $CaCO_3$ and $MgCO_3$ if more slowly dissolved. Supersaturation, rarely incompatible waters.
2. Dark colored, brown to black		
2.1 Compact, brown	Essentially insoluble brown color dissolves on heating. Acid turns yellow. White, insoluble residue.	See 1.1 and 1.2 above for white residue. Brown, iron oxide is corrosion product or precipitate due to oxygen.
2.2 Compact, black	Black material dissolves slowly with evolution of H_2S white, insoluble residue.	See 1.1 and 1.2 above for residue. Black color is iron sulfide-corrosion product, incompatible waters, or both.
2.3 Compact, brown or black	Easily soluble in 4% HCl (dilute 15% 1:4) with gas bubbles. Brown or black color remains.	$CaCO_3$ with iron oxide or iron sulfide coloring matter.
2.4 Soft muck, usually brown or black:		
2.41	Insoluble.	See 1.1 above.
2.42	Dissolves, no bubbles.	See 1.2 above.
2.43	Dissolves, gas bubbles.	See 1.3 above.
2.44	Insoluble, except brown material, yellow solution.	Iron oxide, see 2.1 above.
2.45	Black material dissolves, evolution of H_2S.	Iron sulfide, see 2.2 above.

Note. Discussion of inert residue and organic slime is omitted from the above outline. It should be emphasized that acid-insoluble residue occurs in all scale deposits, sometimes being the major ingredient. Also, "soft muck" deposits may contain all the others, in finely-divided state and their recognition may be difficult due to more or less organic slime.

REFERENCES

1   Leontaritis K (1996) *State of Art Asphaltene Deposition and Modelling, Controlling Hydrates, Waxes and Asphaltenes.* IBC UK Conferences Limited, Aberdeeen, UK

2   Minssieux L (1997) Core damage from crude asphaltene deposition. *Paper SPE 37250* presented at the 1997 SPE International Symposium on Oilfield Chemistry, Houston, Texas

3   Minssieux L (1998) Removal of asphalt deposits by cosolvent squeeze: mechanisms and screening. *Paper SPE 39447* presented at the 1998 SPE Formation Damage Control Conference, Lafayette, Louisiana

4   Leontaritis KJ, Amaefule JO, Charles RE (1992) A systematic approach for the prevention and treatment of formation damage caused by asphaltene deposition. *SPE Production & Facilities,* August 1994, 157–164

5   Moricca G, Trabucchi G (1996) Effective removal of asphaltene deposits from pipelines and treating plants. *Paper SPE 36834* presented at the 1996 SPE European Petroleum Conference, Milan, Italy

6   Garland E (1989) The asphaltic properties of an apparently ordinary crude oil may lead to re-thinking of field exploitation. *Paper SPE 19731* presented at the 64th Annual Technical Conference and Exhibition of the SPE, San Antonio, Texas

7   Izquierdo A, Rivas O (1997) A global approach to asphaltene deposition problem. *Paper SPE 37251* presented at the 1997 SPE International Symposium on Oilfield Chemistry, Houston, Texas

8   Szewczyk V, Thomas M, Behar E (1998) Prediction of volumetric properties and (multi-) phase behaviour of asphaltenic crudes. *Revue de l'Institut Français du Pétrole,* 53, 1, 51–58

9   Dayvault GP, Patterson DE (1989) Solvent and acid stimulation increase production in Los Angeles basin waterflood. *Paper SPE 18816* presented at the 1989 SPE Regional California Meeting, Bakersfield, California

10 Trbovich MG, King GE (1991) Asphaltene deposit removal: Long-lasting treatment with a co-solvent. *Paper SPE 21038* presented at the 1991 SPE International Symposium on Oilfield Chemistry, Anaheim, California

11 Coppel CP, Newberg PL (1972) Field results of solvent stimulation in a low-gravity-oil reservoir. *Journal of Petroleum Technology,* Richardson, TX, October 1972, 1213–1218

12 Groffe P, Volle JL, Ziada A (1995) Application of chemicals in prevention and treatment of asphaltene precipitation in crude oils. *Paper SPE 30128* presented at the 1995 SPE European Formation Damage Conference, The Hague, The Netherlands

13 Bouts MN, Wiersma RJ, Muljs HM, Samuel AJ (1995) An evaluation of new asphaltene inhibitors: laboratory study and field testing. *Journal of Petroleum Technology*, Richardson, TX, September 1995, 782–787

14 Di Lullo AG, Lockart TP, Carniani C, Tambini M (1998) A technico-economic feasibility study of asphaltene inhibition squeeze treatments. *Paper SPE 50656* presented at the 1998 SPE European Petroleum Conference, The Hague, The Netherlands

15 Paterson P (1996) *Design Guidelines for Hydrate Prevention, Controlling Hydrates, Waxes and Asphaltenes.* IBC UK Conferences Limited, Aberdeeen, UK

16 Behar E, Delion AS, Durand JP, Sugier A, Thomas M (1995) Hydrates problem within the framework of multiphase production and transport of crude oils and natural gases. *Revue de l'Institut Français du Pétrole,* 50, 5, 627–640

17 Palermo T (1996) Hydrates control in multiphase flow. New trends, Natural gas: from exploration to utilization. *OAPEC-IFP Joint Workshop,* Rueil-Malmaison, France

U

18 TR Oil Services, *Hydreat. Advanced Hydrate Prevention Technology.* A Clariant Group Company, 17–20

19 Volle JL (1996) Operational aspects of wax deposition from field discovery to the cleaning of a plugged line. *Controlling Hydrates, Waxes and Asphaltenes*, IBC UK Conferences Limited, Aberdeeen, UK

20 Ruffier-Meray V, Calange S, Behar E (1996) Cryspar: A thermodynamic model for waxy crude oils. *Controlling Hydrates, Waxes and Asphaltenes,* IBC UK Conferences Limited, Aberdeeen, UK

21 Lessons in Well Servicing and Workover (1980) *Well Servicing and Repair.* Petroleum Extension Service, Austin, Texas

22 Behar E, Delion AS, Durand JP, Sugier A, Thomas M (1994) Plugging control of production facilities by hydrates. *Annals of the New York Academy of Sciences*, Vol. 715, April 29, 1994

23 Kohler N (1997) Inhibition des dépôts minéraux dans les puits de production. *Note No. 107, Institut Français du Pétrole*, Division Chimie Appliquée, Biotechnologies, Matériaux, April 1997

24 Allen TO, Roberts AP (1997) *Production Operations.* Vol. 2, Ch. 9, OGCI, Tulsa, OK

25 Hinrichen CJ (1998) Preventing scale deposition in oil production facilities: an industry review. *Paper No. 61* presented at Corrosion 98, NACE International, Houston, TX

26 Sloan DE, Jr. (1998) *Offshore Hydrate Engineering Handbook.* ARCO Exploration and Production Technology Company

27 Barrufet MA, Morales G (1999) Evaluation of hydrate formation and inhibition using equations of state. *Petroleum Engineer International*, January 1999, pp. 65–70

28 Emmons DH, Jordan MM (1999) *The Development of Near-Real Time Monitoring of Scale Deposition.* Nalco/Exxon Energy Chemicals, L.P.

29 Thomas FB, Bennion DB (1999) Development and evaluation of paraffin technology: current status. *Journal of Petroleum Technology*, February 1999, pp. 60–61

U

V

Well Servicing

V

Well Servicing

V1 WELL CONTROL OPERATION [1]

The standard **API RP 59** has been designed to serve as a direct field aid in well control and also as a technical source for teaching well control principles. These recommended practices contain current practices for utilization in land, marine, and offshore drilling operations.

The recommended practices are separed into two main systems:

- Blowout preventers at the surface within reach and sight of the driller
- Blowout preventers installed on the seafloor with relatively long choke and kill lines.

In this standard, sections have been prepared to establish practices and procedures pertaining to both surface and subsea blowout preventer installations.

V2 WIRELINE OPERATION [2, 3]

Wireline operation is the basic method of handling a producing well.

2.1 Principle and area of application

Wireline operations are used to work on a producing well by means of a steel cable to enter, run in, set and retrieve the tools and measurement instruments needed for rational production.

Wireline jobs can be classified into three different types:

- Monitoring and cleaning the tubing or the bottom of the hole (inside diameter, corrosion, clogging, sediment top, etc.)
- Measurement operations (bottomhole temperature and pressure recordings, sampling, locating interfaces, production logs, etc.)
- Setting or retrieving tools and operations in the well (setting and pulling subsurface safety valves, bottomhole chokes, plugs, gas lift valves, etc.; actuating circulating sleeves; fishing, perforating).

V

2.2 Dimensions

Dimensions of the more commonly used wireline tools are given in Tables V1 and V2. Diameters 0.082 in. (2.08 mm) and 0.092 in. (2.34 mm) are the most utilized.

Table V1 Diameters.

mm	in.
1.68	0.066
1.83	0.072
2.08	0.082
2.34	0.092
2.67	0.105
2.74	0.108
3.18	0.125
3.25	0.128

Table V2 Standard lengths.

m	ft
3 048	10 000
3 657	12 000
4 572	15 000
5 486	18 000
6 096	20 000
7 620	25 000

2.3 Well-measuring wire [4]

2.3.1 Requirements

Well-measuring shall be in accordance with Table V3 (established for improved plow steel). For well-measuring wire of other materials or coating, refer to supplier for physical properties.

Well-servicing wire shall consist of one continuous piece of wire without brazing or welding of the finished wire. The wire shall be made from the best quality of specified grade of material, shall be of good workmanship, and shall be free from defects which might affect its appearance or service-ability. Coating on well-measuring wire shall be optional with the purchaser.

The minimum torsions for individual bright wire shall be the number of 360° revolutions in a 8-in. (203 mm) length that the wire must withstand before breakage occurs.

Table V3 Requirements for well measuring wire, bright or drawn galvanized carbon steel (improved plow steel) [4].

Wire diameter		Approximate wire weight		Nominal strength		Minimum torsion revolutions
(mm) (± 0.03)	(in.) (± 0.0001)	(kg/m)	(lb/ft)	(kN)	(lbf)	
1.68	0.066	0.018	0.012	3.61	811	32
1.83	0.072	0.021	0.014	4.27	961	29
2.08	0.082	0.027	0.018	5.51	1 239	26
2.34	0.092	0.034	0.023	6.88	1 547	23
2.67	0.105	0.045	0.030	8.74	1 966	20
2.74	0.108	0.048	0.032	9.38	2 109	19
3.18	0.125	0.062	0.042	12.43	2 794	*
3.25	0.128	0.065	0.044	13.01	2 924	*

* Values to be agreed upon between purchaser and manufacturer:
  – Extra improved plow steel: 1.18 nominal strength
  – Extra extra improved plow steel: 1.25 nominal strength.

2.3.2 Testing

A specimen of wire 3 ft (0.91 m) long shall be cut from each coil of well-measuring wire. One section of this specimen shall be tested for elongation and tensile simultaneously. The ultimate elongation shall be measured on a 10-in. (254 mm) length of specimen, at instant of rupture, which must occur within the 10-in. (254 mm) gauge length. When determining elongation, a stress shall be imposed upon the wire equal to 100 000 psi (690 MPa) at which point the extensometer is applied. Directly to the reading of the extensometer shall be added 0.4 percent to allow for the initial elongation ocurring before application of the extensometer.

The remaining section of the 3-ft (0.91 m) test specimen shall be gauged for size and tested for torsional requirements in accordance with Paragr. 2.3.1, above.

If, when making any individual test, the first specimen fails, not more than two additional specimens from the same wire shall be tested. If the average of any two tests shows acceptance, it shall be used as the value to represent the wire.

2.4 Well-measuring strand [4]

V

2.4.1 Construction

- Well-measuring strand shall be bright (uncoated) or drawn-galvanized.
- Well-measuring strand shall be left lay. The lay of the finished strand shall not exceed 19 times the nominal diameter.
- Well-measuring strands may be of various combinations of wires but are commonly furnished in both 1 × 16 (1 6/9) and 1 × 19 (1 6/12) constructions.

2.4.2 Requirements

Well-measuring strands shall conform to the properties listed in Table V4.

Table V4 Requirements for well servicing strand bright or drawn
galvanized carbon steel [4].

Nominal diameter	Minimum diameter		Maximum diameter		Approximate weight		Galvanized improved plow steel		Galvanized extra improved plow steel	
(in.)	(mm)	(in.)	(mm)	(in.)	(kg/m)	(lb/ft)	(kN)	(lbf)	(kN)	(lbf)
3/16	4.775	0.188	5.105	0.201	0.109	0.073	18.7	4 200	20.9	4 700
7/32	5.563	0.219	5.893	0.232	0.149	0.100	26.2	5 900	29.4	6 600
1/4	6.350	0.250	6.731	0.265	0.189	0.127	32.5	7 300	36.5	8 200
5/16	7.950	0.313	8.357	0.329	0.327	0.220	49.4	11 100	55.6	12 500

2.4.3 Testing

When testing finished strands to their breaking strength, suitable sockets or other acceptable means of holding small cords shall be used.

2.4.4 Wire guy strand and structural rope and strand

They shall conform to the following ASTM standards (Table V5).

Table V5 [4].

Galvanized wire guy strand	A-475: Zinc-coated steel wire strand
Aluminized wire guy strand	A-474: Aluminum-coated steel wire strand
Galvanized structural strand	A-586: Zinc-coated steel structural strand
Galvanized structural rope	A-603: Zinc-coated steel structural wire rope

V3 SUBSURFACE SAFETY VALVES [5, 6, 7]

Specifications are defined with Standards ISO 10432 and API spec. 14A, 9th Ed.

A subsurface safety valve (SSSV) is a device whose design function is to prevent uncontrolled well flow when closed. These devices may be installed and retrieved by wireline or pump down methods (wireline retrievable) or be an integral part of the tubing string (tubing retrievable).

3.1 Definitions

- **SSSV:** A subsurface safety valve is a device whose design function is to prevent uncontrolled well flow when closed. These devices may be installed and retrieved by wireline or pump down method (wireline retrievable) or be an integral part of the tubing string (tubing retrievable).
- **SCSSV:** A surface controlled subsurface safety valve is an SSSV controlled from the surface.
- **SSCSV:** A subsurface controlled subsurface safety valve is an SSSV actuated by the characteristics of the well.
- **SV lock:** A safety valve lock is a device attached to or a part of the SSSV that holds the SSSV in place.
- **SVLN:** A safety valve landing nipple is a receptacle in the production string with internal sealing surfaces in which an SSSV may be installed. It may include recesses for locking devices to hold the SSSV in place and may be ported for communication to an outside source for SSSV operation.
- **SSSV equipment:** The subsurface safety valve, safety valve lock, safety valve landing nipple, and all components that establish tolerances and/or clearances which may affect performance or interchangeability of the SSSV equipment.

3.2 Suggestions for ordering subsurface safety valve equipment

In placing orders for subsurface safety valve equipment in accordance with ISO 10432-1, Subpart 2, the operator should specify the following on the purchase order.

- **Specification and edition.**
- **Tubing:** size, weight, grade, connection.
- **SSSV equipment**
  - Type system (SCSSV or SSCSV).
  - Type, model and quantity.
  - Class of service:
    Class 1. Standard service.
    Class 2. Sandy service.
    Class 3. Stress corrosion cracking service.
    Class 4. Weight loss corrosion service.
  - Size: see Tables V6 and V7.

V

– Rated working pressure: the most commonly used are listed below:

MPa	psi
20.7	3 000
34.5	5 000
41.4	6 000
69.0	10 000
103.5	15 000

– Temperature range:
The lowest high temperature rating of any component of the SSSV.
The highest low temperature rating of any component of the SSSV.
– Special features (see API RP 14B).

- **SCSSV**
  - Control system pressure.
  - Setting depth (see API RP 14B).
  - Type control fluid (see API RP 14B).
  - Strength (tubing retrievable only).
- **SSCSV**
  - Orifice size, spring, spacers, dome charge, etc. (see API RP 14B).
- **Other equipment**
  - Safety valve lock.
  - Safety valve landing nipple.

3.3 Standard safety valve landing nipples

They are listed in Table V6.

Table V6 [5-6].

Nominal nipple size		Tubing or casing size		Sealing surface ID	
(mm)	(in.)	(mm)	(in.)	(mm)	(in.)
31.8	1 1/4	42.2	1.66	31.75	1.250
38.1	1 1/2	48.3	1.90	38.10	1.500
44.5	1 3/4	52.4	2.063	41.28	1.625
50.8	2	60.3	2 3/8	47.63	1.875
63.5	2 1/2	73.0	2 7/8	58.72	2.312
76.2	3	88.9	3 1/2	71.42	2.812
88.9	3 1/2	101.6	4	84.15	3.313
101.6	4	114.3	4 1/2	96.82	3.812
114.3	4 1/2	127.0	5	104.78	4.125
127.0	5	139.7	5 1/2	115.87	4.562
152.4	6	177.8	7	151.46	5.963

Reprinted Courtesy of the American Petroleum Institute, 1220 L Street, NW, Washington, DC, 20005.

3.4 Standard outside diameters wireline retrievable subsurface safety valves

They are listed in Table V7.

Table V7 [5–6].

Nominal SSSV size		Standard valve OD	
(mm)	(in.)	(mm)	(in.)
31.8	1 1/4	31.50	1.240
38.1	1 1/2	37.85	1.490
44.5	1 3/4	41.02	1.615
50.8	2	47.37	1.865
63.5	2 1/2	58.47	2.302
76.2	3	71.17	2.802
88.9	3 1/2	83.90	3.303
101.6	4	96.57	3.802
114.3	4 1/2	104.52	4.115
127.0	5	115.62	4.552
152.4	6	151.21	5.953

Reprinted Courtesy of the American Petroleum Institute, 1220 L Street, NW, Washington, DC, 20005.

V4 SNUBBING [8]

4.1 Introduction

Snubbing is the process of running or pulling tubing, drillpipe or other tubulars in the presence of sufficient surface pressure present to cause the tubular to be forced out of the hole. That is, in snubbing the force due to formation pressure's acting to eject the tubular exceeds the buoyed weight of the tubular. As illustrated in Fig. V1, the well force F_w, is a combination of the well pressure force (F_{wp}), buoyant force (F_b) and friction force (F_f).

Stripping is similar to snubbing in that the tubular is being run into or pulled out of the hole under pressure.

Snubbing or stripping operations through rams can be performed at any pressure. Snubbing or stripping operations through a good quality annular preventer are generally limited to pressures less than 14 MPa (2 000 psi). Operations conducted through a stripper rubber or rotating head should be limited to pressures less than 1 750 kPa (250 psi). Although slower, ram-to-ram is the safest procedure for conducting operations under pressure.

Some of the more common snubbing applications are as follows:
• Tripping tubulars under pressure
• Pressure control/well killing operations
• Fishing, milling or drilling under pressure
• Completion operations under pressure.

V

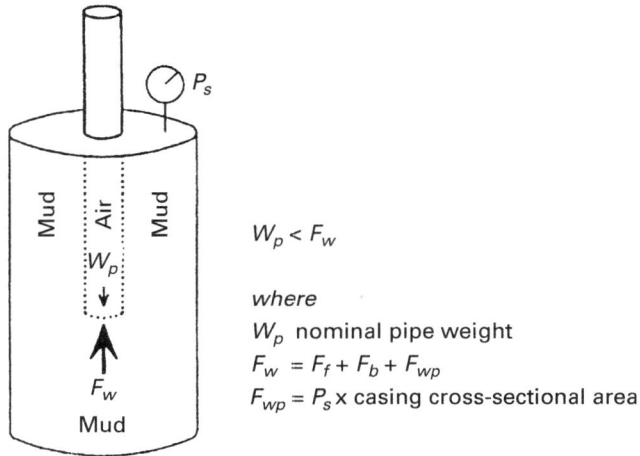

$W_p < F_w$

where

W_p nominal pipe weight

$F_w = F_f + F_b + F_{wp}$

$F_{wp} = P_s \times$ casing cross-sectional area

Figure V1 Snubbing principle [8].

4.2 The snubbing stack

There are many acceptable snubbing stack arrangements. The basic snubbing stack is illustrated in Fig. V2. As illustrated, the lowermost rams are blind safety rams. Above the blind safety rams are the pipe safety rams. Above the pipe safety rams is the bottom snubbing ram, followed by a spacer spool and the upper snubbing ram. Since a ram preventer should not be operated with a pressure differential across the ram, an equalizing loop is required to equalize the pressure across the snubbing rams during the snubbing operation. The pipe safety rams are used only when the snubbing rams become worn and require changing.

4.3 Theoretical considerations

As shown in Fig. V1, snubbing is required when the well force, F_w, exceeds the total weight of the tubular. The snubbing force F_{sn} is equal to the net upward force as illustrated in Eq. V1 and Fig. V1.

$$F_{sn} = W_p L - \left(F_f + F_b + F_{wp} \right) \tag{V1}$$

where

W_p nominal weight of the pipe (lbm/ft)

L length of pipe (ft)

F_f friction force (lbf)

F_b buoyant force (lbf)

F_{wp} well pressure force (lbf).

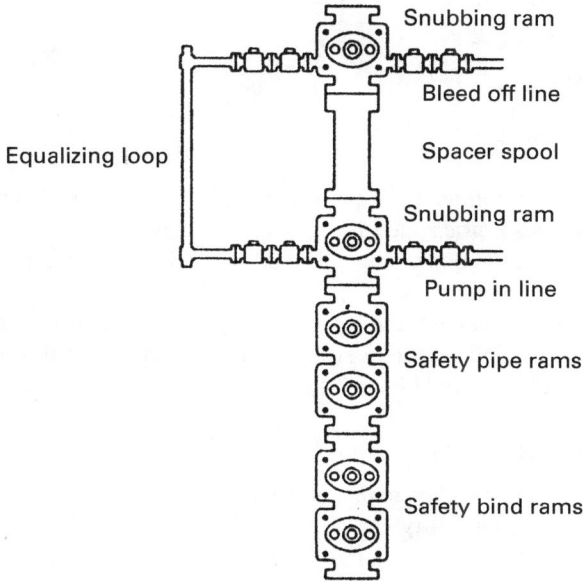

Figure V2 Basic snubbing stack [8].

The well pressure force, F_{wp}, is given by Eq. V2:

$$F_{wp} = 0.7854 D_p^2 P_s \qquad\qquad\qquad (V2)$$

where

P_s surface pressure (psi)

D_p outside diameter of tubular exposed to P_s (in.).

As shown in Eq. V2, the diameter of the pipe within the seal element must be considered. When running a pipe through an annular or stripper, the outside diameter of the connection is the determining variable. When stripping or snubbing pipe from ram to ram, only the pipe body is contained within the seal elements; therefore, the outside diameter of the tube will determine the force required to push the pipe into the well. With drillpipes, there is a significant difference between the diameter of the pipe body and the tool joint.

V

V5 DRILLSTEM TESTING (DST) [9]

5.1 Background [9]

5.1.1 Objective

Drillstem testing provides a method of temporarily completing a well to determine the productive characteristics of a specific zone. As originally conceived, a drillstem test provided primarily an indication of formation content. The pressure chart was available, but served mainly to evaluate tool operation.

Currently, analysis of pressure data in a properly planned and executed DST can provide, at reasonable cost, good data to help evaluate the productivity of the zone, the completion practices, the extent of formation damage, and perhaps the need for stimulation.

5.1.2 Reservoir characteristics

Reservoir characteristics that may be estimated from DST analysis include:
- (a) Average effective permeability
- (b) Reservoir pressure
- (c) Wellbore damage
- (d) Barriers, permeability changes fluid, contacts
- (e) Radius of investigation
- (f) Depletion.

5.1.3 Basics of DST operations [9, 10]

Simply a drillstem test is made by running in the hole on drill pipe a bottom assembly consisting of a packer and a surface operated valve (Fig. V3). The DST valve is closed while the drill string is run, thus, pressure inside the drill pipe is very low compared to hydrostatic mud column pressure. Once on bottom, the packer is set to isolate the desired formation zone from the mud column, and the control valve is opened to allow formation fluids to enter the drill pipe.

After a suitable period, the valve is closed, and a pressure buildup occurs below the valve as formation fluids repressure the area around the wellbore. After a suitable buildup time, the control valve usually is opened again, and the flowing and shut-in periods repeated, to obtain additional data and verification.

5.2 Theory of pressure buildup analysis [9]

Transient pressure analysis of a DST is based on the Horner pressure buildup equation. This equation describes the repressuring of the wellbore area during the shut-in period, as formation fluid moves into the "pressure sink" created by the flowing portion of the DST.

$$p_{ws} = p_i - \frac{162.6q\mu B}{kh}\log_{10}\left(\frac{t_p' - \Delta t'}{\Delta t'}\right) \qquad (V3)$$

where

    p_{ws} measured pressure in the wellbore during buildup (psi)
    t'_p flowing time (min)
    $\Delta t'$ shut-in time (min)
    p_i shut-in reservoir pressure (psi)
    q rate of flow (st bbl/D)
    μ fluid viscosity (cp)
    B formation volume factor (reservoir bbl/stock tank bbl)
    k formation permeability (md)
    h formation thickness (ft).

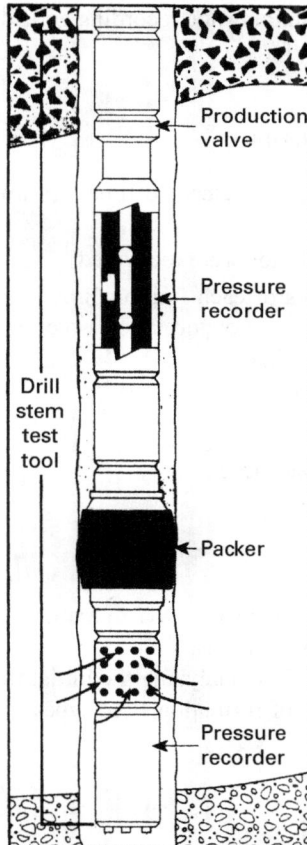

Figure V3 A drillstem test tool records pressure and samples the formation fluid [10].

Conditions which must be assumed during the buildup period for Eq. V3 to be strictly correct are:

(a) radial flow, homogeneous formation
(b) steady-state conditions
(c) infinite reservoir
(d) single-phase flow.

Most of these conditions are met on a typical DST. Steady-state flow is perhaps the condition causing the primary concern, particularly at early shut-in time.

5.3 Recommendations for obtaining good test data [9]

The key to DST evaluation is obtaining and recording good data. The DST must be planned to fit the specific situation.

5.3.1 Recording surface events

Recording surface events, both character and time, are important to chart analysis. For example:

• What types and amounts of fluids were recovered?
• What were the characteristics of each fluid (salinity and perhaps resistivity of water, was it water cushion, mud filtrate or formation water, how much of each; API gravity and gas-oil ratio of recoverd oil, etc.)?
• What size chokes were used?
• When were they changed?
• When did fluid come to the surface?

5.3.2 Time intervals

Time intervals allotted to each of the basic DST operations should ideally be adjusted during the test based on surface observations.

The "closed chamber method" of analyzing the initial flowing period (discussed later) provides very early indications of formation fluid types and rates for use by the on-site supervisor in running the DST.

5.3.3 Initial flowing period

This must be sufficient to relieve the effect of supercharge or overpressure in the formation immediately surrounding the wellbore.

Normally 5 to 20 minutes is sufficient; however, longer times may be desirable in low productivity reservoirs in order to positively differentiate supercharge from depletion.

5.3.4 Initial shut-in period

With no previous experience, the length of initial shut-in period may be based on statistical studies as follows:
- With 30 minute shut-in, 50% of tests reached static reservoir pressure.
- With 45 minute shut-in, 75% of tests reached static reservoir pressure.
- With 60 minute shut-in, 92% of tests reached static reservoir pressure.

5.3.5 Final flow period

This should be at least one hour of good to strong blow. The longer the flow period, the deeper the radius of investigation. If fluid reaches the surface, additional time is desirable to obtain accurate volume gauges and gas-oil ratios. If blow quits, nothing is gained by continuing flow test.

5.3.6 Final shut-in period (FSI period)

This is the most important portion of the DST as far as formation evaluation is concerned. The length of shut-in should be based on events during the flow period.
- If formation fluid surfaces, FSI period should be one-half the flowing time (but never less than 30 minutes).
- With good to strong blow, FSI period should be equal the flowing time (but never less than 45 minutes).
- With poor blow, FSI period should be twice the flowing time (120 minutes, if possible).

V6 WORKOVER PLANNING [11]

Recommendations to repair a well will vary with the defined problem. Some of the workover considerations that influence a job design and its execution are the following:
1. What corrective action is proposed to repair the well?
2. What well fluid is required to control the zone?
3. Can the well be repaired by working through tubing, or is it necessary to remove tubulars?
4. What type of rig equipment is necessary?
5. Are there any special techniques required such as sand control, gas lift or stimulation?
6. What type of downhole completion equipment is required, and is it readily available?
7. Will there be any special logistical problems during the job?

6.1 Workover operations

Table V8 reviews several types of workover operation and reasons for the workover.

V

Table V8 [11].

Reason for workover	Operation
Increase production from current zone or recomplete in new zone	• Reperforation (major) • Stimulation (major) • Put on artificial lift system (routine)
Lower operating costs	• Reduce water production with cement squeeze (major) • Use sand control procedures (major) • Change out lift systems (routine) • Plug and abandon (major)
Return well to its production objective	• Pull and replace wireline equipment (routine) • Pull and replace leaking tubing (routine) • Pull and repair lift equipment (routine)

6.2 Analysis of problem wells

Well problems may be categorized as follows:
- Mechanical (rod breaks, pump failures, seals, packer, or tubing leaks, etc.).
- Restricted producing rate with high flowing bottomhole pressure p_{wf} (plugged: tubing intake, pump, gas-lift valve, etc.).
- Restricted producing rate with low flowing bottomhole pressure p_{wf} (plugged perforations, skin effect, sand movement, etc.).
- Excessive water or gas production.
- Restricted injection rate with higher injection pressures, p_{inj} than normal (formation or injection plugging).
- Restricted injection rate with normal injection pressures, p_{inj} (leaking flow lines or bypass, bad injection pumps).
- Failure of the operational personnel to adequately relate production to economics.

6.3 Workover considerations

After the problem is identified and its solution selected, the workover program is established. Each step of the workover must be considered to optimize the overall operation. While each workover may vary with respect to the actual operation (major or routine), they generally have a common order of doing the work to be performed. This order usually follows these steps.
1. Preparation of the well site for the rig, storage of materials and equipment.
2. Selection of the well servicing fluid.
3. Manipulation of downhole safety equipment, well surface equipment, and lease equipment before initiating workover.
4. Killing the well (circulating, reverse or normal, or bullheading, if necessary).

5. Removal of production equipment (rods, pump, wireline equipment, tubing, packer, etc.).
6. Determine the mechanical condition of the casing, liner, open hole. Run gauge ring, casing scraper, CBL, pipe inspection log, production log; remove scale, paraffin, salt, sand.
7. Run equipement for use in workover operation (bridge plug, packer, workover tubing string, cement retainer, test tools, perforating gun, wireline equipment).
   Note: It is important to measure and record all dimensions and tolerances of tools run and especially their fishing necks.
8. Performance of desired workover operation (major or routine).
9. Examine mechanical condition of production equipment to be run back in the hole (steam clean tubing, clean and inspect all downhole equipment that has not been repaired or replaced).
10. Run production equipment, keeping detailed and accurate records of what was run and where, along with its dimensions and materials.
11. Return well to desired status (production or injection).
12. Monitor results, evaluate economics, write a detailed workover history complete with a well diagram, and see that all items are properly filed in the well file.

Don't try harder, work smarter.

REFERENCES

1   Standard API RP 59 (1987) *Recommended Practices for Well Control Operations*. American Petroleum Institute, Washington, DC
2   Institut Français du Pétrole (1970) *Formulaire du producteur*, Chap. VII. Editions Technip, Paris
3   Perrin D, Caron M, Gaillot G (1999) *Well Completion and Servicing*, Chap. 5. Editions Technip, Paris
4   Standard API Spec. 9A (1995) *Specfication for Wire Rope*. American Petroleum Institute, Washington, DC
5   Standard ISO DIS 10432 (1999) *Standard for Subsurface Safety Valves*. Geneva, Switzerland
6   Standard API Spec. 14A (1994) *Specification for Subsurface Safety Valve Equipment* (8th & 9th Ed.). American Petroleum Institute, Washington, DC
7   Standard API RP 14B (1990) *Recommended Practice for Design, Installation, Repair and Operation of Subsurface Safety Valve Systems* (3rd Ed.). American Petroleum Institute, Washington, DC
8   Grace RD (1995) *Advanced Blowout & Well Control*, Chap. 6. Gulf Publishing Company
9   Allen TO, Roberts AP (1994) *Production Operations*, Vol. 1, Well Completions, Workover, and Stimulation. OGCI, Tulsa, OK, 3–14, 3–27
10 Van Dyke K (1997) *A Primer of Oilwell Service, Workover, and Completion*, Chap. 4, Petroleum Extension Service. The University of Texas at Austin
11 Patton LD, Abbott WA (1985) *Well Completions and Workovers: The Systems Approach*, Chap. 21. Energy Publications, Dallas, TX.

V

W

Cased-Hole Logging and Imaging

W

W

Cased-Hole logging and Imaging

Wireline logs are the most reliable source of information for determining the presence of oil and gas. They are applied throughout the life of the well for various geological, geophysical, petrophysical and engineering applications.

Production logging, cement and corrosion evaluation and nuclear measurements made after casing has been set are increasingly used to identify problems and to monitor well performance.

Referring to Fig. W1, different regions of investigation are selected in which tools are specifically designed to operate:

- **Region 1**
  Various formation evaluation services are used to measure parameters (see Paragr. W1).

- **Region 2**
  Cement evaluation logs inspect (see Paragr. W2).

- **Region 3**
  Casing inspection devices operate (see Paragr. W3).

- **Region 4**
  Production logging tools measure the region inside casing (see Paragr. W4).

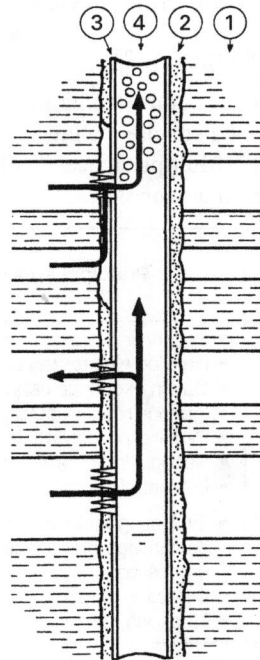

Figure W1 Cased-hole logging environment [37].

W

451

W1 FORMATION EVALUATION [1–5, 33, 37, 38]

Different technologies can be used:
- (a) Natural gamma ray logging
- (b) Pulsed neutron capture logs
- (c) Neutron logging
- (d) Carbon/oxygen logging
- (e) Sonic and acoustic techniques
- (f) Other services.

1.1 Natural gamma ray logging

This tool contains no source and responds only to gamma ray emissions from the downhole environment.

1.1.1 Applications

- Primary depth control log.
- Bed definition.
- Evaluation of shale volume from the gamma ray log.
- Presence of highly radioactive zones.

1.1.2 Tools from Service Companies

Schlumberger [1, 2]

NGS* Natural Gamma Ray Spectrometry Log*. It can be used to help identify clay type and to calculate clay volumes.

MTT* Multiple Tracer Tool*. It can be used to determine the distribution of the materials placed downhole.

Tools	Principal applications	OD (in.)	Maximum pressure (MPa)	Maximum temperature (°C)	Maximum hole size (in.)
NGS	• Reservoir delineation • Detailed well-to-well correlation • Cation exchange capacity studies • Igneous rock recognition • Definition of clay content • Recognition of radioactive minerals • Estimates of potassium and uranium	3 3/8 3 5/8 3 7/8	140 140 140	150 180 150	24 24 24
MTT	• Evaluating hydraulic fractures, squeeze cementing and enhanced oil recovery processes • Detecting zones of lost circulation • Monitoring gravel packing	1 11/16 2	140 140	180 180	

* **Note.** *Placed after the name of a tool, the* * *indicates that it is a trade mark of the Company under which it is listed.*

Baker Atlas [33]

SL* Spectrolog*. It records gamma ray intensity versus borehole depth.
PRISM* Precision Radioactive Isotope Spectral Measurement*. It overcomes the limitations of conventional evaluation methods.

Tools	Principal applications	OD (in.)	Maximum pressure (MPa)	Maximum temperature (°C)	Maximum hole size (in.)
SL	• Define clay content and clay type • Mineral identification • Aid for fracture detection	3.63	140	200 & 260	24
PRISM	• Evaluation of well treatment such as: – hydraulic stimulation – selective cementing – gravel packing	1.69	100	200	9 5/8

Halliburton [38]

CSNG* Compensated Spectral Natural Gamma Log*. It allows to more thoroughly characterize reservoirs and to confidently estimate net pay.
TracerScan* for radioactive surveys.

Tools	Principal applications	OD (in.)	Maximum pressure (MPa)	Maximum temperature (°C)	Maximum hole size (in.)
CSNG	• Characterize reservoirs more thoroughly • Confidently estimate net pay	3 5/8	140	175	20
Tracer-Scan	• Lithology • Fracture evaluation	3 5/8 3 5/8 1 11/16	70 140 140	135 204 175	20 20 20.75

1.2 Pulsed neutron capture logs

They are often run without shutting in production from the well being logged. These tools do not contain chemical neutron sources. Neutrons are generated by the electrical neutron generator named "minitron".

1.2.1 Applications

• Measurement of water saturation, porosity, and presence of gas in the formation.
• Location of gas-oil contact and water-oil contact.

- Correlation with open hole resistivity logs.
- Shale indicator.
- Evaluate changes in saturation due to zones watering out.
- Measure residual oil saturation (ROS).
- Locate and select zones for recompletion.

1.2.2 Tools from Service Companies

Schlumberger [1]

TDT-K*, **TDT-M***, **Dual-Burst* TDT (TDT-P*)** Thermal Decay Time tool*. They can determine both formation sigma and borehole sigma.

RST-A*, **RST-B*** Reservoir Saturation Tool*. They use multiple detectors to separate the signal contributions from the borehole and the formation.

Tools	Principal applications	OD (in.)	Maximum pressure (MPa)	Maximum tempera- ture (°C)	Maximum hole size (in.)
TDT-K TDT-M TDT-P	• Formation sigma • Borehole sigma • Gas detection and discrimination • Porosity analysis • Reservoir and flood monitoring • Through-drillpipe formation evaluation	1 11/16	120	165	
RST	• Same applications as the TDT tools, plus • Oil saturation determination in unknown or fresh formation water (this is only for inelastic mode) • Borehole oil fraction • Sigma measurement	1 11/16 2 1/2	105 105	150 150	7 9 5/8

Baker Atlas [33]

PDK-100* Pulsed Decay (DK)* with 100 counting gates. It measures the macroscopic absorption cross, section, sigma (Σ), of the bulk formation.

W

Tools	Principal applications	OD (in.)	Maximum pressure (MPa)	Maximum tempera- ture (°C)	Maximum hole size (in.)
PDK-100	• Locate gas, oil, and water contacts • Correlate with open and cased-hole logs	1.69	110	180	9 5/8

Halliburton [38]

TMD-L* Thermal Multigate Decay-Lithology tool*. It provides data that helps determine through-casing water saturation estimates, lithology and porosity.

Tools	Principal applications	OD (in.)	Maximum pressure (MPa)	Maximum tempera-ture (°C)	Maximum hole size (in.)
TMD-L	• Reservoir monitoring • Lithology determination • Enhanced oil recovery monitoring • Gas vs. tight determination • Spectral water flow detection	2.125	110	165	9 5/8

1.3 Neutron logging

Neutrons logs are used to evaluate formation porosity, detect gas, and as correlation log between open and cased hole when the gamma ray lacks character. The neutron logs are primarily responsive to the hydrogen density of the pore fluid downhole. Porosity is calculated based on the hydrogen index [37].

1.3.1 Applications

• Evaluation of formation porosity.

• Detection of gas.

1.3.2 Tools from Service Companies

Schlumberger [1]

CNL* Compensated Neutron Log*. It provides lithology and porosity analyses in one trip in the well.
APS* Accelerator Porosity Sonde*.

Tools	Principal applications	OD (in.)	Maximum pressure (MPa)	Maximum tempera-ture (°C)	Maximum hole size (in.)
CNL	• Porosity (thermal neutron porosity) • Clay analysis • Gas detection	2 3/4 3 3/8	175 140	260 200	10 24
APS	• Porosity (epithermal neutron porosity) • Clay • Gas • Saturation by sigma measurement	3 5/8	140	175	22

Baker Atlas [33]

CN * Compensated Neutron Log*. It makes the instrument very desirable in rough or washed out boreholes.

NEU Neutron Log*. It helps distinguish fluids from rock matrix.

Tools	Principal applications	OD (in.)	Maximum pressure (MPa)	Maximum tempera-ture (°C)	Maximum hole size (in.)
CN	• Porosity determination • Locate gas • Identify lithology	2.75 3.63	170 140	200 230	24
NEU	• Indicates porosity • Identify gas zones • Locate fluid contacts	1.38 1.69 3.00 3.63	140 120 170 140	200 260 260 200	10

Halliburton [38]

DSN * Dual Spaced Neutron*.

Tools	Principal applications	OD (in.)	Maximum pressure (MPa)	Maximum tempera-ture (°C)
DSN	• Porosity • Lithology • Production analysis	3 5/8	140	175

1.4 Carbon/oxygen logging

This logging has application in determining the presence of water and oil and their saturations behind casing in formations whose water is fresh or of unknown salinity.

1.4.1 Applications [37]

- Discriminate water/oil contact when salinity of formation water is low or known.
- Evaluate hydrocarbon zones and saturations in fresh, mixed, or unknown water salinity environments.
- Locate water and oil zones in waterfloods where mixed salinities exist between formation and flood waters.

W

- Evaluate saturations in formations behind casings when open hole logs are not available.
- Monitoring of steam and CO_2 flood fronts/breakthrough.

1.4.2 Tools from Service Companies

Schlumberger [1]

GST* Gamma Spectrometry Tool*. It provides information for complete reservoir analysis through casing.

RST* Reservoir Saturation Tool*. It uses multiple detectors to separate the signal contributions from the borehole and the formation.

Tools	Principal applications	OD (in.)	Maximum pressure (MPa)	Maximum temperature (°C)	Maximum hole size (in.)
GST	• Lithology • Porosity • Oil saturation by C/O measurement	3 5/8	140	150	
RST	• Oil saturation determination in unknown or fresh formation water by C/O measurement • Borehole oil fraction • Water flow log • Sigma measurement • Lithology (spectrolith)	1 11/16 2 1/2	105 105	150 150	7 9 5/8

Baker Atlas [33]

RPM* Reservoir Performance Monitor*. It is a pulse neutron tool measuring capture and inelastic gamma ray induced by pulsed neutron, water saturation in unknown water salinity is determined through casing in addition the tool also detect water flow behind casing.

Tools	Principal applications	OD (in.)	Maximum pressure (MPa)	Maximum temperature (°C)	Maximum hole size (in.)
RPM	• Location of hydrocarbons in fresh, brackish, or unknown formation water salinities • Pre-abandonment exploration for bypassed hydrocarbons • Monitor production and reservoir depletion • Monitor enhanced oil recovery projects • Lithology identification and porosity indicator	1 11/16	140	175	9 5/8

W

Halliburton [38]

PSG[*] Pulsed Spectral Gamma Log[*]. It provides information for improved through-casing oil-saturation estimates when formation waters are of low or unknown salinity.
TMD-L[*] Thermal Multigate Decay-Lithology tool[*]. It provides data that helps determine through-casing water saturation estimates.
RMT[*] Reservoir Monitoring Tool[*]. It is a small diameter reservoir monitoring tool delivers high-performance, through-tubing capacity.

Tools	Principal applications	OD (in.)	Maximum pressure (MPa)	Maximum temperature (°C)	Maximum hole size (in.)
PSG	• Reservoir evaluation through casing • Waterflood monitoring	3 3/8	100	150	10
TMD-L	• Reservoir monitoring • Lithology and porosity determination • Enhanced oil recovery monitoring • Gas vs. tight determination • Spectral water flow detection	1 11/16	100	160	16
RMT	• More than double the logging speed • Detailed analysis of the reservoir through a comprehensive reservoir monitoring program • Operating inelastic and capture modes	2 1/8	100	160	9 5/8

1.5 Sonic and acoustic techniques

There are three main types of waves in sonic logging: the compressional, shear, and Stoneley. Acoustic waves propagate from the transmitter to receiver by being refracted at the formation interface in a direction parallel to the wellbore [37].

1.5.1 Applications

Sonic measurements

• Velocity between two receivers.
• Porosity.
• Elastic properties.

Acoustic wave logging

• Early open hole–cased hole comparison.
• Effects of cement coverage.
• Detection of gas in high porosity sands.

1.5.2 Tools from Services Companies

Schlumberger [1]

Array-Sonic* log (monopole only). It digitizes the entire acoustic waveform and discriminates compressional, shear and Stoneley waves.

DSI* Dipole Shear-Sonic Imager*. It combines dipole and monopole technologies to record enhanced measurements of compressional, shear and Stoneley wave-forms in all formations.

Tools	Principal applications	OD (in.)	Maximum pressure (MPa)	Maximum temperature (°C)	Maximum hole size (in.)
Array-Sonic log	• Porosity evaluation • Shear seismic correlation • Thin-bed detection • Fracture prediction and detection • Sand strength analysis • Cement bond logging • Through-casing sonic logging	3 3/8	140	175	18
DSI	• Same applications, plus • Synthetic seismograms • Prediction of rock strength • Porosity and lithology in slow formations • Generation of acoustic hydrocarbon indicators	3 5/8	140	175	17 1/2

Baker Atlas [33]

MAC* Multipole Array Acoustilog*. It determines P- and S-wave velocities in soft formations and measures azimuthal anisotropy.

XMAC* Cross-Multipole Array Acoustilog*. It determines P- and S-wave velocities in soft formations and measures azimuthal anisotropy using the latest acoustilog technology.

Tools	Principal applications	OD (in.)	Maximum pressure (MPa)	Maximum temperature (°C)	Maximum hole size (in.)
MAC	• Synthetic seismograms • Prediction of rock strength • Porosity and lithology in slow formations • Generation of acoustic hydrocarbon	3.63	140	200	17 1/2
XMAC	• Cross-dipole azimuthal anisotropy analysis • Rock mechanical properties • Stoneley-wave permeability analysis • Seismic and log time-depth correlation	3.88	140	200	21

W

Halliburton [38]

FWS[*] Full wave sonic[*]. Using a piezoelectric transmitter and four long-spaced receivers, the FWS system records the entire acoustic wavetrain.

XACT[*] Multipole Acoustic Logging Service[*]. It provides deep-reading full waveform acoustic data in a any downhole environment for enhanced formation evaluation.

Tools	Principal applications	OD (in.)	Maximum pressure (MPa)	Maximum tempera- ture (°C)	Maximum hole size (in.)
FWS	• Porosity evaluation • Shear seismic correlation • Thin-bed detection • Fracture prediction and detection • Sand strength analysis • Cement bond logging • Through-casing sonic logging	3 5/8	140	175	
XACT	• Porosity • Lithology • Fracture evaluation	3 3/8	140	175	18

W2   CEMENT EVALUATION [1–3, 16–19, 33, 38]

2.1   Acoustic cement evaluation surveys

Acoustic cement bond logs measure the loss of acoustic energy as it propagates through casing.

The conventional cement bond log (CBL) is the most common bond log.

2.1.1   CBL-VDL measurement [2]

The cement bond log (CBL), later combined with the variable density log (VDL) waveform, has been for many years the primary way to evaluate cement quality. The principle of the measurement is to record the transit time and attenuation of a 20 kHz acoustic wave after propagation through the borehole fluid and the casing well.

The attenuation rate depends on the cement compressive strength, the casing diameter, the pipe thickness, and the percentage of bonded circumference (Fig. W2). The scheme of figure W2 is only applicable to CBL.

The longer 5-ft spacing is used to record the VDL waveform for better discrimination between casing and formation arrivals. The VDL is generally used to assess the cement to formation bond and helps to detect the presence of channels and the intrusion of gas.

W

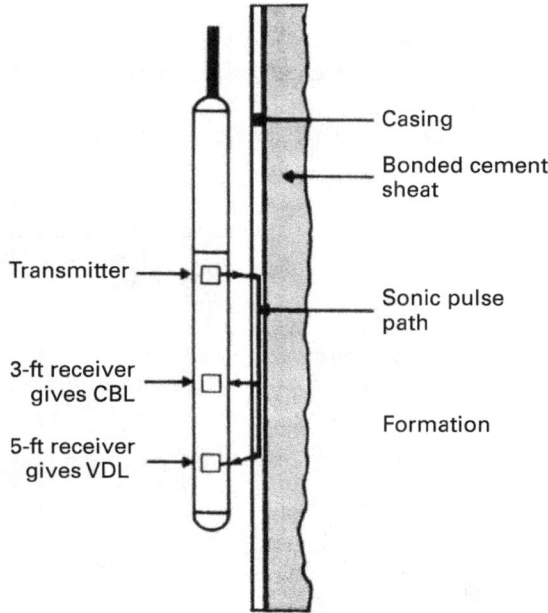

Figure W2 CBL-VDL measurement [2].

2.1.2 Tools from Service Companies

Schlumberger [2, 16]

CBL[*] Cement Bond Log[*]
CBT[*] Cement Bond Tool[*]. It provides a precise axial measurement of cement-to-casing and cement-to-formation bond using high-frequency sound pulses.

Tools	Principal applications	OD (in.)	Maximum pressure (MPa)	Maximum tempera-ture (°C)
CBL	• Evaluation of cement seal in casing-formaton annulus • Location of cement top • Determination of cement quality	3 5/8	140	175
CBT	• Evaluation of cement seal in casing-formaton annulus • Location of cement top • Determination of cement quality	2 3/4	140	175

461

Baker Atlas [33]

BAL* Bond Attenuation Log*. It is a borehole-compensated measurement system which provides direct recording of signal attenuation, downhole calibration, and conventional bond logging measurements.

SBT* Segmented Bond Tool*. It quantitatively measures the cement bond integrity in six angular segments around the casing.

Tools	Principal applications	OD (in.)	Maximum pressure (MPa)	Maximum tempera- ture (°C)	Maximum hole size (in.)
BAL	• Evaluate the effectiveness of the cement bond in the casing-formation annulus by direct attenuation measurements • Locate areas for investigation with tracer surveys and Sonan log for determining channeling in annulus cement • Determine annulus cement strength • Detect presence of microannulus	2 3/4	140	180	9 5/8
SBT	• Provide quantitative analysis of cement bond to the casing in 60° segments around the borehole • Provide 360° evaluation of cement bonding • Identify intervals of uniform bonding and detect cement channels or voids • Provide accurate high-velocity (fast) formation cement evaluation • Orient cement voids or channeling in reference to the low side of the wellbore • Provide qualitative analysis of cement bond to formation.	3 3/8	140	180	16

Halliburton

CBL* Cement Bond Log*.

Tools	Principal applications	OD (in.)	Maximum pressure (MPa)	Maximum tempera- ture (°C)
CBL	• Evaluation of cement seal in casing-formaton annulus • Location of cement top • Determination of cement quality	2 3/4	140	175

Lee Tool/Schlumberger [19]

CMT* Cement Mapping Tool*. It provides a radial map image from 10 evenly-spaced transmitter/receiver pairs.

SCMT* Slim Cement Mapping Tool*. It provides a radial map image from 8 evenly-spaced transmitter/receiver pairs. This tool is a slim (1 11/16 in.) version of the CMT.

Tools	Principal applications	OD (in.)	Maximum pressure (MPa)	Maximum tempera- ture (°C)
CMT	• Cement presence • Integrity of cement	2 7/8 (CMT)	70	150
SCMT	• Identification of cement channeling • Location of areas of stuck casing • Fast formation detection • Logging in horizontal wells; strong centralizers and a light tool optimize response	1 11/16 (SCMT)	70	150

2.2 Ultrasonic pulse echo cement evaluation

This technique provides an estimate of cement coverage and a direct measurement of the cement compressive strength.

2.2.1 Applications [37]

• Directly measure and map the cement distribution around pipe.
• Determine acoustic impedance and possibly compressive strength of annular materials.
• Discriminate gas from liquid behind pipe.
• Assess the presence of gas cut cement.
• Provide an acoustic measurement of the pipe ID, inner wall surface roughness and wall thickness, as well as ovality and tool centralization.

2.2.2 Tools from Service Companies

Schlumberger [1, 17, 18]

CET* Cement Evalution Tool*. It evaluates the quality of cementation in eight directions, 45° apart, with a very fine vertical resolution.

USI* Ultrasonic Imager*. It scans the entire circumference of the wellbore to evaluate the quality of the cement job and determine both internal and external casing corrosion.

Tools	Principal applications	OD (in.)	Maximum pressure (MPa)	Maximum tempera-ture (°C)	Maximum hole size (in.)
CET	• Determination of annulus-cement quality • Identification of channels in cement sheath • Location of cement top • Casing evaluation • Determination of zone isolation.	3 3/8 4	140 140	175 175	7 13 3/8
USI	• Same applications as the CET, plus • Corrosion detection and monitoring (casing ID and wall thickness) • 360° coverage.	3 3/8	140	175	13 3/8

Halliburton [38]

PET* Pulse Echo Tool*. It evaluates the quality of cementation in eight directions, 45° apart, with a very fine vertical resolution.

CAST-V* Circumferential Acoustic Scanning Tool-Visualization*.

Tools	Principal applications	OD (in.)	Maximum pressure (MPa)	Maximum tempera-ture (°C)	Maximum hole size (in.)
PET	• Determination of annulus-cement quality • Identification of channels in cement sheath • Location of cement top • Casing evaluation • Determination of zone isolation.	3 3/8	140	160	12 1/4
CAST-V	• Cement evaluation • Corrosion detection and monitoring	3 5/8	140	175	13 3/8

2.3 Cement-quality logging guidelines [29]

The following guidelines summarize factors that should be considered when cement-quality logs are run and interpreted.

2.3.1 Cement-bond logs

1. Proper centralization is critical in obtaining a useful cement-bond log. In a vertical hole, use at least three strong centralizers, with centralizers directly above and below the transmitter-receiver and one centralizer on the top of the tool string. In deviated wells, pay even more attention to centralization.

W

464

2. Make repeat runs through at least part of the logged section to check for proper tool operation and centralization.

3. Always record a full wave-train display, either a variable density log or an *x-y* (signature) log, because it provides the only information about bonding to the formation and aids in the interpretation of the amplitude log.

4. Consider quantitative interpretation of the amplitude log to be approximate because it is based on empirical correlations and is affected by many factors (e.g., microannulus, eccentering, and fast formations). The amplitude log is best used in conjunction with a full wave-train display.

5. Repeat a cement-bond log with pressure applied at the wellhead when intermediate bonding is indicated to distinguish between a microannulus and poor cement hydraulic integrity.

6. Run the cement-bond log through the cement top whenever possible. A section of log in free pipe allows one to check tool calibration and centralization.

7. Detection of shear waves in the full wave-train display is generally a reliable indication of good bonding to the pipe and to the formation.

8. A transit-time measurement is essential in detecting eccentering and the presence of fast formations.

9. A cement-bond log can clearly indicate regions of free pipe and of very good bonding. Even in these cases, and certainly when intermediate bonding conditions are found, the cement-bond log does not provide conclusive evidence of the ability of the cement to provide hydraulic integrity.

10. An attenuation-ratio log is often preferable to a cement-bond log amplitude measurement because of its lower sensitivity to calibration errors and eccentering. Use the attenuation-ratio log in wells where centralization is difficult.

2.3.2 Ultrasonic-pulse-echo logs

1. Ultrasonic-pulse-echo logs provide an excellent acoustic-caliper log, with a measure of the inside and outside diameters of the casing in some cases.

2. The ultrasonic-pulse-echo log can clearly identify partial contact between the pipe and the cement because of the circumferential picture of cement bonding provided. A channel between the cement and the formation is not normally detectable.

3. The ultrasonic-pulse-echo log has relatively low sensitivity to bonding to the formation.

4. An ultrasonic-pulse-echo log is less sensitive to a liquid-filled microannulus than is a cement-bond log.

5. The presence of gas in a microannulus causes apparently poor bonding on an ultrasonic-pulse-echo log, even though cement integrity may be good.

6. The ultrasonic-pulse-echo is particularly useful when intermediate bonding is indicated by a cement bond because the log can determine whether partial contact is the cause of the intermediate cement-bond log response.

W3 DOWNHOLE CASING INSPECTION [1, 20, 33, 34, 37, 38]

Casing inspection measurements are used to examine the casing or tubing in a well. Vastly differing technologies are used. They are:

- **Mechanical calipers.**
- **Acoustic tools:** they assess the pipe ID, surface roughness, and wall thickness.
- **Electromagnetic tools:** they examine and discriminate the inner from outer pipe surface and may shed light on concentric casing strings.
- **Casing potential profiles:** they foresee electrochemical corrosion and are the basis for cathodic protection.
- **Borehole video cameras:** they provide an actual view of damage or lost equipment downhole. It is developed in Paragr. W6.

3.1 Mechanical calipers

3.1.1 Application

- Examination of the inner pipe surface.

3.1.2 Tools from Service Companies

Schlumberger [1, 20]

MFCT[*] Multifinger Caliper Tool[*]. It is a mechanical casing inspection device using from 36 to 72 fingers.
TGS[*] Tubing Geometry Sonde[*]. It uses a 16-finger caliper to precisely measure variations in the internal diameter of the tubing.

Tools	Principal applications	OD (in.)	Maximum pressure (MPa)	Maximum temperature (°C)	Maximum hole size (in.)
MFCT	• Casing corrosion analysis • Corrosion detection in air or gas drilled holes	3.55 5.4 9.1	140 140 140	175 175 175	7 9 5/8 13 3/8
TGS	• Tubing and through-tubing measurements of corrosion • Corrosion detection in air or gas drilled holes	2 1/8	70	150	13 3/8

Baker Atlas [33]

MFC* Multi-Finger Caliper*. This instrument mechanically measures the inside diameter of downhole tubular goods with two channels of information recorded at the surface.

Tools	Principal applications	OD (in.)	Maximum pressure (MPa)	Maximum tempera- ture (°C)	Maximum hole size (in.)
MFC	• Identify internal metal loss from corrosion, erosion, or mechanical wear • Measure the minimum and maximum internal tubing or casing diameter	3 3/8 5	140 140	180 180	7 9 5/8

Sondex/Halliburton [22, 38]

MIT* Multifinger Imaging System*. A family of five sizes of tools, each with an independant measurement for each finger, covers tubings and casings from 1 3/4 to 13 3/8 inches.

Tools	Principal applications	OD (in.)	Maximum pressure (MPa)	Maximum tempera- ture (°C)	Maximum hole size (in.)
MIT	• Corrosion assessment • Casing wear • Casing deformation • Scale evaluation • Location of holes, cracks and splits.	1 11/16 2 1/8 2 3/4 4 8	100 100 100 100 100	150 150 150 150 150	4 1/2 5 1/2 7 1/2 9 5/8 14

3.2 Acoustic casing inspection tools

The acoustic tools have been introduced in Paragraph 2.2 on ultrasonic pulse echo technology. The tools are:
- The fixed transducer **CET/PET** type.
- The rotating transducer variety **USI** type.
- The new focused transducer **UCI** (Schlumberger), specially designed for casing inspection [41]. It has a better resolution for smaller defects (pressure, temperature) by the focused transducer.

3.3 Electromagnetic casing inspection tools

There are two basic types of electromagnetic casing inspection tools available:
(a) The pad type device
(b) The phase shift type device.

W

467

3.3.1 Applications

- The pad type device detects defects anywhere within the wall, it is hereinafter referred to as the "total wall" test.
- The phase shift type device is used to measure the amount of metal remaining in a casing string. It is best suited for large scale corrosion, vertical splits, holes larger than 2 in. (5 cm) diameter.

3.3.2 Tools from Service Companies for pad type device

Schlumberger [1, 20]

PAL[*] Pipe Analysis Tool (Log)[*]. This tool monitors casing quality and discriminates between internal and external defects.

Tools	Principal applications	OD (in.)	Maximum pressure (MPa)	Maximum temperature (°C)	Maximum hole size (in.)
PAL	• Identification of corrosion damage • Monitor of anticorrosion systems • Evaluation of corrosion progress through periodic logs • Prediction of casing life	4 1/2	140	175	13 3/8

Baker Atlas [33]

DVRT[*] Digital Vertilog[*]. It uses DC flux leakage (FL) measurements to determine the depth of penetration of casing defects.

Tools	Principal applications	OD (in.)	Maximum pressure (MPa)	Maximum temperature (°C)	Maximum hole size (in.)
DVRT	• Detect corrosion and depth of penetration; a base log and periodic surveys can help determine progress of corrosion • Check casing makeup and joint lengths, locate well completion equipment, and detect centralization of primary casing string at the bottom of the next casing string.	3.78 to 7.37	82	140	22

Halliburton [38]

PIT* Pipe Inspection Tool*. PIT* is a cost-effective system for performing these important procedures:

- Determining the need for pipe repair services (such as scrab liners), remedial cementing, or cathodic protection
- Evaluating the economic value of pipe when plug and abandonment or reentry is necessary
- Determining the exact location of perforations or leaks
- Periodic monitoring of the pipe condition in gas storage or injection wells
- Determining the pressure limits for well service operations.

3.3.3 Tools from Service Companies for phase shift type device

Schlumberger [1, 20]

METT* Multifrequency Electromagnetic Thickness Tool*. It uses nondestructive, noncontact induction methods to detect metal loss and changes in casing geometry, regardless of casing fluid type.

Tools	Principal applications	OD (in.)	Maximum pressure (MPa)	Maximum tempera-ture (°C)	Maximum hole size (in.)
METT	• Location of corrosion damage in single or multiple strings • Estimation of life remaining in casing • Evaluation of anticorrosion systems	2 3/4 9 5/8	140 140	175 175	9 5/8 13 3/8

Baker Atlas [33]

DMAG* Digital Magnelog*. It is an electromagnetic multifrequency, multispacing casing inspection service used to detect wall thickness changes in single or multiple casing strings.

Tools	Principal applications	OD (in.)	Maximum pressure (MPa)	Maximum tempera-ture (°C)	Maximum hole size (in.)
DMAG	• Determine joints of casing having different weights or wall thickness • Locate holes greater than 2 in. in diameter	3.63 to 8	140	180	9 5/8 to 13 5/8

W

Halliburton [38]

METG* Multifrequency Electromagnetic Thickness Gauge*.

Tools	Principal applications	OD (in.)	Maximum pressure (MPa)	Maximum tempera- ture (°C)	Maximum hole size (in.)
METG	• Location of corrosion damage in single or multiple strings • Estimation of life remaining in casing • Evaluation of anticorrosion systems	3 3/8 5 1/2	100 100	175 175	7 13 3/8

3.4 Casing potential surveys for cathodic protection

Electrochemical corrosion is a primary cause of casing damage.

3.4.1 Applications

- Location where electrochemical corrosion is occuring in the casing string.
- Cathodic protection of casing.

3.4.2 Tools from Services Companies

Schlumberger [1, 20]

CPET* Corrosion and Protection Evaluation Tool*. It measures casing potential and resistance in cathodically protected wells to evaluate the protection and determine the extent of any corrosion.

Tools	Principal applications	OD (in.)	Maximum pressure (MPa)	Maximum tempera- ture (°C)	Maximum hole size (in.)
CPET	• Evaluate the quality of protection and extent of corrosion in cathodically pro- tected wells • Give rate and location of external casing corrosion in nonprotected wells (current turned off)	3 3/8	140	175	22

W

Baker Atlas [33]

CPP* Casing Potential Profile. It is a wireline service used for determining the existence of active galvanic corrosion.

Tools	Principal applications	OD (in.)	Maximum pressure (MPa)	Maximum temperature (°C)	Maximum hole size (in.)
CPP	• Forecast corrosive damage and determine the depth levels and probable rates • Determine if casing leaks are caused by galvanic action	3.80	70	180	20

W4 PRODUCTION LOGGING [1–3, 11–15, 30, 33, 37]

Production logging services provide information that is vital to the successful completion and optimum recovery of any reservoir.

Measurements that relate to the nature and behavior of fluids and the integrity of the gravel pack and hardware assemblies give the operator pertinent real-time data on producing and injection wells. Many of the sensors can be combined into one tool for simultaneous recording of fluid and pressure data that relate directly to the volume and nature of production at each point of entry downhole. The sensors now included are [2]:

- Temperature control
- Noise logging
- Radioactive tracer logging
- Oxygen activation and other pulsed neutron applications
- Bulk flow rate measurement
- Fluid identification and multiphase flow
- Production logging in horizontal wells.

4.1 Simultaneous production tools

W

Many of these sensors can be combined into one tool and recorded simultaneously to measure fluid entries and exits, standing liquid levels, bottomhole flowing and shut-in pressures, pressure losses in the tubing and the integrity of the gravel pack and hardware assemblies. Since the measurements are made simultaneously, their correlation is unaffected by any well instability that might cause downhole conditions to vary over a period of time.

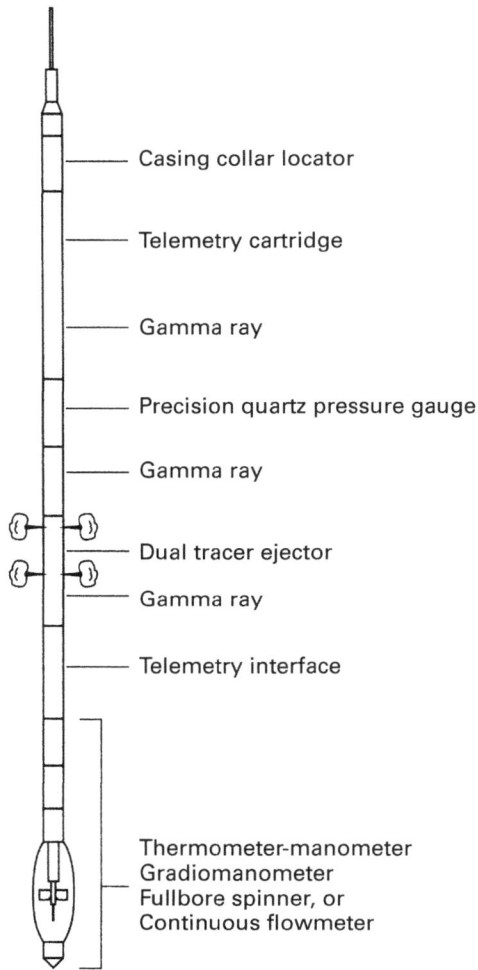

Figure W3 Simultaneous production tool [2].

Figure W3 shows a schematic of the sensors in a typical production logging tool string [1, 2].

4.1.1 Applications

- Evaluation of well performance with respect to the reservoir.
- Analysis of mechanical problems.

4.1.2 Tools from Service Companies

Schlumberger [1]

PLT* Production Logging Tool*. It provides simultaneous measurement from downhole sensors used for the analysis of producing or injection wells.

CPLT* Combinable Production Logging Tool*. It acquires large amounts of production logging data in one trip in the well.

Tools	Principal applications	OD (in.)	Maximum pressure (MPa)	Maximum tempera- ture (°C)	Maximum hole size (in.)
PLT	Measuring of: • Fluid entries and exits • Standing fluid levels • Bottomhole flowing and shut-in pressures • Pressure losses in the tubing and the integrity of the gravel pack and hardware assemblies	1 11/16	140	180	18
CPLT	• Well production and injection monitoring • Production and injection problem diagnosis • Well testing	1 11/16	140	180	

Baker Atlas [33]

SPL* Simultaneous Production Logging*. It allows simultaneous measurement from a large combination of sensors on a single run in producing or injection wells in single, two or three phase flow regimes. It can be run on memory mode or real time surface read-out with telemetry in vertical, deviated or horizontal wells (using coiled tubing conveyance or well tractor).

Tools	Principal applications	OD (in.)	Maximum pressure (MPa)	Maximum tempera- ture (°C)	Maximum hole size (in.)
SPL	• Flow profile determination • Detection of waterhold • Thief zone detection • Flowing and downhole shut-in pressure	1.69 2.125	100 100	175 175	18

W

Halliburton

PLT Production Logging Tool. It provides simultaneous measurement from downhole sensors used for the analysis of producing or injection wells.

Tools	Principal applications	OD (in.)	Maximum pressure (MPa)	Maximum tempera- ture (°C)	Maximum hole size (in.)
PLT	Measuring of: • Fluid entries and exits • Standing fluid levels • Bottomhole flowing and shut-in pressures • Pressure losses in the tubing and the integrity of the gravel pack and hardware assemblies	1 11/16	70	175	18

4.2 New generation production logging

Compact production services acquire cross-sectional flow measurements from collocated sensors. They give a better understanding of the downhole flow regime. An example is given with the basic PS Platform of Schlumberger (Fig. W4).

4.2.1 Applications

*   Maximization of reservoir performance.
*   Assistance to the efficiency of production and injection monitoring operations.

4.2.2 Tools from Services Companies

Schlumberger [8, 12]

PS Platform[*].

Tools	Principal applications	OD (in.)	Maximum pressure (MPa)	Maximum tempera- ture (°C)
PS Platform	• Two-and three-phase flow profiles with redundant hydrocarbon flow estimate • Production logging in horizontal or devi- ated wells • Remedial operation using identification of water and hydrocarbon entry points • Logging in wells with limited surface height • Electric probe measurement • Advanced well testing	1 11/16 no rollers 2 1/8 rollers	100 100	150 150

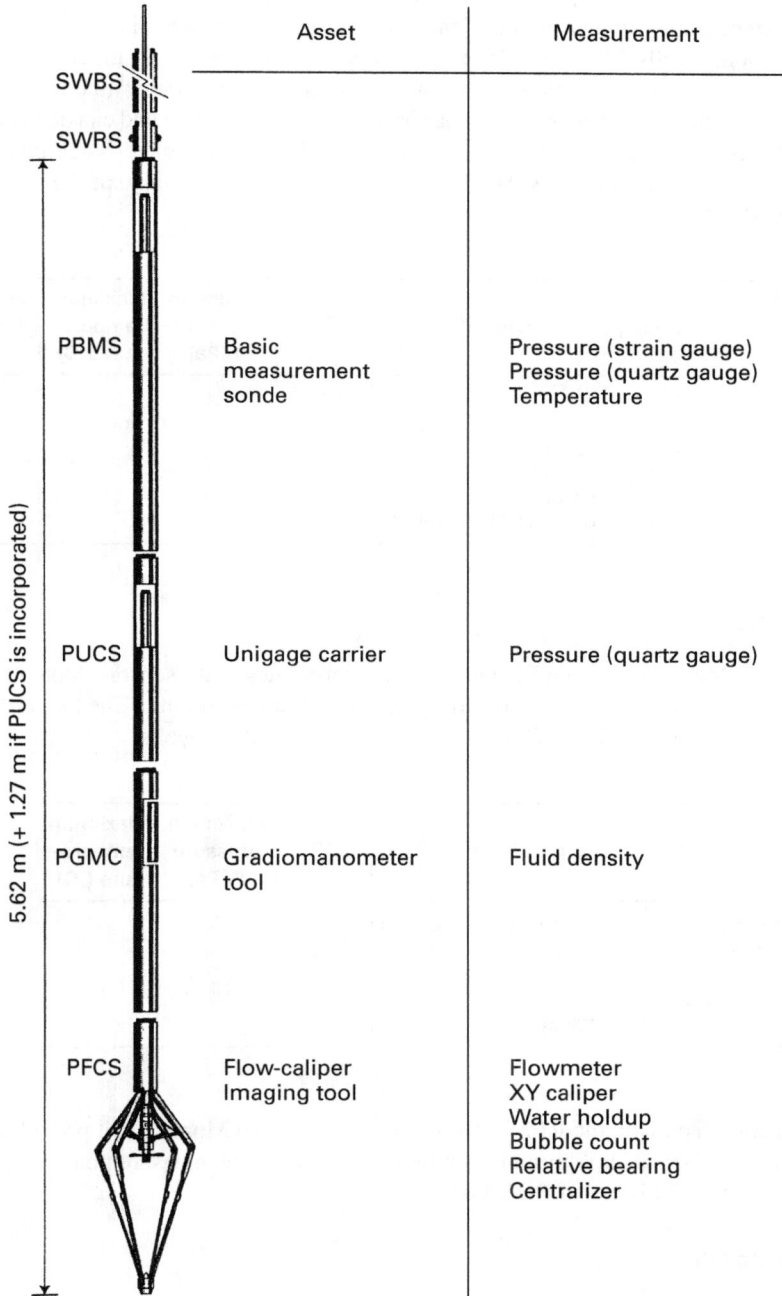

	Asset	Measurement
SWBS		
SWRS		
PBMS	Basic measurement sonde	Pressure (strain gauge) Pressure (quartz gauge) Temperature
PUCS	Unigage carrier	Pressure (quartz gauge)
PGMC	Gradiomanometer tool	Fluid density
PFCS	Flow-caliper Imaging tool	Flowmeter XY caliper Water holdup Bubble count Relative bearing Centralizer

5.62 m (+ 1.27 m if PUCS is incorporated)

Figure W4 The basic PS platform [8].

W

FloView[*] holdup measurement tool facilitates understanding the nature of wellbore flow and significantly improves confidence in the interpretation and diagnosis of problems in all producing wells. There are 4 probe sensors protected inside a caliper measurement section of the tool. The probes use a resistivity measurement to determine the difference between hydrocarbon and water. The measurement is very sensitive and can detect the very oil bubbles of production in a full column of water. The measurement is accurate in high angle wells and horizontal wells. With computer technology, an image of the flow profile is generated at the surface.

Tools	Principal applications	OD (in.)	Maximum pressure (MPa)	Maximum tempera-ture (°C)	Maximum hole size (in.)
FloView	• Accurate water holdup • First oil-entry detection • First water-entry detection for water • Shutoff (at low water cut) • Caliper measurement • Fresh and injected water differentiation	1 11/16	100	150	9

Baker Atlas

MCFM[*] Multicapacitance Flowmeter[*]. It incorporates multi sensors deployed across the borehole for the determination of flow phase in multiphase regime. The tool is designed for horizontal or deviated wells but can also be used in vertical wells.

Tools	Principal applications	OD (in.)	Maximum pressure (MPa)	Maximum tempera-ture (°C)
MCFM	• 3-phase flow rates and 3-phase holdups • Velocity profile • Pressure • Temperature • Caliper measurements	1.7 2.125	103	177

When combined with the **RPM**[*] (Reservoir Performance Monitor[*]), it provides formation saturation, 3-phase holdups and behind casing water flow measurements. The combination RPM-MCFM is called **POLARIS**.

Halliburton [38]

Flow 2000[*] It brings the ultimate combination of technology, efficiency, and value to the wellsite. Integrated Sondex/Halliburton sensors combine reliability with the best sensor technology.

Tools	Principal applications	OD (in.)	Maximum pressure (MPa)	Maximum tempera- ture (°C)
Flow 2000	• Provides fast, high-resolution transient well testing with pressure derivative in real time • Identifies 3-phase flow with low water cuts • Provides fullbore gas holdup • Differential pressure tool • Measures mono- and multiphase flow • Provides strong centralization in vertical, deviated, and horizontal wells	1.688	100	175

4.2.3 Guidelines for running and interpreting spinner flowmeters [29]

A properly run and interpreted spinner-flowmeter log should provide a reliable flow profile for a single-phase well. The following guidelines point out procedures that should be followed to ensure log quality.

Running spinner flowmeters

1. Well conditions must be suitable for a spinner flowmeter. As a minimum, the well fluids should be clean (no entrained solids), and the rate should be as stable as possible.
2. Multiphase-flow effects often render a spinner flowmeter useless. Unless the multiphase-stream flow rate is large, other devices are needed for velocity measurements.
3. A spinner flowmeter should be checked thoroughly on the surface before it is run in the well. The impeller should rotate freely and all electronics should operate as expected.
4. A spinner flowmeter should always be run centralized.
5. Multiple passes, at several different cable speeds and in both up and down directions, should be made across all zones of interest.
6. Stationery readings should be taken at several wellbore locations.
7. A caliper log is needed if the wellbore cross-sectional area is not constant.
8. Repeat runs should always be made to assess well stability and tool performance.

Interpreting spinner-flowmeter logs

1. The multipass method of interpretation should be used whenever significant lengths of wellbore exist without fluid exits.
2. The two-pass interpretation is useful for visually accentuating the spinner-flowmeter response. The multipass method should be applied to confirm two-pass interpretations.
3. Schlumberger's latest interpretation technique **SPRInt** (Single Pass Rate Interpretation) uses PS platform data and requires only two passes in most of the cases (one pass plus verification pass).

W

4. The volumetric flow rate above all zones taking or producing fluid should be calculated from the log and compared with surface flow conditions.

5. Spinner-response characteristics obtained the log interpretation, such as the response slope and threshold velocity, should be close to that predicted by the tool supplier.

6. Stationary measurements are useful, particularly in open-hole completions, where fluid entries may be scattered over the entire completion.

4.3 Water-flow logging

Water-flow logging enables the evaluation of the amount of water flowing in the borehole.

4.3.1 Applications

* Used in producing or water injection wells.
* Accurately detects and quantifies downhole water flow to enhance the planning and improvement of conformance and water management.

4.3.2 Tools from Service Companies

Schlumberger [1, 13, 14]

WFL[*] Water Flow Log[*]. It uses a new oxygen-activation technique to evaluate both upward and downward water flow behind pipe or inside casing.

Tools	Principal applications	OD (in.)	Maximum pressure (MPa)	Maximum tempera- ture (°C)
WFL	• Detection of water flow behind pipe • Water flow inside pipe	1 11/16	120	160

Baker Atlas [33]

WHI[*] Water Holdup Indicator Log[*].

Tools	Principal applications	OD (in.)	Maximum pressure (MPa)	Maximum tempera- ture (°C)	Maximum hole size (in.)
WHI	• Locate water entry into hydrocarbon flow • Locate hydrocarbon/water interface in shut-in well conditions	1.69	100	175	9 5/8

W

Halliburton [38]

SpFL* Spectra Flow Logging Service*. New standard in water-flow technology.

Tools	Principal applications	OD (in.)	Maximum pressure (MPa)	Maximum temperature (°C)	Maximum hole size (in.)
SpFL	• Accurately identify the water source • Evaluate downhole flow patterns • Reduce water disposal costs • Comprehensive water-flow analysis • Precise water velocity measurements	1.69	100	160	16

4.4 Gravel-pack logging

The gravel-pack logging tool uses a gamma ray source and a single gamma ray detector.

4.4.1 Application

• Evaluation of the integrity of gravel pack, to fully understand production or injection profiles.

4.4.2 Tools from Service Companies

Schlumberger [1]

Gravel pack log. This tool is a nuclear measuring device that evaluates the integrity of the gravel pack and identifies the gravel top.

Tools	Principal applications	OD (in.)	Maximum pressure (MPa)	Maximum temperature (°C)
Gravel pack log	• Evaluation of gravel pack before well is placed on production • Repair of gravel pack	1 11/16	120	160

Baker Atlas [33]

PRISM* PRISM Log*. It overcomes the limitations of conventional evaluation methods.

Tools	Principal applications	OD (in.)	Maximum pressure (MPa)	Maximum temperature (°C)	Maximum hole size (in.)
PRISM	• Monitor the effectiveness of gravel-pack procedures for sand control during production in unconsolidated formations	1.69	100	200	9 5/8

W

Halliburton [38]

Gravel-pack log. This tool is a nuclear measuring device that evaluates the integrity of the gravel and identifies the gravel top.

Tools	Principal applications	OD (in.)	Maximum pressure (MPa)	Maximum tempera- ture (°C)
Gravel- pack log	• Evaluation of gravel pack before well is placed on production • Repair of gravel pack	1 11/16	120	160

4.5 Determination of the distribution of the materials placed downhole

The tool discriminates between gamma ray-emitting isotopes placed in the wellbore or isotopes in the formation to identify any downhole process that can be traced with radioactive isotopes.

4.5.1 Applications

- Measure gravel-pack quality by tagging gravel with radioactive materials.
- Detect radioactive tagged acid.
- Measure fracture or propped height with tagged frac fluid or proppant.

4.5.2 Tools from Service Companies

Schlumberger [1]

MTT[*] Multi-Isotope Tracer Tool[*]. Available at the wellsite, it can be used to determine the distribution of the materials placed downhole.

Tools	Principal applications	OD (in.)	Maximum pressure (MPa)	Maximum tempera- ture (°C)
MTT	• Evaluating hydraulic fractures, squeeze cementing and enhanced oil recovery processes • Detecting zones of lost circulation • Monitoring gravel packing	1 11/16 2	140 140	180 180

Baker Atlas [33]

TRL/NFL[*] Tracerlog/Nuclear Flolog[*]. It is very useful in low flow rate producing or injection wells where relatively low flow velocities are recorded.

Tools	Principal applications	OD (in.)	Maximum pressure (MPa)	Maximum tempera-ture (°C)	Maximum hole size (in.)
TRL/NFL	• Location of fractures and etched channels from treating operations • Location of thief zones • Develop cumulative and differential percentage flow profile • Quantitative analysis of injection or production on wells	1.70 1.70	100 125	175 200	20.75 20.75

Halliburton [38]

TracerScan Logging Service[*] Used for radioactive surveys.

Tools	Principal applications	OD (in.)	Maximum pressure (MPa)	Maximum tempera-ture (°C)	Maximum hole size (in.)
Tracer-Scan	• Evaluating hydraulic fractures, squeeze cementing and enhanced oil recovery processes • Detecting zones of lost circulation • Monitoring gravel packing	1 11/16	140	180	20.75

4.5.3 General recommendations for running and interpreting radioactive-tracer logs [29]

1. Whenever possible, tracer-logging tools should be centralized so that tracer is not ejected directly against the casing wall. This is particularly important for velocity-shot logs.

2. A radioactive-tracer log should include a well diagram, a tool sketch, the surface well conditions, and a tabulation of the chart speeds used (e.g., how many seconds per division on time drive). The log should provide **all** information needed for a complete, independent log analysis.

3. Anomalous results, such as the flow rate apparently decreasing then increasing farther downhole, should be presented, not smoothed, on the interpreted log. Such results are either showing a real physical effect or giving a useful indication of log quality.

W

4. Channeling is sometimes indicated by a large difference between the profile determined with the tracer-loss log and that determined by the velocity-shot log. When the tracer-loss log shows a significantly higher flow rate at certain depth locations than the velocity-shot log, flow outside the casing can explain the discrepancy. This interpretation should be applied cautiously, however, because of the inaccuracy of the tracer-loss log.

5. The two-pulse log may be preferable to a tracer-loss log, especially in wellbores with varying cross-sectional areas.

4.6 Temperature logging

Temperature control is the mainstay of logging for fluid movement detection downhole.

4.6.1 Applications

- Location of production sources downhole and zones of injection.
- Assistance in location channels.
- Possible discrimination of gas from liquid entries.
- Evaluation of the height of an induced fracture.
- Location of zones of acid placement.
- Detection of the top of cement.

4.6.2 Tools from Service Companies

Schlumberger [1]

Temperature log. It is normally run with other services.

Tools	Principal applications	OD (in.)	Maximum pressure (MPa)	Maximum tempera- ture (°C)
Tempera- ture log	• Detection of gas entries in air-drilled • Location of lost-circulation zones • Determination of geothermal gradients • Detection of events behind casing/ tubing • Location of cement top	3 5/8	140	175

W

Baker Atlas [33]

TEMP* Temperature Log*. It provides a continuous measurement of borehole fluid temperature.

Tools	Principal applications	OD (in.)	Maximum pressure (MPa)	Maximum tempera-ture (°C)	Maximum hole size (in.)
TEMP	• Locate points of gas entry • Distinguish producing and non-producing zones • Locate tubing and casing leaks • Locate cement tops	3.38 1.69 1.69 1.69 1.69	140 100 125 100 100	200 175 200 315 177	20

Halliburton

Temperature log.

Tools	Principal applications	OD (in.)	Maximum pressure (MPa)	Maximum tempera-ture (°C)
Tempera-ture log	• Locate points of gas entry • Distinguish producing and non-producing zones • Locate tubing and casing leaks • Locate cement tops	1 11/16	140	175

4.6.3 Guidelines for running and interpreting temperature logs [29]

Recommendations for running temperature logs

1. For routine surveys, stabilize injection or production conditions (rate and temperature) for 48 hours before surveying.
2. More information about the flow profile can usually be obtained from a shut-in log than from a flowing temperature log. Whenever possible, run a shut-in log. In many instances, running a succession of shut-in logs will aid in flow profile definition.
3. Permit little or no surface leakoff during an injection survey. Permit no injection or leakoff during a shut-in survey. Even a few barrels of backflow can impair the survey.
4. Check the lubricator grease head to ensure proper pressure balance to prevent injection of large amounts of grease down the tubing. Be sure the temperature instrument is responding properly (is free of grease, etc.).

W

5. Log when entering the hole to record undisturbed temperatures if possible. If forced to log up, use as slow a cable speed as possible.

6. Logging speeds should not exceed about 6 m/min (20 ft/min) with current temperature-sensing instruments. After the continuous survey is finished, make a few stationary measurements to check the response rate.

7. Allow sufficient time between runs on decay-time surveys (successive shut-in logs) for temperature equilibrium to be restored within the wellbore. Typically, 1 to 1 1/2 hours should be allowed between surveys.

Guidelines for interpreting temperature logs

1. A suite of flowing and shut-in logs is generally easier to interpret and more diagnostic than a single log.

2. Long cumulative injection times result in slower temperature recovery after shut-in more vertical smearing of temperature profiles. Injection times exceeding roughly 2 years decrease identification of multiple injection intervals (within a gross injection zone) on shut-in temperature curves.

3. Previous injection intervals significantly affect the temperature shut-in curve for up to 6 months (or even longer) after injection has been terminated in a particular interval; as wells become more mature, distinction between previous and current injection intervals is increasingly difficult without varying the injection water temperature.

4. Identification of current injection intervals in mature wells can be improved by injection of warmer or cooler water a few hours before shut-in.

5. Injection zones as thin as 2 m (6 ft) or less can be identified on the temperature profile.

6. Thief-zone losses of 0.3 m^3/(d·m) or 5 bbl/(D-ft), or more, cause anomalies on the shut-in temperature curve of the same magnitude as major injection intervals. Even thief losses as low as 0.03 m^3/(d·m) or 0.5 bbl/(D-ft) cause sizable anomalies on shut-in curves after sufficient injection time.

7. Wellbore arrangement of tubing, casing, and open hole significantly affects the shut-in temperature curve in noninjection intervals. Added insulation at any depth generally speeds the return to geothermal temperature.

8. Wellbore arrangements, except for shot holes, have little effect on shut-in temperature curves in the injection intervals themselves.

9. Shot and enlarged holes cause anomalies that can be mistaken for injection intervals on the shut-in temperature curves.

10. Large masses of cement in enlarged holes, whether in open hole or behind casing, will cause warm anomalies on late-shut-in-time (24 hours) curves in injection wells and cool anomalies on late shut-in logs from production wells.

11. Backflow at the surface during shut-in will decrease temperature curves.

12. To apply quantitative interpretation methods based on the Ramey equation, reservoir zones must be far enough apart that vertical heat conduction is not significant between zones.

13. Though temperature logs have limited depth resolution in defining a flow profile, they are sometimes the most accurate logs available in multiphase production wells. For this reason, they should be routinely included in the suite of logs run in a multiphase-flow well.

14. A sharp break toward geothermal temperature positively identifies the bottom of the fluid injection interval.

4.7 Multiphase flow

Ideally, a flowmeter measures the bulk total flow rate regardless of the number and distribution of phases present.

4.7.1 Fluid identification

The holdup of a fluid phase is the fraction of volume of that phase present in the flowing stream.

4.7.2 Tools from Service Companies

Schlumberger

FloView* (see Paragr. 4.2, p. 470)
RST* (see Paragr. 1.4, p. 451). For gas hold-up measurement.
NFD* Nuclear Fluid Densimeter*.
Gradiomanometer.
Holdup meter (capacitance tool).

Baker Atlas

POLARIS* It is the combination of **RPM*** (presented in Paragr. 1.4, p. 451) and **MCFM*** (presented in Paragr. 4.2, p. 470).

Halliburton [38]

GHT* Gas Holdup Tool*. This new generation tool measures gas holdup directly and accurately over a cross-sectional volume element of the wellbore.

Tools	Principal applications	OD (in.)	Maximum pressure (MPa)	Maximum tempera-ture (°C)
GHT	• Improve the knowledge of actual downhole flow conditions • Expedite the evaluation of the reservoir and maximize its value and productivity • Avoid costly errors in judgment based on inaccurate flow analysis	1.687	100	175

4.7.3 Guidelines for running and interpreting production logs in multiphase flow [29]

Multiphase flow presents the most difficult conditions faced in production logging. In these conditions, the logger must recognize that tool responses may not represent average properties of the flow stream and that considerable error may be introduced by the interpretation procedures. The following guidelines summarize the most important factors to consider when multiphase-flow production logs are run and interpreted.

1. Multiphase flow may occur in the region of the wellbore being logged in any production well, even if multiphase-flow conditions are not clearly indicated by surface conditions. For example, a well producing only gas at the surface may contain gas and condensate or gas and water at bottomhole conditions. Thus, the logger must be prepared for the prospect of multiphase flow in any production well.

2. Under many multiphase-flow conditions, flow properties will fluctuate with time at any pipe location. This leads to "noisy" log responses that can be difficult to interpret. Stationary measurements are particularly necessary in this type of flow so that log responses can be averaged with time.

3. As in single-phase-flow logging, stable, known flow rates are needed for proper logging measurements and interpretation. Fluid properties at downhole conditions of each phase are also required for interpretation of multiphase-flow logs.

4. Running a suite of logs with the well shut in can be very helpful if operations permit. A shut-in temperature log in multiphase flows has the same advantages as in single-phase flows. A shut-in well is also helpful in calibrating density and capacitance logs because the phases will separate with the well shut in, allowing measurements to be made on the individual phases.

5. Well deviation will often profoundly affect production logs in multiphase flow, particularly at low flow rates. Inclinations as small as $2°$ from vertical lead to a nonuniform distribution of the phases across the pipe cross section, making the responses of spinner-flowmeter, gamma-ray-densitometer, and capacitance logs inaccurate.

6. A temperature log is not greatly affected by multiphase-flow conditions in the wellbore. For this reason, though it can usually be used only in a qualitative way, the temperature log is often the most reliable log in multiphase flow.

7. Spinner-flowmeter logs are not reliable in multiphase flow unless the flow rates are sufficiently high to make the flow similar to a single-phase flow. Flow-concentrating flowmeters provide the best velocity measurements at applicable flow rates (generally less than about 480 to 640 m³/d (3 000 to 4 000 bbl/D).

8. In gravel-pack completions, flow-concentrating flowmeters may yield erroneous results because of fluid bypassing the tool by flowing through the gravel pack.

9. Radioactive-tracer logs are not often used in production wells by some operators because radioactive fluids will be produced to the surface, though their use varies significantly from company to company and region to region. Radioactive-tracer logs offer a possible means to measure the velocity of the continuous phase directly.

10. Density tools are best suited for distinguishing between gas and liquid because of the larger density contrast than that between oil and water. Gamma ray densitometers are hampered by small sample size and the statistical natures of the measurements and of the flow stream. Stationary measurements are advantageous with any density tool to minimize the effect of statistical fluctuations.

11. A capacitance tool is often better than a density tool for distinguishing between oil and water. However, keep in mind that capicitance-tool response is not a linear function of water fraction and that capacitance tool sensitivity is greatly diminished when water is the continuous phase.

12. Quantitative analysis of production logs in multiphase flow requires independent estimation of the slip velocity. Slip velocity can be obtained from laboratory data, from log response above all production zones, or from two-phase-flow correlations. Whatever source is used, the estimation of slip velocity lends considerable uncertainty to any quantitative interpretation of the flow profile.

13. When logs are interpreted in three-phase flow, it is generally assumed that there is no slip between oil and water so that the flow can be treated like gas/liquid two-phase flow. The validity of this assumption is untested.

14. Log-interpretation results in multiphase flows are very sensitive to the holdup measurement (density or capacitance log). The log interpreter should estimate the possible range of these measurements when the accuracy of the interpreted flow profile is considered.

15. Because of the many uncertainties, the more logs run the better, with a combination of temperature, flow-concentrating-flowmeter, density, and capicitance logs preferred. This provides a means of recognizing erroneous or misleading log responses by checking each log for consistency with the other logs.

16. Any other available information about the well or reservoir (reservoir pressure, behavior of offset wells, etc.) should be used to check the validity of the production log interpretation. Unusual conditions indicated by the logs, e.g. apparent downflows, often result from the nonuniform nature of the multiphase-flow stream. Independant information can sometimes distinguish between "real" conditions and artifacts resulting from misleading production logging responses.

W5 OTHER CASED-HOLE SERVICES

5.1 Tools used [1–3, 23–25, 40]

W

- The **Guidance Continuous Tool*** (**GCT***) (Schlumberger) provides a highly accurate, continuous gyroscopic directional measurement in cased holes. The measurement is based on a two-axis gyroscope whose spin axis is horizontally maintened and aligned toward north. The position of the gyro is sensed by an accelerometer and a gyro-axis positional resolver. This information is combined with data from another accelerometer to derive the azimuth and inclination of the hole [1, 2].

487

- The **Noise Logging Tool** (Schlumberger). A typical noise logging tool consists of a transducer that converts sound to an electrical signal. The transducer is designed to respond to sound originating in any direction around the borehole; therefore, it has no directional properties. An amplifier, contained in the tool, transmits the signal up the cable [40].
- The **Correlated Electromagnetic Retrieval Tool*** (**CERT***) (Schlumberger) is a wireline electromagnetic fishing device designed to retrieve metallic junk in cased or open holes. It offers the following advantages over permanent magnet-type fishing tools. [1, 2].
  - The tool is nonmagnetic when the power is off so it will not disrupt navigational instruments during transport to the well.
  - The tool is not activated until fishing depth is reached so that a clean surface is maintained for maximum lifting capacity.
  - Casing collar and gamma ray logs can be run with the CERT tool for positive depth control.
  - A "fish detector" circuit provides an indication of fish contact and/or loss so that the progress can be monitored all the way to the surface.
- The **Well Tractor** (Welltec) is ideal for conveying standard well intervention equipment and tools into horizontal and highly deviated extended reach wells. It is a reliable and extremely fast alternative to the standard pipe or coiled tubing conveyed operations. It operates in live oil and gas producing wells under extreme hostile downhole conditions with high temperature and high well pressure [23].
- The downhole **Tractor "MULE"** by Sondex [24].

5.2 Tool applications and specifications [1–3, 23, 24]

Schlumberger

Tools	Principal applications	OD (in.)	Maximum pressure (MPa)	Maximum temperature (°C)	Maximum hole size (in.)
GCT	• Cluster drilling operations • Exact locations of blowout and relief wells • Precise targets within reservoirs for infill drilling • Locations of old wells for secondary recovery projects	3 3/8 3 5/8	70 140	120 150	10 13 3/8
CERT	• Retrieval of lost metallic objects	5	140	175	

Sondex

Tools	Principal applications	OD (in.)	Maximum pressure (MPa)	Maximum temperature (°C)
MULE	• Tractor for through tubing intervention	2 1/8	100	150

Welltec

Tools	Principal applications	OD (in.)	Maximum pressure (MPa)	Maximum temperature (°C)	Maximum hole size (in.)
Well Tractor	• Perforation • Plug setting • Caliper runs • Drift runs • Sample bailer runs • Fishing • Cement bond logs • Production logging	2 1/8 3 1/8 4 3/4	140	175	6.5 7.7 13.4

W6 IMAGING

The inspection of the inside of casing can be obtained:
- (a) either by an imaging log
- (b) or by a borehole video.

6.1 Imaging log

It is an acoustic device that produces images of the borehole wall. The images that are recorded are the variable density representation of reflected acoustic waves.

6.1.1 Applications

- Presentation of a complete 360° borehole image.
- Location and evaluation of internal casing corrosion, mechanical wear, defects, and perforations.
- Provide borehole geometry.

6.1.2 Tools from Service Companies

Schlumberger

UBI[*] Ultrasonic Borehole Imager[*]. The UBI was developed from USI with a high-resolution transducer for openhole imaging applications. For casing internal geometry measurements where casing resonance is not needed, the more focused UBI transducer gives a much better image resolution [1, 17, 18].

BHTV[*] Borehole Televiewer Tool[*]. It uses a single transducer that rotates at 3 rps and makes 250 measurements at each rotation. Thus, the resolution is essentially limited only by the physical size of the transducer and the data sampling rate [2, 20].

Tools	Principal applications	OD (in.)	Maximum pressure (MPa)	Maximum tempera-ture (°C)	Maximum hole size (in.)
UBI	• Imaging in oil-base muds • Casing internal corrosion and wear • Other detections in open hole	3 3/8	140	175	13 3/8
BHTV	• Inspect the inside of casing • Leaks (holes)	1 3/4 3 3/8	70 70	150 150	7 9 5/8

Baker Atlas [33]

Star Imager[*]**.** It is an integrated acoustic and resistivity borehole imaging service that provides detailed geological and petrophysical evaluation of subsurface formations.

CBIL[*] Circumferential Borehole Imaging Log[*]. It is an acoustic device that produces images of the borehole wall.

Tools	Principal applications	OD (in.)	Maximum pressure (MPa)	Maximum tempera-ture (°C)	Maximum hole size (in.)
Star Imager	• Borehole radius • Borehole fluid velocity • Borehole orientation	5.00	140	175	21
CBIL	• 360° borehole image • Locate and evaluate internal casing corrosion, mechanical wear, defects, and perforations	3.63 3.63	100 140	160 200	

W

Halliburton [38]

CAST-V[*] Circumferential Acoustic Scanning Tool-Visualization[*].

Tools	Principal applications	OD (in.)	Maximum pressure (MPa)	Maximum tempera-ture (°C)	Maximum hole size (in.)
CAST-V	• Internal diameter • Metal thickness • Inner and external wall corrosion • Oriented measurements	3 3/8	140	175	13 3/8

6.2 Borehole video

Borehole videos are the ultimate tools to assess conditions downhole. The problems can be viewed directly without significant interpretation.

6.2.1 Conditions of utilization

The fluid in the well must be clear. Most videos have been run in wells where the borehole was filled with filtered water or blown down with nitrogen to provide a clear fluid for logging. Some reports indicate that wellbore fluids may be suitable if one can see through the produced water at about 15 cm (6 in.) at least.

6.2.2 Tools from Service Companies

Hitwell camera

VSAS and **Slim hole camera** (Camera run on electric wireline).

Tools	Principal applications	OD (in.)	Maximum pressure (MPa)	Maximum tempera-ture (°C)
VSAS	• Diagnose downhole mechanical problems	2 3/8	70	200
Slim hole camera	• Can be run on electric line. No optical cable required • Fishing	1 1/16	70	200

W

DHV/Halliburton [25, 38]

HawkEye Video System*. It is a highly portable downhole video inspection system built for quick deployment to any location worldwide. It transmits a continuous sequence of still camera images from the downhole video camera to the surface on any standard single conductor wireline cable. Video images from downhole arrive at the surface and are updated every 3.5 seconds.

Tools	Principal applications	OD (in.)	Maximum pressure (MPa)	Maximum tempera- ture (°C)
HawkEye Video System DHV	• Ideal for diagnosing downhole mechanical problems where there is a single, still object to be observed and analyzed • Fishing, corrosion, production	1 11/16 2 1/8	70 70	100 150

W7 EVOLUTION OF WIRELINE SERVICE [3, 28]

7.1 Nuclear techniques

In order to make nuclear measurements in cased holes, the instrument needs to go through tubing, as typical cased-hole wirelines are 1 11/16 in. in diameter. Therefore a small diameter crystal is needed.

7.2 Measurements through tubing [3, 28]

- **Smaller-higher efficiency crystals allowed small tools to be used for measurements through casing and tubing**.
  In order for the detector to measure gamma ray energy levels immediately after a high energy neutron, the small diameter crystal must have a very high density (similar to steel) and be "crystal clear" in order to allow the flashes of light to enter the photo multiplier.

- **Old sodium iodide crystal** (crystal density: 3.67 g/cm^3).

- **Technology breakthrough: new GSO crystal**.
  The technical innovation to solve this challenge was developed by Hitachi, in the early 1980's as they were experimenting with a new "rare earth" material Gd_2SiO_5 known as GSO. Schlumberger research worked together with Hitachi to improve the crystal has a density of 6.71 g/cm^3, which is close the density of steel, yet it is "crystal clear". It allowed small diameter cased logging tools to be developed to measure water saturation and porosity behind casing

W

7.3 Production logging horizontal wells

Imaging of flow profile in casing

Schlumberger's latest production logging tool is called **Flagship Tool String** and consists of:

* Spinner(s)
* FloView
* Phase velocity log
* RST
* Water flow log
* Three phase holp-up measurement
* Caliper.

REFERENCES

1   Wireline Services Catalog, Schlumberger Document, SMP-7005
2   Cased Hole Log Interpretation Principles/Applications (1989) Schlumberger Document, SMP-7025
3   Edmundson H (1999) *Introduction to Cased Hole Logging. Production Logging, Evaluation Behind Casing and Reservoir Monitoring.* Journée SAID/AFTP/SPE, Schlumberger Forum, February 2, 1999, Montrouge, France
4   RST* Reservoir Saturation Tool (1993) Schlumberger Document, SMP-9250
5   Silipigno L (1999) *Latest Advances in RST technology: RST-Pro.* Journée SAID/AFTP/SPE, Schlumberger Forum, February 2, 1999, Montrouge, France
6   Catala G (1999) *A New Generation Production Logging Platform.* Journée SAID/AFTP/SPE, Schlumberger Forum, February 2, 1999, Montrouge, France
7   PS Platform. Schlumberger Document, SMP-5257
8   PS Platform (1998) Schlumberger Document, SMP-5693
9   Universal Pressure Platform (1995) Schlumberger Document, SMP-4043
10 Lenn C, Kuchuk FJ, Rounce J, Hook P (1998) *Horizontal Well Performance Evaluation and Fluid Entry Mechanisms.* SPE Annual Technical Conference and Exhibition, New Orleans, Louisiana, SPE 49089
11 Hervé X, Galley J, Ferguson R, Manin Y (1999) *A Step ahead for Production Logging: A Case History in the North Sea.* OMC, Ravenna, Italy
12 Rounce J, Lenn C, Catala G (1999) *Pinpointing Fluid Entries in Producing Wells.* SPE Middle East Oil Show, Bahrain, SPE 53249
13 FloView* (1996) Schlumberger Document, SMP-5602
14 Hervé X, Bouroumeau-Fuseau P, Quin E, Bigno Y (1996) Advanced production logging applications in Dunbar field. *World Oil*, January 1999, 45–50
15 Beresford T (1999) *A new age of Production Logging Pulsed Neutron Diagnostic for Multilaterals Wells.* Journée SAID/AFTP/SPE, Schlumberger Forum, February 2, 1999, Montrouge, France
16 Cement Evaluation Services (1989) Schlumberger Document, SMP-5098
17 Ultrasonic Imaging, USI*, UBI* (1993) Schlumberger Document, SMP-9230

W

18 USI* UltraSonic Imager (1991) Schlumberger Document, M-092047

19 Cement Mapping Tools. Lee Tool Document

20 Corrosion Evaluation (1992) Schlumberger Document, SMP-9110

21 Broughton D (1999) *Internal Inspection of Casing and Tubing Using Multi-Finger Calipers.* Journée SAID/AFTP/SPE, Schlumberger Forum, February 2, 1999, Montrouge, France

22 Multifinger Imaging Tools. Sondex Document

23 Well Tractor. Welltec Document

24 Searight T (1999) *Deployment of Intervention Tools in Cased Hole Horizontal Wells.* Journée SAID/AFTP/SPE, Schlumberger Forum, February 2, 1999, Montrouge, France

25 HawkEye Video System. DHV Document

26 Hitwell Video, Inc.

27 Van Derspek AM (1999) *Neural and Identification of flow regime usingBond Spectra of flow generated sound.* Journée SAID/AFTP/SPE, Schlumberger Forum, February 2, 1999, Montrouge, France

28 Henry KC (1998) *Advancements in Wireline Technology Past, Present and Future.* 12th Offshore South East Asia Conference and Exhibition, 2-4 December 1998, Singapore

29 Hill AD (1990) *Production Logging. Theoretical and Interpretive Elements.* Henry L. Doherty Memorial Fund of AIME, Society of Petroleum Engineers, Richardson, TX

30 Laurent P, Singer J, Dueso A, Douglas A (1997) *Evaluation and Monitoring of Existing Wells.* 1997 Well Evaluation Conference, Produced by A J Torre and Associates, Houston, Texas

31 Davies J, van Dillewijn J, Hervé X, Kusaka K (1997) *Spinners Run While Perforating.* 1997 Offshore Europe Conference, SPE 38549, Aberdeen, Scotland

32 Bamforth S, Besson C, Stephson K, Whittaker C, Brown G, Catala G, Rouault G, Théron B, Conort G, Lenn C, Roscoe B (1996) Revitalizing production logging. *Oilfield Review,* Schlumberger, Winter 1996, 44–60

33 Western Atlas (1997) Services Catalog

34 Monrose H, Boyer S (1992) Casing corrosion: origin and detection, *The Log Analyst,* Nov.-Dec. 1992, 507-517

35 Rouillac D (1994) *Cement Evaluation Logging Handbook.* Editions Technip, Paris

36 Desbrandes R (1985) *Encyclopedia of Well Logging.* Editions Technip, Paris

37 Smolen JJ (1996) *Cased Hole and Production Log Evaluation.* PennWell Books

38 Vallat D (1999) *Logging & Perforating. Casedhole. Reservoir Monitoring.* Halliburton Company

39 Bigelow EL (1999) A practical approach to the interpretation of emet bond logs. *Journal of Petroleum Technology,* July 1985

40 Schlumberger (January 1997) *Wireline and testing,* Chapter G: Other sensors

41 Schlumberger (September 1996) *Corrosion Evaluation using Wireline Logging Techniques,* Section 8: The Ultrasonic Imaging Platform.

W

X

Financial Formulas for Investment Decisions

X

Financial Formulas for Investment Decisions

X1 BASIC FORMULAS

- **Future Value of a sum S**, invested at the beginning of year 1, at a **simple interest rate** of i, for n years (interest revenues are distributed every year):

$$FV = S(1 + ni) \qquad \text{(X1)}$$

- **Future Value of a sum S**, invested at the beginning of year 1 at an **annually compounded interest rate** of i for n years (interest revenues are reinvested to earn interest in subsequent periods):

$$FV = S(1 + i)^n \qquad \text{(X2)}$$

- **Future Value P** [1] of a **yearly installment C** invested at the beginning of every year for n years, at an annually compounded interest rate of i:

$$P = \sum_{k=1}^{n} C(1 + i)^k$$

$$P = C(1 + i)\frac{(1 + i)^n - 1}{i}$$

---

[1] We need to calculate the sum of the finite geometric series $Z = C\,(1 + x + x^2 + \ldots + x^{n-1})$.
Multiplying both sides by x, we have $Zx = C\,(x + x^2 + \ldots + x^n)$.
Subtracting (X3) from (X4) gives us:

$$Z\,(x - 1) = C\,(x^n - 1) \rightarrow \quad Z = C\frac{x_n - 1}{x - 1}$$

Now let $x = (1 + i)$, then $P = Zx = C(1 + i)\ \frac{(1 + i)^n - 1}{i}$.

X

Subsequently, the yearly installment C required to obtain an asset P after n years is:

$$C = \frac{P}{(1+i)} \frac{i}{(1+i)^n - 1} \qquad (X3)$$

- **Yearly installment C necessary to reimburse** over n years **a debt D** (incurred at the beginning of year 1 at a compounded interest rate of i):

$$D = \sum_{k=1}^{n} \frac{C}{(1+i)^k} = \frac{C}{i} \frac{(1+i)^n - 1}{(1+i)^n} = \frac{C}{i}\left(1 - \frac{1}{(1+i)^n}\right) \qquad (X4)$$

$$C = D \frac{i(1+i)^n}{(1+i)^n - 1} \qquad (X5)$$

The values of multiplying factors in the paragraph are given as functions of i and n in Tables X1 and X2.

X2 PRESENT VALUE AND THE OPPORTUNITY COST OF CAPITAL

As a rule, investors are always interested in evaluating with accuracy the present value of future cash flows generated by the projects they wish to invest in.

The present value of a sum C one year from now must be less than C. In fact, *a dollar today is always worth more than a dollar tomorrow*, because the dollar today can be invested to start earning interest immediately. The present value of a delayed payoff is found by multiplying the payoff by a discount factor which is less than 1.

- **Present value** of a project that **produces a cash flow C, n years from now**:

$$PV = \frac{C}{(1+r)^n} \qquad (X6)$$

where $1/(1 + r)^n$ is the discount factor .

r is referred to as the **discount rate**, or the **opportunity cost of capital**. The cost of capital is the return foregone by **not** investing in securities.

The value for r is often taken as **the annual interest rate of money invested over n years**.

Table X3 gives the discount factors for rates ranging from 1 to 30%, and for years between 1 and 30.

- **Present value of an asset producing cash flows C_1 to C_n over n years** (discounted cash flow formula):

$$PV = \frac{C}{1 + r_1} + \frac{C_2}{(1+r_2)^2} + \ldots + \frac{C_n}{(1+r_n)^n}$$

or

$$PV = \sum_{t=1}^{n} \frac{C_t}{(1+r_t)^t} \qquad (X7)$$

However, it is often assumed that **the term structure of interest rates is "flat"**, i.e., that **the discount rate r remains the same** regardless of the date of the payoff. Then:

$$PV = \sum_{t=1}^{n} \frac{C_t}{(1+r)^t}$$

- **Present value PV[2] of an asset producing yearly cash flows C over n years, with a constant discount rate r:**

$$PV = \sum_{t=1}^{n} \frac{C}{(1+r)^t}$$

$$PV = \frac{C}{r}\left(1 - \frac{1}{(1+r)^n}\right) \tag{X8}$$

X3 PRESENT VALUE APPLIED TO OILFIELD PRODUCTION

3.1 Case 1. Present value of an oilfield generating a fixed yearly cash flow C, occurring at half year

$$PV = \sum_{t=1}^{n} C\frac{(1+r)^{1/2}}{(1+r)^t}$$

$$PV = nFC \tag{X9}$$

where $$F = \frac{(1+r)^{1/2}}{nr}\left(1 - \frac{1}{(1+r)^n}\right) \tag{X10}$$

and $C = qc$, with
  q yearly field output
  c cash flow per output unit.

---

(2) In footnote (1), we showed that the sum of the finite geometric series $W = C(x + x^2 + ... + x^n)$ is:

$$W = Cx\frac{x^n - 1}{x - 1}$$

Now let $x = 1/(1 + r)$, then:

$$PV = W = C\frac{1}{(1+r)}\frac{1/(1+r)^n - 1}{1/(1+r) - 1} = \frac{C}{r}\frac{(1+r)^n - 1}{(1+r)^n} = \frac{C}{r}\left(1 - \frac{1}{(1+r)^n}\right)$$

3.2 Case 2a. Present value of an oilfield generating a decreasing output, using a continuously compounded interest rate

The present value of cash flows produced over n years can be written as follows:

$$PV = \int_{t=0}^{n} \frac{C_t}{(1+r)^t}\, dt = C \int_{t=0}^{n} \frac{(1-d)^t}{(1+r)^t}\, dt$$

$$PV = C \int_{t=0}^{n} e^{-t(b+j)}\, dt$$

Finally: $PV = nFC$

where $F = \dfrac{1 - e^{-n(b+j)}}{n(b+j)}$ (X11)   or $nF = \dfrac{1}{\ln\left(\dfrac{1+r}{1-d}\right)}\left(1 - \dfrac{1}{\left(\dfrac{1+r}{1-d}\right)^n}\right)$ (X12)

with
- q_0 initial annual output
- c cash flow per output unit
- d decreasing output rate
- $C = q_0 c$ cash flow in the first year
- $C_t = q_0 c\,(1-d)^t$ cash flow in year t
- j continuously compounded interest rate $e^j = 1 + r$
- b continuous decline in output $e^{-b} = 1 - d$.

F can be calculated using the function $1 - e^{-x}/x$.
Table X5 gives the values of e^{nj} for nj ranging from 0.01 to 3.99.

nF can be extrapolated from Table X4, using the discount rate s such that $1 + s = (1+r)/(1-d)$. Table X4 then gives the values of nF for $n = 1$ to 30, and $s = 1\%$ to 30%.

3.3 Case 2b. Present value of an oilfield generating a constant output, using a continuously compounded interest rate

$$PV = nFC$$

where $F = \dfrac{1 - e^{-nj}}{nj}$ (X13)   or $nF = \dfrac{1}{\ln(1+r)}\left(1 - \dfrac{1}{(1+r)^n}\right)$ (X14)

Table X4 gives the values of nF for $n = 1$ to 30 and $r = 1\%$ to 30%.

X4 MAKING WISE INVESTMENT DECISIONS NET PRESENT VALUE AND ALTERNATIVE INVESTMENT CRITERIA

First, the investor must **forecast the cash flows** generated by the project over its economic life. Second, he must determine the appropriate **opportunity cost of capital** to discount the future cash flows of the project.

The opportunity-cost concept makes sense only if assets of equivalent risk are compared. An investor should therefore identify financial assets with risks equivalent to the project under consideration, estimate the expected rate of return on these assets, and use this rate as opportunity cost.

To analyse possible investment projects, an investor will use different criteria, in order to:
- Determine which project is acceptable, and which is not
- Choose among different solutions for a same project
- Select the best among various acceptable projects.

4.1 Net Present Value

The Net Present Value of a project producing cash flows C_1 to C_n over n years equals the Present Value minus the required initial investment I (or minus the present value of the successive outlays, if the project requires outlays in two or more periods):

$$NPV = \sum_{t=1}^{n} \frac{C_t}{(1+r)^t} - I \qquad (X15)$$

or, **in case of oilfield production**:

$$NPV = nFC - I$$

The *NPV* rule states that a project is worth undertaking when the net present value is positive, i.e., when the project is worth more than it costs.

4.2 Profitability index

The Profitability index is the ratio between the net present value of the project and the required investment I. This rule is used as a complement to the *NPV* rule when funds to be invested are limited. The investor then selects the projects that offer the highest net present value per dollar of initial outlay.

$$\text{Profitability index} = \frac{NPV}{I} \qquad (X16)$$

X

Note. The ratio of present value to initial investment is known as the benefit-cost ratio. To calculate it, simply add 1.0 to the profitability index.

4.3 Payback period

The **payback period** of a project is found by counting the number of years it takes before the cumulative discounted cash flow equals the initial investment I.

The payback period n is such that:

$$PV = I$$

$$0 = \sum_{t=1}^{n} \frac{C_t}{(1+r)^t} - I \qquad (X17)$$

When the yearly cash flow is a constant, Eq. X4 gives:

$$I = \frac{C}{r}\left(1 - \frac{1}{(1+r)^n}\right) \qquad (X18)$$

Knowing I/C and r, n can be extrapolated from Table X2.

In case of oilfield production, $I = nFC$. Therefore:

$$nF = \frac{I}{C}$$

In Cases 2a and 2b, knowing I/C and r, n can be extrapolated from Table X4, which gives nF as a function of n and r.

Before selecting projects on the basis of the payback period, an investor should remember that the payback rule ignores all cash flows after the cutoff date. He will then avoid to reject good long-lived projects in favor of poor short-lived ones.

4.4 Internal Rate of Return

The Internal Rate of Return *(IRR)* is defined as the rate of discount at which a project would have zero *NPV*. The *IRR* rule states that an investment can be accepted if it offers an *IRR* in excess of the opportunity cost of capital.

For a project lasting n years, *IRR* is the discount rate R such that:

$$0 = \sum_{t=1}^{n} \frac{C_t}{(1+R)^t} - I \qquad (X19)$$

When the yearly cash flow is a constant, Eq. X4 gives:

$$I = \frac{C}{R}\left(1 - \frac{1}{(1+R)^n}\right) \tag{X20}$$

Knowing I/C, the internal rate of return R can be extrapolated from Table X2.

In case of an oilfield or a well producing for n years, *IRR* is the discount rate R such that:

$$I = \frac{1 - e^{-n(b+J)}}{(b+J)}C = nFC \tag{X21}$$

where

$e^J = 1 + R$ or $J = \ln(1 + R)$
$e^{-b} = 1 - d$ or $b = -\ln(1 - d)$
$C = q_0 c$ cash flow in the first year.

Knowing I/C, the internal rate of return R can be extrapolated from Table X4 which gives nF as a function or n and r.

Using *IRR* as a criterion of decision, an investor should be careful of:

- Multiple rates of return: it happens when there is more than one change in the signs of the cash flows, and the project may then have several *IRR*s or none at all.
- Differences in short-term and long-term interest rates: when the opportunity cost of capital significantly varies with the date of the cash flow, there is no simple yardstick for evaluating the *IRR* of a project.
- Short lived projects with relatively little up-front investment: the *IRR* may be high, but such projects may not add much to the value of the company.

REFERENCES

1   Pike R, Neale B (1999) *Corporate Finance and Investment.* Prentice Hall
2   Brealey RA, Myers SC (2000) *Principles of Corporate Finance.* McGraw-Hill
3   Babusiaux D (1990) *Décision d'investissement et calcul économique dans l'entreprise.* Editions Technip, Paris.

X

Table X1 Annuity tables: *i* interest rate per year, *n* number of years.

n	2% $(1+i)^n$	2% $\frac{i(1+i)^n}{(1+i)^n-1}$	3% $\frac{i}{(1+i)^n-1}$	3% $(1+i)^n$	3% $\frac{i(1+i)^n}{(1+i)^n-1}$	4% $\frac{i}{(1+i)^n-1}$	4% $(1+i)^n$	4% $\frac{i(1+i)^n}{(1+i)^n-1}$	5% $\frac{i}{(1+i)^n-1}$	5% $(1+i)^n$	5% $\frac{i(1+i)^n}{(1+i)^n-1}$	6% $\frac{i}{(1+i)^n-1}$	6% $(1+i)^n$	6% $\frac{i(1+i)^n}{(1+i)^n-1}$	7% $(1+i)^n$	7% $\frac{i}{(1+i)^n-1}$	7% $\frac{i(1+i)^n}{(1+i)^n-1}$
1	1.0200	1.02000	1.00000	1.0300	1.03000	1.00000	1.0400	1.04000	1.00000	1.0500	1.05000	1.00000	1.0600	1.06000	1.0700	1.00000	1.07000
2	1.0404	0.51505	0.49261	1.0609	0.52261	0.49020	1.0816	0.53020	0.48780	1.1025	0.53780	0.48544	1.1236	0.54544	1.1449	0.48309	0.55309
3	1.0612	0.34675	0.32353	1.0927	0.35353	0.32035	1.1249	0.36035	0.31721	1.1576	0.36721	0.31411	1.1910	0.37411	1.2250	0.31105	0.38105
4	1.0824	0.26262	0.23903	1.1255	0.26903	0.23549	1.1699	0.27549	0.23201	1.2155	0.28201	0.22859	1.2625	0.28859	1.3108	0.22523	0.29523
5	1.1041	0.21216	0.18835	1.1593	0.21835	0.18463	1.2167	0.22463	0.18097	1.2763	0.23097	0.17740	1.3382	0.23740	1.4026	0.17389	0.24389
6	1.1262	0.17853	0.15460	1.1941	0.18460	0.15076	1.2653	0.19076	0.14702	1.3401	0.19702	0.14336	1.4185	0.20336	1.5007	0.13980	0.20980
7	1.1487	0.15451	0.13051	1.2299	0.16051	0.12661	1.3159	0.16661	0.12282	1.4071	0.17282	0.11914	1.5036	0.17914	1.6058	0.11555	0.18555
8	1.1717	0.13651	0.11246	1.2668	0.14246	0.10853	1.3686	0.14853	0.10472	1.4775	0.15472	0.10104	1.5938	0.16104	1.7182	0.09747	0.16747
9	1.1951	0.12252	0.09843	1.3048	0.12843	0.09449	1.4233	0.13449	0.09069	1.5513	0.14069	0.08702	1.6895	0.14702	1.8385	0.08349	0.15349
10	1.2190	0.11133	0.08723	1.3439	0.11723	0.08329	1.4802	0.12329	0.07950	1.6289	0.12950	0.07587	1.7908	0.13587	1.9672	0.07238	0.14238
11	1.2434	0.10218	0.07808	1.3842	0.10808	0.07415	1.5395	0.11415	0.07039	1.7103	0.12039	0.06679	1.8983	0.12679	2.1049	0.06336	0.13336
12	1.2682	0.09456	0.07046	1.4258	0.10046	0.06655	1.6010	0.10655	0.06283	1.7959	0.11283	0.05928	2.0122	0.11928	2.2522	0.05590	0.12590
13	1.2936	0.08812	0.06403	1.4685	0.09403	0.06014	1.6651	0.10014	0.05646	1.8856	0.10646	0.05296	2.1329	0.11296	2.4098	0.04965	0.11965
14	1.3195	0.08260	0.05853	1.5126	0.08853	0.05467	1.7317	0.09467	0.05102	1.9799	0.10102	0.04758	2.2609	0.10758	2.5785	0.04434	0.11434
15	1.3459	0.07783	0.05377	1.5580	0.08377	0.04994	1.8009	0.08994	0.04634	2.0789	0.09634	0.04296	2.3966	0.10296	2.7590	0.03979	0.10979
16	1.3728	0.07365	0.04961	1.6047	0.07961	0.04582	1.8730	0.08582	0.04227	2.1829	0.09227	0.03895	2.5404	0.09895	2.9522	0.03586	0.10586
17	1.4002	0.06997	0.04595	1.6528	0.07595	0.04220	1.9479	0.08220	0.03870	2.2920	0.08870	0.03544	2.6928	0.09544	3.1588	0.03243	0.10243
18	1.4282	0.06670	0.04271	1.7024	0.07271	0.03899	2.0258	0.07899	0.03555	2.4066	0.08555	0.03236	2.8543	0.09236	3.3799	0.02941	0.09941
19	1.4568	0.06378	0.03981	1.7535	0.06981	0.03614	2.1068	0.07614	0.03275	2.5270	0.08275	0.02962	3.0256	0.08962	3.6165	0.02675	0.09675
20	1.4859	0.06116	0.03722	1.8061	0.06722	0.03358	2.1911	0.07358	0.03024	2.6533	0.08024	0.02718	3.2071	0.08718	3.8697	0.02439	0.09439
21	1.5157	0.05878	0.03487	1.8603	0.06487	0.03128	2.2788	0.07128	0.02800	2.7860	0.07800	0.02500	3.3996	0.08500	4.1406	0.02229	0.09229
22	1.5460	0.05663	0.03275	1.9161	0.06275	0.02920	2.3699	0.06920	0.02597	2.9253	0.07597	0.02305	3.6035	0.08305	4.4304	0.02041	0.09041
23	1.5769	0.05467	0.03081	1.9736	0.06081	0.02731	2.4647	0.06731	0.02414	3.0715	0.07414	0.02128	3.8197	0.08128	4.7405	0.01871	0.08871
24	1.6084	0.05287	0.02905	2.0328	0.05905	0.02559	2.5633	0.06559	0.02247	3.2251	0.07247	0.01968	4.0489	0.07968	5.0724	0.01719	0.08719
25	1.6406	0.05122	0.02743	2.0938	0.05743	0.02401	2.6658	0.06401	0.02095	3.3864	0.07095	0.01823	4.2919	0.07823	5.4274	0.01581	0.08581
26	1.6734	0.04970	0.02594	2.1566	0.05594	0.02257	2.7725	0.06257	0.01956	3.5557	0.06956	0.01690	4.5494	0.07690	5.8074	0.01456	0.08456
27	1.7069	0.04829	0.02456	2.2213	0.05456	0.02124	2.8834	0.06124	0.01829	3.7335	0.06829	0.01570	4.8223	0.07570	6.2139	0.01343	0.08343
28	1.7410	0.04699	0.02329	2.2879	0.05329	0.02001	2.9987	0.06001	0.01712	3.9201	0.06712	0.01459	5.1117	0.07459	6.6488	0.01239	0.08239
29	1.7758	0.04578	0.02211	2.3566	0.05211	0.01888	3.1187	0.05888	0.01605	4.1161	0.06605	0.01358	5.4184	0.07358	7.1143	0.01145	0.08145
30	1.8114	0.04465	0.02102	2.4273	0.05102	0.01783	3.2434	0.05783	0.01505	4.3219	0.06505	0.01265	5.7435	0.07265	7.6123	0.01059	0.08059

Table X1   Annuity tables: i interest rate per year, n number of years.

n	8% $(1+i)^n$	8% $\frac{i}{(1+i)^n-1}$	8% $\frac{i(1+i)^n}{(1+i)^n-1}$	9% $(1+i)^n$	9% $\frac{i}{(1+i)^n-1}$	9% $\frac{i(1+i)^n}{(1+i)^n-1}$	10% $(1+i)^n$	10% $\frac{i}{(1+i)^n-1}$	10% $\frac{i(1+i)^n}{(1+i)^n-1}$	11% $(1+i)^n$	11% $\frac{i}{(1+i)^n-1}$	11% $\frac{i(1+i)^n}{(1+i)^n-1}$	12% $(1+i)^n$	12% $\frac{i}{(1+i)^n-1}$	12% $\frac{i(1+i)^n}{(1+i)^n-1}$	13% $(1+i)^n$	13% $\frac{i}{(1+i)^n-1}$	13% $\frac{i(1+i)^n}{(1+i)^n-1}$
1	1.0800	1.00000	1.08000	1.0900	1.00000	1.09000	1.1000	1.00000	1.10000	1.1100	1.00000	1.11000	1.1200	1.00000	1.12000	1.1300	1.00000	1.13000
2	1.1664	0.48077	0.56077	1.1881	0.47847	0.56847	1.2100	0.47619	0.57619	1.2321	0.47393	0.58393	1.2544	0.47170	0.59170	1.2769	0.46948	0.59948
3	1.2597	0.30803	0.38803	1.2950	0.30505	0.39505	1.3310	0.30211	0.40211	1.3676	0.29921	0.40921	1.4049	0.29635	0.41635	1.4429	0.29352	0.42352
4	1.3605	0.22192	0.30192	1.4116	0.21867	0.30867	1.4641	0.21547	0.31547	1.5181	0.21233	0.32233	1.5735	0.20923	0.32923	1.6305	0.20619	0.33619
5	1.4693	0.17046	0.25046	1.5386	0.16709	0.25709	1.6105	0.16380	0.26380	1.6851	0.16057	0.27057	1.7623	0.15741	0.27741	1.8424	0.15431	0.28431
6	1.5869	0.13632	0.21632	1.6771	0.13292	0.22292	1.7716	0.12961	0.22961	1.8704	0.12638	0.23638	1.9738	0.12323	0.24323	2.0820	0.12015	0.25015
7	1.7138	0.11207	0.19207	1.8280	0.10869	0.19869	1.9487	0.10541	0.20541	2.0762	0.10222	0.21222	2.2107	0.09912	0.21912	2.3526	0.09611	0.22611
8	1.8509	0.09401	0.17401	1.9926	0.09067	0.18067	2.1436	0.08744	0.18744	2.3045	0.08432	0.19432	2.4760	0.08130	0.20130	2.6584	0.07839	0.20839
9	1.9990	0.08008	0.16008	2.1719	0.07680	0.16680	2.3579	0.07364	0.17364	2.5580	0.07060	0.18060	2.7731	0.06768	0.18768	3.0040	0.06487	0.19487
10	2.1589	0.06903	0.14903	2.3674	0.06582	0.15582	2.5937	0.06275	0.16275	2.8394	0.05980	0.16980	3.1058	0.05698	0.17698	3.3946	0.05429	0.18429
11	2.3316	0.06008	0.14008	2.5804	0.05695	0.14695	2.8531	0.05396	0.15396	3.1518	0.05112	0.16112	3.4785	0.04842	0.16842	3.8359	0.04584	0.17584
12	2.5182	0.05270	0.13270	2.8127	0.04965	0.13965	3.1384	0.04676	0.14676	3.4985	0.04403	0.15403	3.8960	0.04144	0.16144	4.3345	0.03899	0.16899
13	2.7196	0.04652	0.12652	3.0658	0.04357	0.13357	3.4523	0.04078	0.14078	3.8833	0.03815	0.14815	4.3635	0.03568	0.15568	4.8980	0.03335	0.16335
14	2.9372	0.04130	0.12130	3.3417	0.03843	0.12843	3.7975	0.03575	0.13575	4.3104	0.03323	0.14323	4.8871	0.03087	0.15087	5.5348	0.02867	0.15867
15	3.1722	0.03683	0.11683	3.6425	0.03406	0.12406	4.1772	0.03147	0.13147	4.7846	0.02907	0.13907	5.4736	0.02682	0.14682	6.2543	0.02474	0.15474
16	3.4259	0.03298	0.11298	3.9703	0.03030	0.12030	4.5950	0.02782	0.12782	5.3109	0.02552	0.13552	6.1304	0.02339	0.14339	7.0673	0.02143	0.15143
17	3.7000	0.02963	0.10963	4.3276	0.02705	0.11705	5.0545	0.02466	0.12466	5.8951	0.02247	0.13247	6.8660	0.02046	0.14046	7.9861	0.01861	0.14861
18	3.9960	0.02670	0.10670	4.7171	0.02421	0.11421	5.5599	0.02193	0.12193	6.5436	0.01984	0.12984	7.6900	0.01794	0.13794	9.0243	0.01620	0.14620
19	4.3157	0.02413	0.10413	5.1417	0.02173	0.11173	6.1159	0.01955	0.11955	7.2633	0.01756	0.12756	8.6128	0.01576	0.13576	10.1974	0.01413	0.14413
20	4.6610	0.02185	0.10185	5.6044	0.01955	0.10955	6.7275	0.01746	0.11746	8.0623	0.01558	0.12558	9.6463	0.01388	0.13388	11.5231	0.01235	0.14235
21	5.0338	0.01983	0.09983	6.1088	0.01762	0.10762	7.4002	0.01562	0.11562	8.9492	0.01384	0.12384	10.8038	0.01224	0.13224	13.0211	0.01081	0.14081
22	5.4365	0.01803	0.09803	6.6586	0.01590	0.10590	8.1403	0.01401	0.11401	9.9336	0.01231	0.12231	12.1003	0.01081	0.13081	14.7138	0.00948	0.13948
23	5.8715	0.01642	0.09642	7.2579	0.01438	0.10438	8.9543	0.01257	0.11257	11.0263	0.01097	0.12097	13.5523	0.00956	0.12956	16.6266	0.00832	0.13832
24	6.3412	0.01498	0.09498	7.9111	0.01302	0.10302	9.8497	0.01130	0.11130	12.2392	0.00979	0.11979	15.1786	0.00846	0.12846	18.7881	0.00731	0.13731
25	6.8485	0.01368	0.09368	8.6231	0.01181	0.10181	10.8347	0.01017	0.11017	13.5855	0.00874	0.11874	17.0001	0.00750	0.12750	21.2305	0.00643	0.13643
26	7.3964	0.01251	0.09251	9.3992	0.01072	0.10072	11.9182	0.00916	0.10916	15.0799	0.00781	0.11781	19.0401	0.00665	0.12665	23.9905	0.00565	0.13565
27	7.9881	0.01145	0.09145	10.2451	0.00973	0.09973	13.1100	0.00826	0.10826	16.7386	0.00699	0.11699	21.3249	0.00590	0.12590	27.1093	0.00498	0.13498
28	8.6271	0.01049	0.09049	11.1671	0.00885	0.09885	14.4210	0.00745	0.10745	18.5799	0.00626	0.11626	23.8839	0.00524	0.12524	30.6335	0.00439	0.13439
29	9.3173	0.00962	0.08962	12.1722	0.00806	0.09806	15.8631	0.00673	0.10673	20.6237	0.00561	0.11561	26.7499	0.00466	0.12466	34.6158	0.00387	0.13387
30	10.0627	0.00883	0.08883	13.2677	0.00734	0.09734	17.4494	0.00608	0.10608	22.8923	0.00502	0.11502	29.9599	0.00414	0.12414	39.1159	0.00341	0.13341

X

Table X1 Annuity tables: *i* interest rate per year, *n* number of years.

n	14% $(1+i)^n$	14% $\frac{i}{(1+i)^n-1}$	14% $\frac{i(1+i)^n}{(1+i)^n-1}$	15% $(1+i)^n$	15% $\frac{i}{(1+i)^n-1}$	15% $\frac{i(1+i)^n}{(1+i)^n-1}$	16% $(1+i)^n$	16% $\frac{i}{(1+i)^n-1}$	16% $\frac{i(1+i)^n}{(1+i)^n-1}$	17% $(1+i)^n$	17% $\frac{i}{(1+i)^n-1}$	17% $\frac{i(1+i)^n}{(1+i)^n-1}$	18% $(1+i)^n$	18% $\frac{i}{(1+i)^n-1}$	18% $\frac{i(1+i)^n}{(1+i)^n-1}$	19% $(1+i)^n$	19% $\frac{i}{(1+i)^n-1}$	19% $\frac{i(1+i)^n}{(1+i)^n-1}$
1	1.1400	1.00000	1.14000	1.1500	1.00000	1.15000	1.1600	1.00000	1.16000	1.1700	1.00000	1.17000	1.1800	1.00000	1.18000	1.1900	1.00000	1.19000
2	1.2996	0.46729	0.60729	1.3225	0.46512	0.61512	1.3456	0.46296	0.62296	1.3689	0.46083	0.63083	1.3924	0.45872	0.63872	1.4161	0.45662	0.64662
3	1.4815	0.29073	0.43073	1.5209	0.28798	0.43798	1.5609	0.28526	0.44526	1.6016	0.28257	0.45257	1.6430	0.27992	0.45992	1.6852	0.27731	0.46731
4	1.6890	0.20320	0.34320	1.7490	0.20027	0.35027	1.8106	0.19738	0.35738	1.8739	0.19453	0.36453	1.9388	0.19174	0.37174	2.0053	0.18899	0.37899
5	1.9254	0.15128	0.29128	2.0114	0.14832	0.29832	2.1003	0.14541	0.30541	2.1924	0.14256	0.31256	2.2878	0.13978	0.31978	2.3864	0.13705	0.32705
6	2.1950	0.11716	0.25716	2.3131	0.11424	0.26424	2.4364	0.11139	0.27139	2.5652	0.10861	0.27861	2.6996	0.10591	0.28591	2.8398	0.10327	0.29327
7	2.5023	0.09319	0.23319	2.6600	0.09036	0.24036	2.8262	0.08761	0.24761	3.0012	0.08495	0.25495	3.1855	0.08236	0.26236	3.3793	0.07985	0.26985
8	2.8526	0.07557	0.21557	3.0590	0.07285	0.22285	3.2784	0.07022	0.23022	3.5115	0.06769	0.23769	3.7589	0.06524	0.24524	4.0214	0.06289	0.25289
9	3.2519	0.06217	0.20217	3.5179	0.05957	0.20957	3.8030	0.05708	0.21708	4.1084	0.05469	0.22469	4.4355	0.05239	0.23239	4.7854	0.05019	0.24019
10	3.7072	0.05171	0.19171	4.0456	0.04925	0.19925	4.4114	0.04690	0.20690	4.8068	0.04466	0.21466	5.2338	0.04251	0.22251	5.6947	0.04047	0.23047
11	4.2262	0.04339	0.18339	4.6524	0.04107	0.19107	5.1173	0.03886	0.19886	5.6240	0.03676	0.20676	6.1759	0.03478	0.21478	6.7767	0.03289	0.22289
12	4.8179	0.03667	0.17667	5.3503	0.03448	0.18448	5.9360	0.03241	0.19241	6.5801	0.03047	0.20047	7.2876	0.02863	0.20863	8.0642	0.02890	0.21690
13	5.4924	0.03116	0.17116	6.1528	0.02911	0.17911	6.8858	0.02718	0.18718	7.6987	0.02538	0.19538	8.5994	0.02369	0.20369	9.5964	0.02210	0.21210
14	6.2613	0.02661	0.16661	7.0757	0.02469	0.17469	7.9875	0.02290	0.18290	9.0075	0.02123	0.19123	10.1472	0.01968	0.19968	11.4198	0.01823	0.20823
15	7.1379	0.02281	0.16281	8.1371	0.02102	0.17102	9.2655	0.01936	0.17936	10.5387	0.01782	0.18782	11.9737	0.01640	0.19640	13.5895	0.01509	0.20509
16	8.1372	0.01962	0.15962	9.3576	0.01795	0.16795	10.7480	0.01641	0.17641	12.3303	0.01500	0.18500	14.1290	0.01371	0.19371	16.1715	0.01252	0.20252
17	9.2765	0.01692	0.15692	10.7613	0.01537	0.16537	12.4677	0.01395	0.17395	14.4265	0.01266	0.18266	16.6722	0.01149	0.19149	19.2441	0.01041	0.20041
18	10.5752	0.01462	0.15462	12.3755	0.01319	0.16319	14.4625	0.01188	0.17188	16.8790	0.01071	0.18071	19.6733	0.00964	0.18964	22.9005	0.00868	0.19868
19	12.0557	0.01266	0.15266	14.2318	0.01134	0.16134	16.7765	0.01014	0.17014	19.7484	0.00907	0.17907	23.2144	0.00810	0.18810	27.2516	0.00724	0.19724
20	13.7435	0.01099	0.15099	16.3665	0.00976	0.15976	19.4608	0.00867	0.16867	23.1056	0.00769	0.17769	27.3930	0.00682	0.18682	32.4294	0.00605	0.19605
21	15.6676	0.00954	0.14954	18.8215	0.00842	0.15842	22.5745	0.00742	0.16742	27.0336	0.00653	0.17653	32.3238	0.00575	0.18575	38.5910	0.00505	0.19505
22	17.8610	0.00830	0.14830	21.6447	0.00727	0.15727	26.1864	0.00635	0.16635	31.6293	0.00555	0.17555	38.1421	0.00485	0.18485	45.9233	0.00423	0.19423
23	20.3616	0.00723	0.14723	24.8915	0.00628	0.15628	30.3762	0.00545	0.16545	37.0062	0.00472	0.17472	45.0076	0.00409	0.18409	54.6487	0.00354	0.19354
24	23.2122	0.00630	0.14630	28.6252	0.00543	0.15543	35.2364	0.00467	0.16467	43.2973	0.00402	0.17402	53.1090	0.00345	0.18345	65.0320	0.00297	0.19297
25	26.4619	0.00550	0.14550	32.9190	0.00470	0.15470	40.8742	0.00401	0.16401	50.6578	0.00342	0.17342	62.6686	0.00292	0.18292	77.3881	0.00249	0.19249
26	30.1666	0.00480	0.14480	37.8568	0.00407	0.15407	47.4141	0.00345	0.16345	59.2697	0.00292	0.17292	73.9490	0.00247	0.18247	92.0918	0.00209	0.19209
27	34.3899	0.00419	0.14419	43.5353	0.00353	0.15353	55.0004	0.00296	0.16296	69.3455	0.00249	0.17249	87.2598	0.00209	0.18209	109.5893	0.00175	0.19175
28	39.2045	0.00366	0.14366	50.0656	0.00306	0.15306	63.8004	0.00255	0.16255	81.1342	0.00212	0.17212	102.9666	0.00177	0.18177	130.4112	0.00147	0.19147
29	44.6931	0.00320	0.14320	57.5755	0.00265	0.15265	74.0085	0.00219	0.16219	94.9271	0.00181	0.17181	121.5005	0.00149	0.18149	155.1893	0.00123	0.19123
30	50.9502	0.00280	0.14280	66.2118	0.00230	0.15230	85.8499	0.00189	0.16189	111.0647	0.00154	0.17154	143.3706	0.00126	0.18126	184.6753	0.00103	0.19103

Table X1   Annuity tables: *i* interest rate per year, *n* number of years.

i	20%			21%			22%			23%			24%			25%		
n	$(1+i)^n$	$\frac{i}{(1+i)^n-1}$	$\frac{i(1+i)^n}{(1+i)^n-1}$	$(1+i)^n$	$\frac{i}{(1+i)^n-1}$	$\frac{i(1+i)^n}{(1+i)^n-1}$	$(1+i)^n$	$\frac{i}{(1+i)^n-1}$	$\frac{i(1+i)^n}{(1+i)^n-1}$	$(1+i)^n$	$\frac{i}{(1+i)^n-1}$	$\frac{i(1+i)^n}{(1+i)^n-1}$	$(1+i)^n$	$\frac{i}{(1+i)^n-1}$	$\frac{i(1+i)^n}{(1+i)^n-1}$	$(1+i)^n$	$\frac{i}{(1+i)^n-1}$	$\frac{i(1+i)^n}{(1+i)^n-1}$
1	1.2000	1.00000	1.20000	1.2100	1.00000	1.21000	1.2200	1.00000	1.22000	1.2300	1.00000	1.23000	1.2400	1.00000	1.24000	1.2500	1.00000	1.25000
2	1.4400	0.45455	0.65455	1.4641	0.45249	0.66249	1.4884	0.45045	0.67045	1.5129	0.44843	0.67843	1.5376	0.44643	0.68643	1.5625	0.44444	0.69444
3	1.7280	0.27473	0.47473	1.7716	0.27218	0.48218	1.8158	0.26966	0.48966	1.8609	0.26717	0.49717	1.9066	0.26472	0.50472	1.9531	0.26230	0.51230
4	2.0736	0.18629	0.38629	2.1436	0.18363	0.39363	2.2153	0.18102	0.40102	2.2889	0.17845	0.40845	2.3642	0.17593	0.41593	2.4414	0.17344	0.42344
5	2.4883	0.13438	0.33438	2.5937	0.13177	0.34177	2.7027	0.12921	0.34921	2.8153	0.12670	0.35670	2.9316	0.12425	0.36425	3.0518	0.12185	0.37185
6	2.9860	0.10071	0.30071	3.1384	0.09820	0.30820	3.2973	0.09576	0.31576	3.4628	0.09339	0.32339	3.6352	0.09107	0.33107	3.8147	0.08882	0.33882
7	3.5832	0.07742	0.27742	3.7975	0.07507	0.28507	4.0227	0.07278	0.29278	4.2593	0.07057	0.30057	4.5077	0.06842	0.30842	4.7684	0.06634	0.31634
8	4.2998	0.06061	0.26061	4.5950	0.05841	0.26841	4.9077	0.05630	0.27630	5.2389	0.05426	0.28426	5.5895	0.05229	0.29229	5.9605	0.05040	0.30040
9	5.1598	0.04808	0.24808	5.5599	0.04605	0.25605	5.9874	0.04411	0.26411	6.4439	0.04225	0.27225	6.9310	0.04047	0.28047	7.4506	0.03876	0.28876
10	6.1917	0.03852	0.23852	6.7275	0.03667	0.24667	7.3046	0.03489	0.25489	7.9259	0.03321	0.26321	8.5944	0.03160	0.27160	9.3132	0.03007	0.28007
11	7.4301	0.03110	0.23110	8.1403	0.02941	0.23941	8.9117	0.02781	0.24781	9.7489	0.02629	0.25629	10.6571	0.02485	0.26485	11.6415	0.02349	0.27349
12	8.9161	0.02526	0.22526	9.8497	0.02373	0.23373	10.8722	0.02228	0.24228	11.9912	0.02093	0.25093	13.2148	0.01965	0.25965	14.5519	0.01845	0.26845
13	10.6993	0.02062	0.22062	11.9182	0.01923	0.22923	13.2641	0.01794	0.23794	14.7491	0.01673	0.24673	16.3863	0.01560	0.25560	18.1899	0.01454	0.26454
14	12.8392	0.01689	0.21689	14.4210	0.01565	0.22565	16.1822	0.01449	0.23449	18.1414	0.01342	0.24342	20.3191	0.01242	0.25242	22.7374	0.01150	0.26150
15	15.4070	0.01388	0.21388	17.4494	0.01277	0.22277	19.7423	0.01174	0.23174	22.3140	0.01079	0.24079	25.1956	0.00992	0.24992	28.4217	0.00912	0.25912
16	18.4884	0.01144	0.21144	21.1138	0.01044	0.22044	24.0856	0.00953	0.22953	27.4462	0.00870	0.23870	31.2426	0.00794	0.24794	35.5271	0.00724	0.25724
17	22.1861	0.00944	0.20944	25.5477	0.00855	0.21855	29.3844	0.00775	0.22775	33.7588	0.00702	0.23702	38.7408	0.00636	0.24636	44.4089	0.00576	0.25576
18	26.6233	0.00781	0.20781	30.9127	0.00702	0.21702	35.8490	0.00631	0.22631	41.5233	0.00568	0.23568	48.0386	0.00510	0.24510	55.5112	0.00459	0.25459
19	31.9480	0.00646	0.20646	37.4043	0.00577	0.21577	43.7358	0.00515	0.22515	51.0737	0.00459	0.23459	59.5679	0.00410	0.24410	69.3889	0.00366	0.25366
20	38.3376	0.00536	0.20536	45.2593	0.00474	0.21474	53.3576	0.00420	0.22420	62.8206	0.00372	0.23372	73.8641	0.00329	0.24329	86.7362	0.00292	0.25292
21	46.0051	0.00444	0.20444	54.7637	0.00391	0.21391	65.0963	0.00343	0.22343	77.2694	0.00302	0.23302	91.5915	0.00265	0.24265	108.4202	0.00233	0.25233
22	55.2061	0.00369	0.20369	66.2641	0.00322	0.21322	79.4175	0.00281	0.22281	95.0413	0.00245	0.23245	113.5735	0.00213	0.24213	135.5253	0.00186	0.25186
23	66.2474	0.00307	0.20307	80.1795	0.00265	0.21265	96.8894	0.00229	0.22229	116.9008	0.00198	0.23198	140.8312	0.00172	0.24172	169.4066	0.00148	0.25148
24	79.4968	0.00255	0.20255	97.0172	0.00219	0.21219	118.2050	0.00188	0.22188	143.7880	0.00161	0.23161	174.6306	0.00138	0.24138	211.7582	0.00119	0.25119
25	95.3962	0.00212	0.20212	117.3909	0.00180	0.21180	144.2101	0.00154	0.22154	176.8593	0.00131	0.23131	216.5420	0.00111	0.24111	264.6978	0.00095	0.25095
26	114.4755	0.00176	0.20176	142.0429	0.00149	0.21149	175.9364	0.00126	0.22126	217.5369	0.00106	0.23106	268.5121	0.00090	0.24090	330.8722	0.00076	0.25076
27	137.3706	0.00147	0.20147	171.8719	0.00123	0.21123	214.6424	0.00103	0.22103	267.5704	0.00086	0.23086	332.9550	0.00072	0.24072	413.5903	0.00061	0.25061
28	164.8447	0.00122	0.20122	207.9651	0.00101	0.21101	261.8637	0.00084	0.22084	329.1115	0.00070	0.23070	412.8642	0.00058	0.24058	516.9879	0.00048	0.25048
29	197.8136	0.00102	0.20102	251.6377	0.00084	0.21084	319.4737	0.00069	0.22069	404.8072	0.00057	0.23057	511.9516	0.00047	0.24047	646.2349	0.00039	0.25039
30	237.3763	0.00085	0.20085	304.4816	0.00069	0.21069	389.7579	0.00057	0.22057	497.9129	0.00046	0.23046	634.8199	0.00038	0.24038	807.7936	0.00031	0.25031

X

Table X2 Annuity tables $\frac{1}{r}\left(1 - \frac{1}{(1+r)^n}\right)$ = present value of \$1 per year for each of n years.

n	0.5%	1.0%	1.5%	2.0%	2.5%	3.0%	3.5%	4.0%	4.5%	5.0%	5.5%	6.0%	6.5%	7.0%	7.5%	8.0%	8.5%	9.0%	9.5%	10.0%
1	0.9950	0.9901	0.9852	0.9804	0.9756	0.9709	0.9662	0.9615	0.9569	0.9524	0.9479	0.9434	0.9390	0.9346	0.9302	0.9259	0.9217	0.9174	0.9132	0.9091
2	1.9851	1.9704	1.9559	1.9416	1.9274	1.9135	1.8997	1.8861	1.8727	1.8594	1.8463	1.8334	1.8206	1.8080	1.7956	1.7833	1.7711	1.7591	1.7473	1.7355
3	2.9702	2.9410	2.9122	2.8839	2.8560	2.8286	2.8016	2.7751	2.7490	2.7232	2.6979	2.6730	2.6485	2.6243	2.6005	2.5771	2.5540	2.5313	2.5089	2.4869
4	3.9505	3.9020	3.8544	3.8077	3.7620	3.7171	3.6731	3.6299	3.5875	3.5460	3.5052	3.4651	3.4258	3.3872	3.3493	3.3121	3.2756	3.2397	3.2045	3.1699
5	4.9259	4.8534	4.7826	4.7135	4.6458	4.5797	4.5151	4.4518	4.3900	4.3295	4.2703	4.2124	4.1557	4.1002	4.0459	3.9927	3.9406	3.8897	3.8397	3.7908
6	5.8964	5.7955	5.6972	5.6014	5.5081	5.4172	5.3286	5.2421	5.1579	5.0757	4.9955	4.9173	4.8410	4.7665	4.6938	4.6229	4.5536	4.4859	4.4198	4.3553
7	6.8621	6.7282	6.5982	6.4720	6.3494	6.2303	6.1145	6.0021	5.8927	5.7864	5.6830	5.5824	5.4845	5.3893	5.2966	5.2064	5.1185	5.0330	4.9496	4.8684
8	7.8230	7.6517	7.4859	7.3255	7.1701	7.0197	6.8740	6.7327	6.5959	6.4632	6.3346	6.2098	6.0888	5.9713	5.8573	5.7466	5.6392	5.5348	5.4334	5.3349
9	8.7791	8.5660	8.3605	8.1622	7.9709	7.7861	7.6077	7.4353	7.2688	7.1078	6.9522	6.8017	6.6561	6.5152	6.3789	6.2469	6.1191	5.9952	5.8753	5.7590
10	9.7304	9.4713	9.2222	8.9826	8.7521	8.5302	8.3166	8.1109	7.9127	7.7217	7.5376	7.3601	7.1888	7.0236	6.8641	6.7101	6.5613	6.4177	6.2788	6.1446
11	10.6770	10.3676	10.0711	9.7868	9.5142	9.2526	9.0016	8.7605	8.5289	8.3064	8.0925	7.8869	7.6890	7.4987	7.3154	7.1390	6.9690	6.8052	6.6473	6.4951
12	11.6189	11.2551	10.9075	10.5753	10.2578	9.9540	9.6633	9.3851	9.1186	8.8633	8.6185	8.3838	8.1587	7.9427	7.7353	7.5361	7.3447	7.1607	6.9838	6.8137
13	12.5562	12.1337	11.7315	11.3484	10.9832	10.6350	10.3027	9.9856	9.6829	9.3936	9.1171	8.8527	8.5997	8.3577	8.1258	7.9038	7.6910	7.4869	7.2912	7.1034
14	13.4887	13.0037	12.5434	12.1062	11.6909	11.2961	10.9205	10.5631	10.2228	9.8986	9.5896	9.2950	9.0138	8.7455	8.4892	8.2442	8.0101	7.7862	7.5719	7.3667
15	14.4166	13.8651	13.3432	12.8493	12.3814	11.9379	11.5174	11.1184	10.7395	10.3797	10.0376	9.7122	9.4027	9.1079	8.8271	8.5595	8.3042	8.0607	7.8282	7.6061
16	15.3399	14.7179	14.1313	13.5777	13.0550	12.5611	12.0941	11.6523	11.2340	10.8378	10.4622	10.1059	9.7678	9.4466	9.1415	8.8514	8.5753	8.3126	8.0623	7.8237
17	16.2586	15.5623	14.9076	14.2919	13.7122	13.1661	12.6513	12.1657	11.7072	11.2741	10.8646	10.4773	10.1106	9.7632	9.4340	9.1216	8.8252	8.5436	8.2760	8.0216
18	17.1728	16.3983	15.6726	14.9920	14.3534	13.7535	13.1897	12.6593	12.1600	11.6896	11.2461	10.8276	10.4325	10.0591	9.7060	9.3719	9.0555	8.7556	8.4713	8.2014
19	18.0824	17.2260	16.4262	15.6785	14.9789	14.3238	13.7098	13.1339	12.5933	12.0853	11.6077	11.1581	10.7347	10.3356	9.9591	9.6036	9.2677	8.9501	8.6496	8.3649
20	18.9874	18.0456	17.1686	16.3514	15.5892	14.8775	14.2124	13.5903	13.0079	12.4622	11.9504	11.4699	11.0185	10.5940	10.1945	9.8181	9.4633	9.1285	8.8124	8.5136
21	19.8880	18.8570	17.9001	17.0112	16.1845	15.4150	14.6980	14.0292	13.4047	12.8212	12.2752	11.7641	11.2850	10.8355	10.4135	10.0168	9.6436	9.2922	8.9611	8.6487
22	20.7841	19.6604	18.6208	17.6580	16.7654	15.9369	15.1671	14.4511	13.7844	13.1630	12.5832	12.0416	11.5352	11.0612	10.6172	10.2007	9.8098	9.4424	9.0969	8.7715
23	21.6757	20.4558	19.3309	18.2922	17.3321	16.4436	15.6204	14.8568	14.1478	13.4886	12.8750	12.3034	11.7701	11.2722	10.8067	10.3711	9.9629	9.5802	9.2209	8.8832
24	22.5629	21.2434	20.0304	18.9139	17.8850	16.9355	16.0584	15.2470	14.4955	13.7986	13.1517	12.5504	11.9907	11.4693	10.9830	10.5288	10.1041	9.7066	9.3341	8.9847
25	23.4456	22.0232	20.7196	19.5235	18.4244	17.4131	16.4815	15.6221	14.8282	14.0939	13.4139	12.7834	12.1979	11.6536	11.1469	10.6748	10.2342	9.8226	9.4376	9.0770
26	24.3240	22.7952	21.3986	20.1210	18.9506	17.8768	16.8904	15.9828	15.1466	14.3752	13.6625	13.0032	12.3924	11.8258	11.2995	10.8100	10.3541	9.9290	9.5320	9.1609
27	25.1980	23.5596	22.0676	20.7069	19.4640	18.3270	17.2854	16.3296	15.4513	14.6430	13.8981	13.2105	12.5750	11.9867	11.4414	10.9352	10.4646	10.0266	9.6183	9.2372
28	26.0677	24.3164	22.7267	21.2813	19.9649	18.7641	17.6670	16.6631	15.7429	14.8981	14.1214	13.4062	12.7465	12.1371	11.5734	11.0511	10.5665	10.1161	9.6971	9.3066
29	26.9330	25.0658	23.3761	21.8444	20.4535	19.1885	18.0358	16.9837	16.0219	15.1411	14.3331	13.5907	12.9075	12.2777	11.6962	11.1584	10.6603	10.1983	9.7690	9.3696
30	27.7941	25.8077	24.0158	22.3965	20.9303	19.6004	18.3920	17.2920	16.2889	15.3725	14.5337	13.7648	13.0587	12.4090	11.8104	11.2578	10.7468	10.2737	9.8347	9.4269

X

Table X2 Annuity tables $\frac{1}{r}\left(1 - \frac{1}{(1+r)^n}\right)$ = present value of $1 per year for each of n years.

Interest rate per year

n	10.5%	11.0%	11.5%	12.0%	12.5%	13.0%	13.5%	14.0%	14.5%	15.0%	15.5%	16.0%	16.5%	17.0%	17.5%	18.0%	18.5%	19.0%	19.5%	20.0%
1	0.9050	0.9009	0.8969	0.8929	0.8889	0.8850	0.8811	0.8772	0.8734	0.8696	0.8658	0.8621	0.8584	0.8547	0.8511	0.8475	0.8439	0.8403	0.8368	0.8333
2	1.7240	1.7125	1.7012	1.6901	1.6790	1.6681	1.6573	1.6467	1.6361	1.6257	1.6154	1.6052	1.5952	1.5852	1.5754	1.5656	1.5560	1.5465	1.5371	1.5278
3	2.4651	2.4437	2.4226	2.4018	2.3813	2.3612	2.3413	2.3216	2.3023	2.2832	2.2644	2.2459	2.2276	2.2096	2.1918	2.1743	2.1570	2.1399	2.1231	2.1065
4	3.1359	3.1024	3.0696	3.0373	3.0056	2.9745	2.9438	2.9137	2.8841	2.8550	2.8263	2.7982	2.7705	2.7432	2.7164	2.6901	2.6641	2.6386	2.6135	2.5887
5	3.7429	3.6959	3.6499	3.6048	3.5606	3.5172	3.4747	3.4331	3.3922	3.3522	3.3129	3.2743	3.2365	3.1993	3.1629	3.1272	3.0921	3.0576	3.0238	2.9906
6	4.2922	4.2305	4.1703	4.1114	4.0538	3.9975	3.9425	3.8887	3.8360	3.7845	3.7341	3.6847	3.6365	3.5892	3.5429	3.4976	3.4532	3.4098	3.3672	3.3255
7	4.7893	4.7122	4.6370	4.5638	4.4923	4.4226	4.3546	4.2883	4.2236	4.1604	4.0988	4.0386	3.9798	3.9224	3.8663	3.8115	3.7580	3.7057	3.6546	3.6046
8	5.2392	5.1461	5.0556	4.9676	4.8820	4.7988	4.7177	4.6389	4.5621	4.4873	4.4145	4.3436	4.2745	4.2072	4.1415	4.0776	4.0152	3.9544	3.8950	3.8372
9	5.6463	5.5370	5.4311	5.3282	5.2285	5.1317	5.0377	4.9464	4.8577	4.7716	4.6879	4.6065	4.5275	4.4506	4.3758	4.3030	4.2322	4.1633	4.0963	4.0310
10	6.0148	5.8892	5.7678	5.6502	5.5364	5.4262	5.3195	5.2161	5.1159	5.0188	4.9246	4.8332	4.7446	4.6586	4.5751	4.4941	4.4154	4.3389	4.2647	4.1925
11	6.3482	6.2065	6.0697	5.9377	5.8102	5.6869	5.5679	5.4527	5.3414	5.2337	5.1295	5.0286	4.9310	4.8364	4.7448	4.6560	4.5699	4.4865	4.4056	4.3271
12	6.6500	6.4924	6.3406	6.1944	6.0535	5.9176	5.7867	5.6603	5.5383	5.4206	5.3069	5.1971	5.0910	4.9884	4.8892	4.7932	4.7004	4.6105	4.5235	4.4392
13	6.9230	6.7499	6.5835	6.4235	6.2698	6.1218	5.9794	5.8424	5.7103	5.5831	5.4605	5.3423	5.2283	5.1183	5.0121	4.9095	4.8104	4.7147	4.6222	4.5327
14	7.1702	6.9819	6.8013	6.6282	6.4620	6.3025	6.1493	6.0021	5.8606	5.7245	5.5935	5.4675	5.3462	5.2293	5.1167	5.0081	4.9033	4.8023	4.7047	4.6106
15	7.3938	7.1909	6.9967	6.8109	6.6329	6.4624	6.2989	6.1422	5.9918	5.8474	5.7087	5.5755	5.4474	5.3242	5.2057	5.0916	4.9817	4.8759	4.7738	4.6755
16	7.5962	7.3792	7.1719	6.9740	6.7848	6.6039	6.4308	6.2651	6.1063	5.9542	5.8084	5.6685	5.5342	5.4053	5.2814	5.1624	5.0479	4.9377	4.8317	4.7296
17	7.7794	7.5488	7.3291	7.1196	6.9198	6.7291	6.5469	6.3729	6.2064	6.0472	5.8947	5.7487	5.6088	5.4746	5.3459	5.2223	5.1037	4.9897	4.8801	4.7746
18	7.9451	7.7016	7.4700	7.2497	7.0398	6.8399	6.6493	6.4674	6.2938	6.1280	5.9695	5.8178	5.6728	5.5339	5.4008	5.2732	5.1508	5.0333	4.9205	4.8122
19	8.0952	7.8393	7.5964	7.3658	7.1465	6.9380	6.7395	6.5504	6.3701	6.1982	6.0342	5.8775	5.7277	5.5845	5.4475	5.3162	5.1905	5.0700	4.9544	4.8435
20	8.2309	7.9633	7.7098	7.4694	7.2414	7.0248	6.8189	6.6231	6.4368	6.2593	6.0902	5.9288	5.7748	5.6278	5.4872	5.3527	5.2241	5.1009	4.9828	4.8696
21	8.3538	8.0751	7.8115	7.5620	7.3256	7.1016	6.8889	6.6870	6.4950	6.3125	6.1387	5.9731	5.8153	5.6648	5.5210	5.3837	5.2524	5.1268	5.0065	4.8913
22	8.4649	8.1757	7.9027	7.6446	7.4006	7.1695	6.9506	6.7429	6.5459	6.3587	6.1807	6.0113	5.8501	5.6964	5.5498	5.4099	5.2763	5.1486	5.0264	4.9094
23	8.5656	8.2664	7.9845	7.7184	7.4672	7.2297	7.0049	6.7921	6.5903	6.3988	6.2170	6.0442	5.8799	5.7234	5.5743	5.4321	5.2964	5.1668	5.0430	4.9245
24	8.6666	8.3481	8.0578	7.7843	7.5264	7.2829	7.0528	6.8351	6.6291	6.4338	6.2485	6.0726	5.9055	5.7465	5.5951	5.4509	5.3134	5.1822	5.0569	4.9371
25	8.7390	8.4217	8.1236	7.8431	7.5790	7.3300	7.0950	6.8729	6.6629	6.4641	6.2758	6.0971	5.9274	5.7662	5.6129	5.4669	5.3278	5.1951	5.0685	4.9476
26	8.8136	8.4881	8.1826	7.8957	7.6258	7.3717	7.1321	6.9061	6.6925	6.4906	6.2994	6.1182	5.9463	5.7831	5.6280	5.4804	5.3399	5.2060	5.0783	4.9563
27	8.8811	8.5478	8.2355	7.9426	7.6674	7.4086	7.1649	6.9352	6.7184	6.5135	6.3198	6.1364	5.9625	5.7975	5.6408	5.4919	5.3501	5.2151	5.0864	4.9636
28	8.9422	8.6016	8.2830	7.9844	7.7043	7.4412	7.1937	6.9607	6.7409	6.5335	6.3375	6.1520	5.9764	5.8099	5.6518	5.5016	5.3588	5.2228	5.0932	4.9697
29	8.9974	8.6501	8.3255	8.0218	7.7372	7.4701	7.2191	6.9830	6.7606	6.5509	6.3528	6.1656	5.9883	5.8204	5.6611	5.5098	5.3661	5.2292	5.0989	4.9747
30	9.0474	8.6938	8.3637	8.0552	7.7664	7.4957	7.2415	7.0027	6.7778	6.5660	6.3661	6.1772	5.9986	5.8294	5.6690	5.5168	5.3722	5.2347	5.1037	4.9789

Table X2 Annuity tables $\dfrac{1}{r}\left(1 - \dfrac{1}{(1+r)^n}\right)$ = present value of \$1 per year for each of n years.

Interest rate per year

n	20.5%	21.0%	21.5%	22.0%	22.5%	23.0%	23.5%	24.0%	24.5%	25.0%	25.5%	26.0%	26.5%	27.0%	27.5%	28.0%	28.5%	29.0%	29.5%	30.0%
1	0.8299	0.8264	0.8230	0.8197	0.8163	0.8130	0.8097	0.8065	0.8032	0.8000	0.7968	0.7937	0.7905	0.7874	0.7843	0.7813	0.7782	0.7752	0.7722	0.7692
2	1.5186	1.5095	1.5004	1.4915	1.4827	1.4740	1.4654	1.4568	1.4484	1.4400	1.4317	1.4235	1.4154	1.4074	1.3995	1.3916	1.3838	1.3761	1.3685	1.3609
3	2.0901	2.0739	2.0580	2.0422	2.0267	2.0114	1.9962	1.9813	1.9666	1.9520	1.9376	1.9234	1.9094	1.8956	1.8819	1.8684	1.8551	1.8420	1.8290	1.8161
4	2.5644	2.5404	2.5169	2.4936	2.4708	2.4483	2.4261	2.4043	2.3828	2.3616	2.3407	2.3202	2.2999	2.2800	2.2603	2.2410	2.2219	2.2031	2.1845	2.1662
5	2.9580	2.9260	2.8945	2.8636	2.8333	2.8035	2.7742	2.7454	2.7171	2.6893	2.6619	2.6351	2.6087	2.5827	2.5571	2.5320	2.5073	2.4830	2.4591	2.4356
6	3.2847	3.2446	3.2054	3.1669	3.1292	3.0923	3.0560	3.0205	2.9856	2.9514	2.9179	2.8850	2.8527	2.8210	2.7899	2.7594	2.7294	2.7000	2.6711	2.6427
7	3.5557	3.5079	3.4612	3.4155	3.3708	3.3270	3.2842	3.2423	3.2013	3.1611	3.1218	3.0833	3.0456	3.0087	2.9725	2.9370	2.9023	2.8682	2.8348	2.8021
8	3.7807	3.7256	3.6718	3.6193	3.5680	3.5179	3.4690	3.4212	3.3745	3.3289	3.2843	3.2407	3.1981	3.1564	3.1157	3.0758	3.0368	2.9986	2.9613	2.9247
9	3.9674	3.9054	3.8451	3.7863	3.7290	3.6731	3.6186	3.5655	3.5137	3.4631	3.4138	3.3657	3.3187	3.2728	3.2280	3.1842	3.1415	3.0997	3.0589	3.0190
10	4.1223	4.0541	3.9877	3.9232	3.8604	3.7993	3.7398	3.6819	3.6254	3.5705	3.5170	3.4648	3.4140	3.3644	3.3161	3.2689	3.2229	3.1781	3.1343	3.0915
11	4.2509	4.1769	4.1051	4.0354	3.9677	3.9018	3.8379	3.7757	3.7152	3.6564	3.5992	3.5435	3.4893	3.4365	3.3851	3.3351	3.2863	3.2388	3.1925	3.1473
12	4.3576	4.2784	4.2017	4.1274	4.0552	3.9852	3.9173	3.8514	3.7873	3.7251	3.6647	3.6059	3.5489	3.4933	3.4393	3.3868	3.3357	3.2859	3.2374	3.1903
13	4.4461	4.3624	4.2813	4.2028	4.1267	4.0530	3.9816	3.9124	3.8452	3.7801	3.7169	3.6555	3.5959	3.5381	3.4818	3.4272	3.3741	3.3224	3.2722	3.2233
14	4.5196	4.4317	4.3467	4.2646	4.1851	4.1082	4.0337	3.9616	3.8918	3.8241	3.7585	3.6949	3.6331	3.5733	3.5152	3.4587	3.4039	3.3507	3.2990	3.2487
15	4.5806	4.4890	4.4006	4.3152	4.2327	4.1530	4.0759	4.0013	3.9291	3.8593	3.7916	3.7261	3.6626	3.6010	3.5413	3.4834	3.4272	3.3726	3.3197	3.2682
16	4.6312	4.5364	4.4449	4.3567	4.2716	4.1894	4.1100	4.0333	3.9591	3.8874	3.8180	3.7509	3.6858	3.6228	3.5618	3.5026	3.4453	3.3896	3.3356	3.2832
17	4.6732	4.5755	4.4814	4.3908	4.3034	4.2190	4.1377	4.0591	3.9832	3.9099	3.8391	3.7705	3.7042	3.6400	3.5779	3.5177	3.4594	3.4028	3.3480	3.2948
18	4.7080	4.6079	4.5115	4.4187	4.3293	4.2431	4.1601	4.0799	4.0026	3.9279	3.8558	3.7861	3.7187	3.6536	3.5905	3.5294	3.4703	3.4130	3.3575	3.3037
19	4.7370	4.6346	4.5362	4.4415	4.3504	4.2627	4.1782	4.0967	4.0182	3.9424	3.8692	3.7985	3.7302	3.6642	3.6004	3.5386	3.4788	3.4210	3.3649	3.3105
20	4.7610	4.6567	4.5565	4.4603	4.3677	4.2786	4.1929	4.1103	4.0306	3.9539	3.8798	3.8083	3.7393	3.6726	3.6081	3.5458	3.4855	3.4271	3.3706	3.3158
21	4.7809	4.6750	4.5733	4.4756	4.3818	4.2916	4.2047	4.1212	4.0407	3.9631	3.8883	3.8161	3.7465	3.6792	3.6142	3.5514	3.4907	3.4319	3.3750	3.3198
22	4.7974	4.6900	4.5871	4.4882	4.3933	4.3021	4.2144	4.1300	4.0487	3.9705	3.8951	3.8223	3.7522	3.6844	3.6190	3.5558	3.4947	3.4356	3.3783	3.3230
23	4.8111	4.7025	4.5984	4.4985	4.4027	4.3106	4.2222	4.1371	4.0552	3.9764	3.9005	3.8273	3.7567	3.6885	3.6228	3.5592	3.4978	3.4384	3.3810	3.3254
24	4.8225	4.7128	4.6077	4.5070	4.4104	4.3176	4.2285	4.1428	4.0604	3.9811	3.9047	3.8312	3.7602	3.6918	3.6257	3.5619	3.5002	3.4406	3.3830	3.3272
25	4.8320	4.7213	4.6154	4.5139	4.4166	4.3232	4.2336	4.1474	4.0646	3.9849	3.9082	3.8342	3.7630	3.6943	3.6280	3.5640	3.5021	3.4423	3.3845	3.3286
26	4.8398	4.7284	4.6217	4.5196	4.4217	4.3278	4.2377	4.1511	4.0679	3.9879	3.9109	3.8367	3.7652	3.6963	3.6298	3.5656	3.5036	3.4437	3.3857	3.3297
27	4.8463	4.7342	4.6270	4.5243	4.4259	4.3316	4.2411	4.1542	4.0706	3.9903	3.9131	3.8387	3.7670	3.6979	3.6312	3.5669	3.5047	3.4447	3.3867	3.3305
28	4.8517	4.7390	4.6312	4.5281	4.4293	4.3346	4.2438	4.1566	4.0728	3.9923	3.9148	3.8402	3.7684	3.6991	3.6323	3.5679	3.5056	3.4455	3.3874	3.3312
29	4.8562	4.7430	4.6348	4.5312	4.4321	4.3371	4.2460	4.1585	4.0745	3.9938	3.9162	3.8414	3.7695	3.7001	3.6332	3.5687	3.5063	3.4461	3.3879	3.3317
30	4.8599	4.7463	4.6377	4.5338	4.4344	4.3391	4.2478	4.1601	4.0759	3.9950	3.9173	3.8424	3.7703	3.7009	3.6339	3.5693	3.5069	3.4466	3.3884	3.3321

Table X3 Discount factor $\dfrac{1}{(1+r)^n}$ **= present value of \$1 to be received after n years.**

Interest rate per year

r / n	0.5%	1.0%	1.5%	2.0%	2.5%	3.0%	3.5%	4.0%	4.5%	5.0%	5.5%	6.0%	6.5%	7.0%	7.5%	8.0%	8.5%	9.0%	9.5%	10.0%
1	0.9950	0.9901	0.9852	0.9804	0.9756	0.9709	0.9662	0.9615	0.9569	0.9524	0.9479	0.9434	0.9390	0.9346	0.9302	0.9259	0.9217	0.9174	0.9132	0.9091
2	0.9901	0.9803	0.9707	0.9612	0.9518	0.9426	0.9335	0.9246	0.9157	0.9070	0.8985	0.8900	0.8817	0.8734	0.8653	0.8573	0.8495	0.8417	0.8340	0.8264
3	0.9851	0.9706	0.9563	0.9423	0.9286	0.9151	0.9019	0.8890	0.8763	0.8638	0.8516	0.8396	0.8278	0.8163	0.8050	0.7938	0.7829	0.7722	0.7617	0.7513
4	0.9802	0.9610	0.9422	0.9238	0.9060	0.8885	0.8714	0.8548	0.8386	0.8227	0.8072	0.7921	0.7773	0.7629	0.7488	0.7350	0.7216	0.7084	0.6956	0.6830
5	0.9754	0.9515	0.9283	0.9057	0.8839	0.8626	0.8420	0.8219	0.8025	0.7835	0.7651	0.7473	0.7299	0.7130	0.6966	0.6806	0.6650	0.6499	0.6352	0.6209
6	0.9705	0.9420	0.9145	0.8880	0.8623	0.8375	0.8135	0.7903	0.7679	0.7462	0.7252	0.7050	0.6853	0.6663	0.6480	0.6302	0.6129	0.5963	0.5801	0.5645
7	0.9657	0.9327	0.9010	0.8706	0.8413	0.8131	0.7860	0.7599	0.7348	0.7107	0.6874	0.6651	0.6435	0.6227	0.6028	0.5835	0.5649	0.5470	0.5298	0.5132
8	0.9609	0.9235	0.8877	0.8535	0.8207	0.7894	0.7594	0.7307	0.7032	0.6768	0.6516	0.6274	0.6042	0.5820	0.5607	0.5403	0.5207	0.5019	0.4838	0.4665
9	0.9561	0.9143	0.8746	0.8368	0.8007	0.7664	0.7337	0.7026	0.6729	0.6446	0.6176	0.5919	0.5674	0.5439	0.5216	0.5002	0.4799	0.4604	0.4418	0.4241
10	0.9513	0.9053	0.8617	0.8203	0.7812	0.7441	0.7089	0.6756	0.6439	0.6139	0.5854	0.5584	0.5327	0.5083	0.4852	0.4632	0.4423	0.4224	0.4035	0.3855
11	0.9466	0.8963	0.8489	0.8043	0.7621	0.7224	0.6849	0.6496	0.6162	0.5847	0.5549	0.5268	0.5002	0.4751	0.4513	0.4289	0.4076	0.3875	0.3685	0.3505
12	0.9419	0.8874	0.8364	0.7885	0.7436	0.7014	0.6618	0.6246	0.5897	0.5568	0.5260	0.4970	0.4697	0.4440	0.4199	0.3971	0.3757	0.3555	0.3365	0.3186
13	0.9372	0.8787	0.8240	0.7730	0.7254	0.6810	0.6394	0.6006	0.5643	0.5303	0.4986	0.4688	0.4410	0.4150	0.3906	0.3677	0.3463	0.3262	0.3073	0.2897
14	0.9326	0.8700	0.8118	0.7579	0.7077	0.6611	0.6178	0.5775	0.5400	0.5051	0.4726	0.4423	0.4141	0.3878	0.3633	0.3405	0.3191	0.2992	0.2807	0.2633
15	0.9279	0.8613	0.7999	0.7430	0.6905	0.6419	0.5969	0.5553	0.5167	0.4810	0.4479	0.4173	0.3888	0.3624	0.3380	0.3152	0.2941	0.2745	0.2563	0.2394
16	0.9233	0.8528	0.7880	0.7284	0.6736	0.6232	0.5767	0.5339	0.4945	0.4581	0.4246	0.3936	0.3651	0.3387	0.3144	0.2919	0.2711	0.2519	0.2341	0.2176
17	0.9187	0.8444	0.7764	0.7142	0.6572	0.6050	0.5572	0.5134	0.4732	0.4363	0.4024	0.3714	0.3428	0.3166	0.2925	0.2703	0.2499	0.2311	0.2138	0.1978
18	0.9141	0.8360	0.7649	0.7002	0.6412	0.5874	0.5384	0.4936	0.4528	0.4155	0.3815	0.3503	0.3219	0.2959	0.2720	0.2502	0.2303	0.2120	0.1952	0.1799
19	0.9096	0.8277	0.7536	0.6864	0.6255	0.5703	0.5202	0.4746	0.4333	0.3957	0.3616	0.3305	0.3022	0.2765	0.2531	0.2317	0.2122	0.1945	0.1783	0.1635
20	0.9051	0.8195	0.7425	0.6730	0.6103	0.5537	0.5026	0.4564	0.4146	0.3769	0.3427	0.3118	0.2838	0.2584	0.2354	0.2145	0.1956	0.1784	0.1628	0.1486
21	0.9006	0.8114	0.7315	0.6598	0.5954	0.5375	0.4856	0.4388	0.3968	0.3589	0.3249	0.2942	0.2665	0.2415	0.2190	0.1987	0.1803	0.1637	0.1487	0.1351
22	0.8961	0.8034	0.7207	0.6468	0.5809	0.5219	0.4692	0.4220	0.3797	0.3418	0.3079	0.2775	0.2502	0.2257	0.2037	0.1839	0.1662	0.1502	0.1358	0.1228
23	0.8916	0.7954	0.7100	0.6342	0.5667	0.5067	0.4533	0.4057	0.3634	0.3256	0.2919	0.2618	0.2349	0.2109	0.1895	0.1703	0.1531	0.1378	0.1240	0.1117
24	0.8872	0.7876	0.6995	0.6217	0.5529	0.4919	0.4380	0.3901	0.3477	0.3101	0.2767	0.2470	0.2206	0.1971	0.1763	0.1577	0.1412	0.1264	0.1133	0.1015
25	0.8828	0.7798	0.6892	0.6095	0.5394	0.4776	0.4231	0.3751	0.3327	0.2953	0.2622	0.2330	0.2071	0.1842	0.1640	0.1460	0.1301	0.1160	0.1034	0.0923
26	0.8784	0.7720	0.6790	0.5976	0.5262	0.4637	0.4088	0.3607	0.3184	0.2812	0.2486	0.2198	0.1945	0.1722	0.1525	0.1352	0.1199	0.1064	0.0945	0.0839
27	0.8740	0.7644	0.6690	0.5859	0.5134	0.4502	0.3950	0.3468	0.3047	0.2678	0.2356	0.2074	0.1826	0.1609	0.1419	0.1252	0.1105	0.0976	0.0863	0.0763
28	0.8697	0.7568	0.6591	0.5744	0.5009	0.4371	0.3817	0.3335	0.2916	0.2551	0.2233	0.1956	0.1715	0.1504	0.1320	0.1159	0.1019	0.0895	0.0788	0.0693
29	0.8653	0.7493	0.6494	0.5631	0.4887	0.4243	0.3687	0.3207	0.2790	0.2429	0.2117	0.1846	0.1610	0.1406	0.1228	0.1073	0.0939	0.0822	0.0719	0.0630
30	0.8610	0.7419	0.6398	0.5521	0.4767	0.4120	0.3563	0.3083	0.2670	0.2314	0.2006	0.1741	0.1512	0.1314	0.1142	0.0994	0.0865	0.0754	0.0657	0.0573

Table X3 Discount factor $\dfrac{1}{(1+r)^n}$ = present value of \$1 to be received after n years.

Interest rate per year

n	10.5%	11.0%	11.5%	12.0%	12.5%	13.0%	13.5%	14.0%	14.5%	15.0%	15.5%	16.0%	16.5%	17.0%	17.5%	18.0%	18.5%	19.0%	19.5%	20.0%
1	0.9050	0.9009	0.8969	0.8929	0.8889	0.8850	0.8811	0.8772	0.8734	0.8696	0.8658	0.8621	0.8584	0.8547	0.8511	0.8475	0.8439	0.8403	0.8368	0.8333
2	0.8190	0.8116	0.8044	0.7972	0.7901	0.7831	0.7763	0.7695	0.7628	0.7561	0.7496	0.7432	0.7368	0.7305	0.7243	0.7182	0.7121	0.7062	0.7003	0.6944
3	0.7412	0.7312	0.7214	0.7118	0.7023	0.6931	0.6839	0.6750	0.6662	0.6575	0.6490	0.6407	0.6324	0.6244	0.6164	0.6086	0.6010	0.5934	0.5860	0.5787
4	0.6707	0.6587	0.6470	0.6355	0.6243	0.6133	0.6026	0.5921	0.5818	0.5718	0.5619	0.5523	0.5429	0.5337	0.5246	0.5158	0.5071	0.4987	0.4904	0.4823
5	0.6070	0.5935	0.5803	0.5674	0.5549	0.5428	0.5309	0.5194	0.5081	0.4972	0.4865	0.4761	0.4660	0.4561	0.4465	0.4371	0.4280	0.4190	0.4104	0.4019
6	0.5493	0.5346	0.5204	0.5066	0.4933	0.4803	0.4678	0.4556	0.4438	0.4323	0.4212	0.4104	0.4000	0.3898	0.3800	0.3704	0.3612	0.3521	0.3434	0.3349
7	0.4971	0.4817	0.4667	0.4523	0.4385	0.4251	0.4121	0.3996	0.3876	0.3759	0.3647	0.3538	0.3433	0.3332	0.3234	0.3139	0.3048	0.2959	0.2874	0.2791
8	0.4499	0.4339	0.4186	0.4039	0.3897	0.3762	0.3631	0.3506	0.3385	0.3269	0.3158	0.3050	0.2947	0.2848	0.2752	0.2660	0.2572	0.2487	0.2405	0.2326
9	0.4071	0.3909	0.3754	0.3606	0.3464	0.3329	0.3199	0.3075	0.2956	0.2843	0.2734	0.2630	0.2530	0.2434	0.2342	0.2255	0.2170	0.2090	0.2012	0.1938
10	0.3684	0.3522	0.3367	0.3220	0.3079	0.2946	0.2819	0.2697	0.2582	0.2472	0.2367	0.2267	0.2171	0.2080	0.1994	0.1911	0.1832	0.1756	0.1684	0.1615
11	0.3334	0.3173	0.3020	0.2875	0.2737	0.2607	0.2483	0.2366	0.2255	0.2149	0.2049	0.1954	0.1864	0.1778	0.1697	0.1619	0.1546	0.1476	0.1409	0.1346
12	0.3018	0.2858	0.2708	0.2567	0.2433	0.2307	0.2188	0.2076	0.1969	0.1869	0.1774	0.1685	0.1600	0.1520	0.1444	0.1372	0.1304	0.1240	0.1179	0.1122
13	0.2731	0.2575	0.2429	0.2292	0.2163	0.2042	0.1928	0.1821	0.1720	0.1625	0.1536	0.1452	0.1373	0.1299	0.1229	0.1163	0.1101	0.1042	0.0987	0.0935
14	0.2471	0.2320	0.2178	0.2046	0.1922	0.1807	0.1698	0.1597	0.1502	0.1413	0.1330	0.1252	0.1179	0.1110	0.1046	0.0985	0.0929	0.0876	0.0826	0.0779
15	0.2236	0.2090	0.1954	0.1827	0.1709	0.1599	0.1496	0.1401	0.1312	0.1229	0.1152	0.1079	0.1012	0.0949	0.0890	0.0835	0.0784	0.0736	0.0691	0.0649
16	0.2024	0.1883	0.1752	0.1631	0.1519	0.1415	0.1318	0.1229	0.1146	0.1069	0.0997	0.0930	0.0869	0.0811	0.0758	0.0708	0.0661	0.0618	0.0578	0.0541
17	0.1832	0.1696	0.1572	0.1456	0.1350	0.1252	0.1162	0.1078	0.1001	0.0929	0.0863	0.0802	0.0746	0.0693	0.0645	0.0600	0.0558	0.0520	0.0484	0.0451
18	0.1658	0.1528	0.1409	0.1300	0.1200	0.1108	0.1023	0.0946	0.0874	0.0808	0.0747	0.0691	0.0640	0.0592	0.0549	0.0508	0.0471	0.0437	0.0405	0.0376
19	0.1500	0.1377	0.1264	0.1161	0.1067	0.0981	0.0902	0.0829	0.0763	0.0703	0.0647	0.0596	0.0549	0.0506	0.0467	0.0431	0.0398	0.0367	0.0339	0.0313
20	0.1358	0.1240	0.1134	0.1037	0.0948	0.0868	0.0794	0.0728	0.0667	0.0611	0.0560	0.0514	0.0471	0.0433	0.0397	0.0365	0.0335	0.0308	0.0284	0.0261
21	0.1229	0.1117	0.1017	0.0926	0.0843	0.0768	0.0700	0.0638	0.0582	0.0531	0.0485	0.0443	0.0405	0.0370	0.0338	0.0309	0.0283	0.0259	0.0237	0.0217
22	0.1112	0.1007	0.0912	0.0826	0.0749	0.0680	0.0617	0.0560	0.0508	0.0462	0.0420	0.0382	0.0347	0.0316	0.0288	0.0262	0.0239	0.0218	0.0199	0.0181
23	0.1006	0.0907	0.0818	0.0738	0.0666	0.0601	0.0543	0.0491	0.0444	0.0402	0.0364	0.0329	0.0298	0.0270	0.0245	0.0222	0.0202	0.0183	0.0166	0.0151
24	0.0911	0.0817	0.0734	0.0659	0.0592	0.0532	0.0479	0.0431	0.0388	0.0349	0.0315	0.0284	0.0256	0.0231	0.0208	0.0188	0.0170	0.0154	0.0139	0.0126
25	0.0824	0.0736	0.0658	0.0588	0.0526	0.0471	0.0422	0.0378	0.0339	0.0304	0.0273	0.0245	0.0220	0.0197	0.0177	0.0160	0.0144	0.0129	0.0116	0.0105
26	0.0746	0.0663	0.0590	0.0525	0.0468	0.0417	0.0372	0.0331	0.0296	0.0264	0.0236	0.0211	0.0189	0.0169	0.0151	0.0135	0.0121	0.0109	0.0097	0.0087
27	0.0675	0.0597	0.0529	0.0469	0.0416	0.0369	0.0327	0.0291	0.0258	0.0230	0.0204	0.0182	0.0162	0.0144	0.0129	0.0115	0.0102	0.0091	0.0081	0.0073
28	0.0611	0.0538	0.0475	0.0419	0.0370	0.0326	0.0288	0.0255	0.0226	0.0200	0.0177	0.0157	0.0139	0.0123	0.0109	0.0097	0.0086	0.0077	0.0068	0.0061
29	0.0553	0.0485	0.0426	0.0374	0.0329	0.0289	0.0254	0.0224	0.0197	0.0174	0.0153	0.0135	0.0119	0.0105	0.0093	0.0082	0.0073	0.0064	0.0057	0.0051
30	0.0500	0.0437	0.0382	0.0334	0.0292	0.0256	0.0224	0.0196	0.0172	0.0151	0.0133	0.0116	0.0102	0.0090	0.0079	0.0070	0.0061	0.0054	0.0048	0.0042

Table X3   Discount factor $\dfrac{1}{(1+r)^n}$ = present value of \$1 to be received after n years.

Interest rate per year

n \ r	20.5%	21.0%	21.5%	22.0%	22.5%	23.0%	23.5%	24.0%	24.5%	25.0%	25.5%	26.0%	26.5%	27.0%	27.5%	28.0%	28.5%	29.0%	29.5%	30.0%
1	0.8299	0.8264	0.8230	0.8197	0.8163	0.8130	0.8097	0.8065	0.8032	0.8000	0.7968	0.7937	0.7905	0.7874	0.7843	0.7813	0.7782	0.7752	0.7722	0.7692
2	0.6887	0.6830	0.6774	0.6719	0.6664	0.6610	0.6556	0.6504	0.6452	0.6400	0.6349	0.6299	0.6249	0.6200	0.6151	0.6104	0.6056	0.6009	0.5963	0.5917
3	0.5715	0.5645	0.5575	0.5507	0.5440	0.5374	0.5309	0.5245	0.5182	0.5120	0.5059	0.4999	0.4940	0.4882	0.4825	0.4768	0.4713	0.4658	0.4605	0.4552
4	0.4743	0.4665	0.4589	0.4514	0.4441	0.4369	0.4299	0.4230	0.4162	0.4096	0.4031	0.3968	0.3905	0.3844	0.3784	0.3725	0.3668	0.3611	0.3556	0.3501
5	0.3936	0.3855	0.3777	0.3700	0.3625	0.3552	0.3481	0.3411	0.3343	0.3277	0.3212	0.3149	0.3087	0.3027	0.2968	0.2910	0.2854	0.2799	0.2746	0.2693
6	0.3266	0.3186	0.3108	0.3033	0.2959	0.2888	0.2818	0.2751	0.2685	0.2621	0.2559	0.2499	0.2440	0.2383	0.2328	0.2274	0.2221	0.2170	0.2120	0.2072
7	0.2711	0.2633	0.2558	0.2486	0.2416	0.2348	0.2282	0.2218	0.2157	0.2097	0.2039	0.1983	0.1929	0.1877	0.1826	0.1776	0.1729	0.1682	0.1637	0.1594
8	0.2250	0.2176	0.2106	0.2038	0.1972	0.1909	0.1848	0.1789	0.1732	0.1678	0.1625	0.1574	0.1525	0.1478	0.1432	0.1388	0.1345	0.1304	0.1264	0.1226
9	0.1867	0.1799	0.1733	0.1670	0.1610	0.1552	0.1496	0.1443	0.1391	0.1342	0.1295	0.1249	0.1206	0.1164	0.1123	0.1084	0.1047	0.1011	0.0976	0.0943
10	0.1549	0.1486	0.1426	0.1369	0.1314	0.1262	0.1212	0.1164	0.1118	0.1074	0.1032	0.0992	0.0953	0.0916	0.0881	0.0847	0.0815	0.0784	0.0754	0.0725
11	0.1286	0.1228	0.1174	0.1122	0.1073	0.1026	0.0981	0.0938	0.0898	0.0859	0.0822	0.0787	0.0753	0.0721	0.0691	0.0662	0.0634	0.0607	0.0582	0.0558
12	0.1067	0.1015	0.0966	0.0920	0.0876	0.0834	0.0794	0.0757	0.0721	0.0687	0.0655	0.0625	0.0596	0.0568	0.0542	0.0517	0.0493	0.0471	0.0450	0.0429
13	0.0885	0.0839	0.0795	0.0754	0.0715	0.0678	0.0643	0.0610	0.0579	0.0550	0.0522	0.0496	0.0471	0.0447	0.0425	0.0404	0.0384	0.0365	0.0347	0.0330
14	0.0735	0.0693	0.0655	0.0618	0.0584	0.0551	0.0521	0.0492	0.0465	0.0440	0.0416	0.0393	0.0372	0.0352	0.0333	0.0316	0.0299	0.0283	0.0268	0.0254
15	0.0610	0.0573	0.0539	0.0507	0.0476	0.0448	0.0422	0.0397	0.0374	0.0352	0.0331	0.0312	0.0294	0.0277	0.0261	0.0247	0.0233	0.0219	0.0207	0.0195
16	0.0506	0.0474	0.0443	0.0415	0.0389	0.0364	0.0341	0.0320	0.0300	0.0281	0.0264	0.0248	0.0233	0.0218	0.0205	0.0193	0.0181	0.0170	0.0160	0.0150
17	0.0420	0.0391	0.0365	0.0340	0.0317	0.0296	0.0276	0.0258	0.0241	0.0225	0.0210	0.0197	0.0184	0.0172	0.0161	0.0150	0.0141	0.0132	0.0123	0.0116
18	0.0349	0.0323	0.0300	0.0279	0.0259	0.0241	0.0224	0.0208	0.0194	0.0180	0.0168	0.0156	0.0145	0.0135	0.0126	0.0118	0.0110	0.0102	0.0095	0.0089
19	0.0289	0.0267	0.0247	0.0229	0.0212	0.0196	0.0181	0.0168	0.0156	0.0144	0.0134	0.0124	0.0115	0.0107	0.0099	0.0092	0.0085	0.0079	0.0074	0.0068
20	0.0240	0.0221	0.0203	0.0187	0.0173	0.0159	0.0147	0.0135	0.0125	0.0115	0.0106	0.0098	0.0091	0.0084	0.0078	0.0072	0.0066	0.0061	0.0057	0.0053
21	0.0199	0.0183	0.0167	0.0154	0.0141	0.0129	0.0119	0.0109	0.0100	0.0092	0.0085	0.0078	0.0072	0.0066	0.0061	0.0056	0.0052	0.0048	0.0044	0.0040
22	0.0165	0.0151	0.0138	0.0126	0.0115	0.0105	0.0096	0.0088	0.0081	0.0074	0.0068	0.0062	0.0057	0.0052	0.0048	0.0044	0.0040	0.0037	0.0034	0.0031
23	0.0137	0.0125	0.0113	0.0103	0.0094	0.0086	0.0078	0.0071	0.0065	0.0059	0.0054	0.0049	0.0045	0.0041	0.0037	0.0034	0.0031	0.0029	0.0026	0.0024
24	0.0114	0.0103	0.0093	0.0085	0.0077	0.0070	0.0063	0.0057	0.0052	0.0047	0.0043	0.0039	0.0035	0.0032	0.0029	0.0027	0.0024	0.0022	0.0020	0.0018
25	0.0094	0.0085	0.0077	0.0069	0.0063	0.0057	0.0051	0.0046	0.0042	0.0038	0.0034	0.0031	0.0028	0.0025	0.0023	0.0021	0.0019	0.0017	0.0016	0.0014
26	0.0078	0.0070	0.0063	0.0057	0.0051	0.0046	0.0041	0.0037	0.0034	0.0030	0.0027	0.0025	0.0022	0.0020	0.0018	0.0016	0.0015	0.0013	0.0012	0.0011
27	0.0065	0.0058	0.0052	0.0047	0.0042	0.0037	0.0033	0.0030	0.0027	0.0024	0.0022	0.0019	0.0018	0.0016	0.0014	0.0013	0.0011	0.0010	0.0009	0.0008
28	0.0054	0.0048	0.0043	0.0038	0.0034	0.0030	0.0027	0.0024	0.0022	0.0019	0.0017	0.0015	0.0014	0.0012	0.0011	0.0010	0.0009	0.0008	0.0007	0.0006
29	0.0045	0.0040	0.0035	0.0031	0.0028	0.0025	0.0022	0.0020	0.0017	0.0015	0.0014	0.0012	0.0011	0.0010	0.0009	0.0008	0.0007	0.0006	0.0006	0.0005
30	0.0037	0.0033	0.0029	0.0026	0.0023	0.0020	0.0018	0.0016	0.0014	0.0012	0.0011	0.0010	0.0009	0.0008	0.0007	0.0006	0.0005	0.0005	0.0004	0.0004

$$nF = \frac{1}{\ln(1+r)}\left(1 - \frac{1}{(1+r)^n}\right) \quad \text{or} \quad nF = \frac{1-e^{-nj}}{j}$$

Table X4 nF = present value of \$1 per year received in a continuous stream for each of n years (discounted at an annually compounded rate r).

n	0.5%	1.0%	1.5%	2.0%	2.5%	3.0%	3.5%	4.0%	4.5%	5.0%	5.5%	6.0%	6.5%	7.0%	7.5%	8.0%	8.5%	9.0%	9.5%	10.0%
$j = \ln(1+r)$	0.0050	0.0100	0.0149	0.0198	0.0247	0.0296	0.0344	0.0392	0.0440	0.0488	0.0535	0.0583	0.0630	0.0677	0.0723	0.0770	0.0816	0.0862	0.0908	0.0953
1	0.998	0.995	0.993	0.990	0.988	0.985	0.983	0.981	0.978	0.976	0.974	0.971	0.969	0.967	0.965	0.962	0.960	0.958	0.956	0.954
2	1.990	1.980	1.971	1.961	1.951	1.942	1.933	1.924	1.914	1.906	1.897	1.888	1.879	1.871	1.862	1.854	1.845	1.837	1.829	1.821
3	2.978	2.956	2.934	2.913	2.892	2.871	2.850	2.830	2.810	2.791	2.771	2.752	2.734	2.715	2.697	2.679	2.661	2.644	2.626	2.609
4	3.960	3.921	3.883	3.846	3.809	3.773	3.737	3.702	3.668	3.634	3.601	3.568	3.536	3.504	3.473	3.443	3.413	3.383	3.354	3.326
5	4.938	4.878	4.818	4.760	4.704	4.648	4.594	4.540	4.488	4.437	4.387	4.338	4.289	4.242	4.196	4.150	4.106	4.062	4.019	3.977
6	5.911	5.824	5.740	5.657	5.577	5.498	5.421	5.346	5.273	5.202	5.132	5.063	4.997	4.931	4.868	4.805	4.744	4.685	4.627	4.570
7	6.879	6.762	6.648	6.536	6.428	6.323	6.221	6.121	6.024	5.930	5.838	5.748	5.661	5.576	5.493	5.412	5.333	5.256	5.181	5.108
8	7.843	7.690	7.542	7.398	7.259	7.124	6.994	6.867	6.743	6.623	6.507	6.394	6.285	6.178	6.074	5.974	5.876	5.780	5.688	5.597
9	8.801	8.609	8.423	8.244	8.070	7.902	7.740	7.583	7.431	7.284	7.142	7.004	6.870	6.741	6.615	6.494	6.376	6.261	6.150	6.042
10	9.755	9.519	9.291	9.072	8.861	8.658	8.461	8.272	8.089	7.913	7.743	7.579	7.420	7.267	7.118	6.975	6.836	6.702	6.573	6.447
11	10.704	10.419	10.146	9.884	9.633	9.391	9.158	8.935	8.719	8.512	8.313	8.121	7.936	7.758	7.586	7.421	7.261	7.107	6.958	6.815
12	11.648	11.311	10.989	10.681	10.385	10.103	9.831	9.572	9.322	9.083	8.853	8.633	8.421	8.218	8.022	7.834	7.653	7.478	7.311	7.149
13	12.588	12.194	11.819	11.461	11.120	10.794	10.482	10.184	9.899	9.627	9.366	9.116	8.876	8.647	8.427	8.216	8.013	7.819	7.632	7.453
14	13.522	13.069	12.637	12.227	11.836	11.465	11.111	10.773	10.451	10.144	9.851	9.571	9.304	9.048	8.804	8.570	8.346	8.131	7.926	7.729
15	14.453	13.934	13.443	12.977	12.536	12.116	11.718	11.339	10.979	10.637	10.311	10.001	9.705	9.423	9.154	8.897	8.652	8.418	8.194	7.980
16	15.378	14.791	14.237	13.713	13.218	12.749	12.305	11.884	11.485	11.107	10.747	10.406	10.082	9.774	9.480	9.201	8.935	8.681	8.439	8.209
17	16.299	15.640	15.019	14.434	13.883	13.363	12.871	12.407	11.969	11.554	11.161	10.789	10.436	10.101	9.783	9.482	9.195	8.923	8.663	8.416
18	17.216	16.480	15.790	15.141	14.532	13.959	13.419	12.911	12.432	11.979	11.553	11.149	10.768	10.407	10.066	9.742	9.435	9.144	8.868	8.605
19	18.128	17.312	16.549	15.835	15.165	14.538	13.948	13.395	12.875	12.385	11.994	11.490	11.080	10.693	10.328	9.983	9.656	9.347	9.054	8.777
20	19.035	18.136	17.297	16.514	15.783	15.100	14.460	13.860	13.298	12.771	12.276	11.811	11.373	10.961	10.572	10.206	9.860	9.533	9.225	8.932
21	19.938	18.951	18.034	17.181	16.386	15.645	14.954	14.308	13.704	13.139	12.610	12.114	11.648	11.210	10.799	10.412	10.048	9.704	9.380	9.074
22	20.836	19.759	18.760	17.834	16.974	16.175	15.431	14.738	14.092	13.489	12.926	12.399	11.906	11.444	11.011	10.604	10.221	9.861	9.522	9.203
23	21.730	20.558	19.475	18.475	17.548	16.689	15.892	15.152	14.464	13.823	13.226	12.669	12.149	11.662	11.207	10.781	10.381	10.005	9.652	9.320
24	22.619	21.349	20.180	19.102	18.108	17.188	16.338	15.550	14.819	14.141	13.510	12.923	12.376	11.866	11.390	10.945	10.528	10.137	9.771	9.427
25	23.504	22.133	20.875	19.718	18.654	17.673	16.768	15.932	15.159	14.443	13.780	13.163	12.590	12.057	11.560	11.096	10.663	10.258	9.879	9.524
26	24.385	22.909	21.559	20.322	19.187	18.144	17.184	16.300	15.485	14.732	14.035	13.389	12.791	12.235	11.718	11.237	10.788	10.369	9.978	9.612
27	25.261	23.677	22.233	20.913	19.706	18.601	17.586	16.654	15.796	15.006	14.277	13.603	12.979	12.402	11.865	11.367	10.903	10.471	10.068	9.692
28	26.133	24.438	22.897	21.493	20.213	19.044	17.974	16.994	16.094	15.268	14.506	13.804	13.156	12.557	12.002	11.487	11.009	10.565	10.151	9.765
29	27.000	25.191	23.551	22.062	20.708	19.475	18.350	17.321	16.380	15.517	14.724	13.994	13.323	12.703	12.129	11.599	11.107	10.651	10.226	9.831
30	27.863	25.937	24.196	22.620	21.191	19.893	18.712	17.636	16.653	15.754	14.930	14.174	13.479	12.838	12.248	11.702	11.197	10.729	10.295	9.891

Interest rate per year

X

$$nF = \frac{1}{\ln(1+r)}\left(1 - \frac{1}{(1+r)^n}\right) \quad \text{or} \quad nF = \frac{1-e^{-nj}}{j}$$

Table X4 nF = present value of \$1 per year received in a continuous stream for each of n years (discounted at an annually compounded rate r).

Interest rate per year

r	10.5%	11.0%	11.5%	12.0%	12.5%	13.0%	13.5%	14.0%	14.5%	15.0%	15.5%	16.0%	16.5%	17.0%	17.5%	18.0%	18.5%	19.0%	19.5%	20.0%
$j=\ln(1+r)$	0.0998	0.1044	0.1089	0.1133	0.1178	0.1222	0.1266	0.1310	0.1354	0.1398	0.1441	0.1484	0.1527	0.1570	0.1613	0.1655	0.1697	0.1740	0.1781	0.1823
n																				
1	0.952	0.950	0.947	0.945	0.943	0.941	0.939	0.937	0.935	0.933	0.931	0.929	0.927	0.925	0.924	0.922	0.920	0.918	0.916	0.914
2	1.813	1.805	1.797	1.790	1.782	1.774	1.767	1.759	1.752	1.745	1.738	1.730	1.723	1.716	1.710	1.703	1.696	1.689	1.683	1.676
3	2.592	2.576	2.559	2.543	2.527	2.512	2.496	2.481	2.465	2.450	2.436	2.421	2.407	2.392	2.378	2.365	2.351	2.337	2.324	2.311
4	3.298	3.270	3.243	3.216	3.190	3.164	3.138	3.113	3.088	3.064	3.040	3.016	2.993	2.970	2.948	2.925	2.904	2.882	2.861	2.840
5	3.936	3.896	3.856	3.817	3.779	3.741	3.704	3.668	3.633	3.598	3.563	3.530	3.497	3.464	3.432	3.401	3.370	3.340	3.310	3.281
6	4.514	4.459	4.406	4.353	4.302	4.252	4.203	4.155	4.108	4.062	4.017	3.972	3.929	3.886	3.845	3.804	3.764	3.724	3.686	3.648
7	5.037	4.967	4.899	4.832	4.768	4.704	4.642	4.582	4.523	4.465	4.409	4.354	4.300	4.247	4.196	4.145	4.096	4.048	4.000	3.954
8	5.510	5.424	5.341	5.260	5.181	5.104	5.029	4.956	4.885	4.816	4.748	4.682	4.618	4.555	4.494	4.434	4.376	4.319	4.264	4.209
9	5.938	5.836	5.738	5.642	5.549	5.458	5.371	5.285	5.202	5.121	5.042	4.966	4.891	4.819	4.748	4.680	4.613	4.547	4.484	4.422
10	6.325	6.208	6.093	5.983	5.876	5.772	5.671	5.573	5.478	5.386	5.297	5.210	5.126	5.044	4.965	4.887	4.812	4.739	4.668	4.599
11	6.676	6.542	6.412	6.287	6.166	6.049	5.936	5.826	5.720	5.617	5.517	5.421	5.327	5.237	5.149	5.063	4.981	4.900	4.822	4.747
12	6.993	6.843	6.699	6.559	6.424	6.294	6.169	6.048	5.931	5.818	5.708	5.603	5.500	5.401	5.305	5.213	5.123	5.036	4.951	4.870
13	7.280	7.115	6.955	6.802	6.654	6.512	6.375	6.242	6.115	5.992	5.874	5.759	5.649	5.542	5.439	5.339	5.243	5.150	5.059	4.972
14	7.540	7.359	7.185	7.018	6.858	6.704	6.556	6.413	6.276	6.144	6.017	5.894	5.776	5.662	5.552	5.446	5.344	5.245	5.150	5.058
15	7.776	7.579	7.392	7.212	7.039	6.874	6.715	6.563	6.416	6.276	6.141	6.010	5.885	5.765	5.649	5.537	5.429	5.326	5.225	5.129
16	7.988	7.778	7.577	7.385	7.201	7.024	6.856	6.694	6.539	6.390	6.248	6.111	5.979	5.853	5.731	5.614	5.502	5.393	5.289	5.188
17	8.181	7.957	7.743	7.539	7.344	7.158	6.980	6.809	6.646	6.490	6.341	6.197	6.060	5.928	5.801	5.679	5.562	5.450	5.342	5.238
18	8.355	8.118	7.892	7.676	7.471	7.275	7.089	6.910	6.740	6.577	6.421	6.272	6.129	5.992	5.861	5.735	5.614	5.498	5.386	5.279
19	8.513	8.263	8.025	7.799	7.584	7.380	7.185	6.999	6.822	6.652	6.491	6.336	6.188	6.047	5.911	5.782	5.657	5.538	5.423	5.313
20	8.656	8.394	8.145	7.909	7.685	7.472	7.269	7.077	6.893	6.718	6.551	6.391	6.239	6.094	5.954	5.821	5.694	5.571	5.454	5.342
21	8.785	8.511	8.253	8.007	7.775	7.554	7.344	7.145	6.955	6.775	6.603	6.439	6.283	6.134	5.991	5.855	5.724	5.600	5.480	5.366
22	8.902	8.618	8.349	8.095	7.854	7.626	7.410	7.205	7.010	6.824	6.648	6.480	6.320	6.168	6.022	5.883	5.751	5.623	5.502	5.385
23	9.008	8.713	8.435	8.173	7.925	7.690	7.468	7.257	7.057	6.868	6.687	6.516	6.353	6.197	6.049	5.908	5.773	5.643	5.520	5.402
24	9.104	8.799	8.513	8.243	7.988	7.747	7.519	7.303	7.099	6.905	6.721	6.546	6.380	6.222	6.072	5.928	5.791	5.660	5.535	5.416
25	9.190	8.877	8.582	8.305	8.043	7.797	7.564	7.344	7.135	6.938	6.750	6.573	6.404	6.244	6.091	5.945	5.807	5.674	5.548	5.427
26	9.269	8.947	8.645	8.360	8.093	7.841	7.603	7.379	7.167	6.966	6.776	6.596	6.424	6.262	6.107	5.960	5.820	5.686	5.559	5.437
27	9.340	9.010	8.700	8.410	8.137	7.880	7.638	7.410	7.194	6.991	6.798	6.615	6.442	6.277	6.121	5.973	5.831	5.696	5.568	5.445
28	9.404	9.066	8.751	8.454	8.176	7.915	7.669	7.437	7.219	7.012	6.817	6.632	6.457	6.291	6.133	5.983	5.840	5.705	5.575	5.452
29	9.462	9.118	8.796	8.494	8.211	7.946	7.696	7.461	7.240	7.031	6.833	6.647	6.470	6.302	6.143	5.992	5.848	5.712	5.581	5.457
30	9.515	9.164	8.836	8.529	8.242	7.973	7.720	7.482	7.258	7.047	6.848	6.659	6.481	6.312	6.152	6.000	5.855	5.718	5.587	5.462

X

$$nF = \frac{1}{\ln(1+r)}\left(1 - \frac{1}{(1+r)^n}\right) \quad \text{or} \quad nF = \frac{1-e^{-nj}}{j}$$

Table X4 nF = present value of \$1 per year received in a continuous stream for each of n years (discounted at an annually compounded rate r).

r	20.5%	21.0%	21.5%	22.0%	22.5%	23.0%	23.5%	24.0%	24.5%	25.0%	25.5%	26.0%	26.5%	27.0%	27.5%	28.0%	28.5%	29.0%	29.5%	30.0%
n \ $j=\ln(1+r)$	0.1865	0.1906	0.1947	0.1989	0.2029	0.2070	0.2111	0.2151	0.2191	0.2231	0.2271	0.2311	0.2351	0.2390	0.2429	0.2469	0.2508	0.2546	0.2585	0.2624
1	0.912	0.910	0.909	0.907	0.905	0.903	0.902	0.900	0.898	0.896	0.895	0.893	0.891	0.889	0.888	0.886	0.884	0.883	0.881	0.880
2	1.669	1.663	1.657	1.650	1.644	1.638	1.631	1.625	1.619	1.613	1.607	1.601	1.596	1.590	1.584	1.578	1.573	1.567	1.562	1.556
3	2.298	2.285	2.272	2.259	2.247	2.235	2.223	2.211	2.199	2.187	2.175	2.164	2.153	2.141	2.130	2.119	2.108	2.098	2.087	2.077
4	2.819	2.799	2.779	2.759	2.739	2.720	2.701	2.682	2.664	2.646	2.628	2.610	2.593	2.576	2.559	2.542	2.525	2.509	2.493	2.477
5	3.252	3.223	3.196	3.168	3.141	3.115	3.089	3.063	3.038	3.013	2.989	2.964	2.941	2.917	2.895	2.872	2.850	2.828	2.806	2.785
6	3.611	3.574	3.539	3.504	3.469	3.436	3.402	3.370	3.338	3.307	3.276	3.246	3.216	3.187	3.158	3.130	3.102	3.075	3.048	3.022
7	3.909	3.865	3.821	3.779	3.737	3.696	3.657	3.617	3.579	3.542	3.505	3.469	3.433	3.399	3.365	3.331	3.299	3.266	3.235	3.204
8	4.156	4.104	4.054	4.004	3.956	3.909	3.862	3.817	3.773	3.730	3.687	3.646	3.605	3.566	3.527	3.489	3.451	3.415	3.379	3.344
9	4.361	4.302	4.245	4.189	4.134	4.081	4.029	3.978	3.928	3.880	3.833	3.786	3.741	3.697	3.654	3.612	3.570	3.530	3.491	3.452
10	4.532	4.466	4.403	4.340	4.280	4.221	4.164	4.108	4.053	4.000	3.948	3.898	3.849	3.801	3.754	3.708	3.663	3.619	3.577	3.535
11	4.673	4.602	4.532	4.465	4.399	4.335	4.273	4.213	4.154	4.096	4.041	3.986	3.934	3.882	3.832	3.783	3.735	3.689	3.643	3.599
12	4.790	4.713	4.639	4.566	4.496	4.428	4.361	4.297	4.234	4.173	4.114	4.057	4.001	3.946	3.893	3.841	3.791	3.742	3.694	3.648
13	4.888	4.806	4.727	4.650	4.575	4.503	4.433	4.365	4.299	4.235	4.173	4.112	4.054	3.997	3.941	3.887	3.835	3.784	3.734	3.686
14	4.968	4.882	4.799	4.718	4.640	4.564	4.491	4.420	4.351	4.284	4.220	4.157	4.096	4.036	3.979	3.923	3.869	3.816	3.765	3.715
15	5.036	4.945	4.858	4.774	4.693	4.614	4.538	4.464	4.393	4.324	4.257	4.192	4.129	4.068	4.009	3.951	3.895	3.841	3.788	3.737
16	5.091	4.998	4.907	4.820	4.736	4.655	4.576	4.500	4.426	4.355	4.286	4.220	4.155	4.092	4.032	3.973	3.916	3.860	3.806	3.754
17	5.137	5.041	4.948	4.858	4.771	4.687	4.607	4.529	4.453	4.381	4.310	4.242	4.176	4.112	4.050	3.990	3.932	3.875	3.821	3.767
18	5.176	5.076	4.981	4.889	4.800	4.714	4.632	4.552	4.475	4.401	4.329	4.259	4.192	4.127	4.064	4.003	3.944	3.887	3.831	3.778
19	5.207	5.106	5.008	4.914	4.823	4.736	4.652	4.571	4.492	4.417	4.344	4.273	4.205	4.139	4.075	4.014	3.954	3.896	3.840	3.785
20	5.234	5.130	5.030	4.935	4.842	4.754	4.668	4.586	4.506	4.430	4.356	4.284	4.215	4.149	4.084	4.022	3.961	3.903	3.846	3.791
21	5.256	5.150	5.049	4.952	4.858	4.768	4.681	4.598	4.518	4.440	4.365	4.293	4.223	4.156	4.091	4.028	3.967	3.908	3.851	3.796
22	5.274	5.167	5.064	4.966	4.871	4.780	4.692	4.608	4.527	4.448	4.373	4.300	4.230	4.162	4.096	4.033	3.972	3.913	3.855	3.800
23	5.289	5.181	5.077	4.977	4.881	4.789	4.701	4.616	4.534	4.455	4.379	4.306	4.235	4.167	4.101	4.037	3.975	3.916	3.858	3.802
24	5.301	5.192	5.087	4.986	4.890	4.797	4.708	4.622	4.540	4.460	4.384	4.310	4.239	4.170	4.104	4.040	3.978	3.918	3.860	3.804
25	5.312	5.201	5.095	4.994	4.897	4.803	4.714	4.627	4.544	4.464	4.388	4.314	4.242	4.173	4.107	4.042	3.980	3.920	3.862	3.806
26	5.320	5.209	5.102	5.000	4.902	4.808	4.718	4.631	4.548	4.468	4.391	4.316	4.245	4.175	4.109	4.044	3.982	3.922	3.864	3.807
27	5.328	5.216	5.108	5.005	4.907	4.813	4.722	4.635	4.551	4.471	4.393	4.318	4.247	4.177	4.110	4.046	3.983	3.923	3.865	3.808
28	5.334	5.221	5.113	5.010	4.911	4.816	4.725	4.637	4.554	4.473	4.395	4.320	4.248	4.179	4.112	4.047	3.984	3.924	3.866	3.809
29	5.338	5.225	5.117	5.013	4.914	4.819	4.727	4.640	4.555	4.474	4.397	4.322	4.249	4.180	4.113	4.048	3.985	3.925	3.866	3.810
30	5.343	5.229	5.120	5.016	4.916	4.821	4.729	4.641	4.557	4.476	4.398	4.323	4.250	4.181	4.113	4.048	3.986	3.925	3.867	3.810

X

Table X5 Values of e^{nj} = future value of $1 invested at a continuously compounded rate j for n years.

n j	0.00	0.01	0.02	0.03	0.04	0.05	0.06	0.07	0.08	0.09
0.00	1.0000	1.0101	1.0202	1.0305	1.0408	1.0513	1.0618	1.0725	1.0833	1.0942
0.10	1.1052	1.1163	1.1275	1.1388	1.1503	1.1618	1.1735	1.1853	1.1972	1.2092
0.20	1.2214	1.2337	1.2461	1.2586	1.2712	1.2840	1.2969	1.3100	1.3231	1.3364
0.30	1.3499	1.3634	1.3771	1.3910	1.4049	1.4191	1.4333	1.4477	1.4623	1.4770
0.40	1.4918	1.5068	1.5220	1.5373	1.5527	1.5683	1.5841	1.6000	1.6161	1.6323
0.50	1.6487	1.6653	1.6820	1.6989	1.7160	1.7333	1.7507	1.7683	1.7860	1.8040
0.60	1.8221	1.8404	1.8589	1.8776	1.8965	1.9155	1.9348	1.9542	1.9739	1.9937
0.70	2.0138	2.0340	2.0544	2.0751	2.0959	2.1170	2.1383	2.1598	2.1815	2.2034
0.80	2.2255	2.2479	2.2705	2.2933	2.3164	2.3396	2.3632	2.3869	2.4109	2.4351
0.90	2.4596	2.4843	2.5093	2.5345	2.5600	2.5857	2.6117	2.6379	2.6645	2.6912
1.00	2.7183	2.7456	2.7732	2.8011	2.8292	2.8577	2.8864	2.9154	2.9447	2.9743
1.10	3.0042	3.0344	3.0649	3.0957	3.1268	3.1582	3.1899	3.2220	3.2544	3.2871
1.20	3.3201	3.3535	3.3872	3.4212	3.4556	3.4903	3.5254	3.5609	3.5966	3.6328
1.30	3.6693	3.7062	3.7434	3.7810	3.8190	3.8574	3.8962	3.9354	3.9749	4.0149
1.40	4.0552	4.0960	4.1371	4.1787	4.2207	4.2631	4.3060	4.3492	4.3929	4.4371
1.50	4.4817	4.5267	4.5722	4.6182	4.6646	4.7115	4.7588	4.8066	4.8550	4.9037
1.60	4.9530	5.0028	5.0531	5.1039	5.1552	5.2070	5.2593	5.3122	5.3656	5.4195
1.70	5.4739	5.5290	5.5845	5.6407	5.6973	5.7546	5.8124	5.8709	5.9299	5.9895
1.80	6.0496	6.1104	6.1719	6.2339	6.2965	6.3598	6.4237	6.4883	6.5535	6.6194
1.90	6.6859	6.7531	6.8210	6.8895	6.9588	7.0287	7.0993	7.1707	7.2427	7.3155
2.00	7.3891	7.4633	7.5383	7.6141	7.6906	7.7679	7.8460	7.9248	8.0045	8.0849
2.10	8.1662	8.2482	8.3311	8.4149	8.4994	8.5849	8.6711	8.7583	8.8463	8.9352
2.20	9.0250	9.1157	9.2073	9.2999	9.3933	9.4877	9.5831	9.6794	9.7767	9.8749
2.30	9.974	10.074	10.176	10.278	10.381	10.486	10.591	10.697	10.805	10.913
2.40	11.023	11.134	11.246	11.359	11.473	11.588	11.705	11.822	11.941	12.061
2.50	12.182	12.305	12.429	12.554	12.680	12.807	12.936	13.066	13.197	13.330
2.60	13.464	13.599	13.736	13.874	14.013	14.154	14.296	14.440	14.585	14.732
2.70	14.880	15.029	15.180	15.333	15.487	15.643	15.800	15.959	16.119	16.281
2.80	16.445	16.610	16.777	16.945	17.116	17.288	17.462	17.637	17.814	17.993
2.90	18.174	18.357	18.541	18.728	18.916	19.106	19.298	19.492	19.688	19.886
3.00	20.086	20.287	20.491	20.697	20.905	21.115	21.328	21.542	21.758	21.977
3.10	22.198	22.421	22.646	22.874	23.104	23.336	23.571	23.807	24.047	24.288
3.20	24.533	24.779	25.028	25.280	25.534	25.790	26.050	26.311	26.576	26.843
3.30	27.113	27.385	27.660	27.938	28.219	28.503	28.789	29.079	29.371	29.666
3.40	29.964	30.265	30.569	30.877	31.187	31.500	31.817	32.137	32.460	32.786
3.50	33.115	33.448	33.784	34.124	34.467	34.813	35.163	35.517	35.874	36.234
3.60	36.598	36.966	37.338	37.713	38.092	38.475	38.861	39.252	39.646	40.045
3.70	40.447	40.854	41.264	41.679	42.098	42.521	42.948	43.380	43.816	44.256
3.80	44.701	45.150	45.604	46.063	46.525	46.993	47.465	47.942	48.424	48.911
3.90	49.402	49.899	50.400	50.907	51.419	51.935	52.457	52.985	53.517	54.055

X

Y

List of Standards for Petroleum Production

Y

List of Standards for Petroleum Production

Y1 INTRODUCTION

1.1 Need for ISO standards

The objective of harmonized standards is to provide transparent and unambiguous requirements which are identical for all countries. ISO (International Organization for Standardization) is a worldwide federation of national standardization organizations, and is the standardization body responsible for the development of standards for global use.

Since the oil and gas industry is operating on a global level the European part of this industry was advised, through the OGP (formely E & P forum), to develop ISO standards.

The intention is that these ISO standards will be accepted by CEN (Comité Européen de Normalisation) for European harmonization. The Vienna agreement between ISO and CEN is in existence to facilitate the process.

From the above, the need for ISO standards is apparent for a globally operating line of business.

1.2 Conformity assessment and certification

In the oil and gas industry API documents (American Petroleum Institute) are often used in commercial contracts, and the documents ensure that form, fit and function are defined. In addition API run a licensing program which enables manufacturers to demonstrate that products with an associated mark conform to the applicable API specification, and that its organization has a quality management system conform to API Q1 in force.

In some cases additional certification has been required and is mandatory practice. Within the European infra-structure certification principles have been developed in conjunction with standardization. The origin has been the principles developed by the ISO for this purpose.

Y

1.3 API documents

Three types of document are mainly published.

*   **API Bulletin**

    API Bulletins are published to provide minimum performance properties, formulas and calculations on which the design of equipment may be based.

*   **API Specification**

    API Specifications are published as aids to the procurement of standardized equipment and materials, as well as instructions to manufacturers of equipment or materials covered by an API Specification. The specifications are not intended to obviate the need for sound engineering, nor to inhibit in any way anyone from purchasing or producing products to other specifications.

*   **API Recommended Practices (RP)**

    API Recommended Practices are published to facilitate the broad availability of proven, sound engineering and operating practices. The Recommended Practices are not intended to obviate the need for applying sound judgement as to when and where the Recommended Practices should be utilized.

The following paragraphs list the numbers of API documents and ISO standards used for well production.

Y2 TUBULAR GOODS

Title	API	ISO	Agreement CEN
Thread Compounds for Casing, Tubing and Line Pipe	RP 5A3 1st Ed., 1996	13 678 Pub.: 2000	
Field Inspection of New Casing, Tubing, and Plain End Drill Pipe	RP 5A5 7th Ed., 2003	15 463 Pub.: 2003	yes
Threading, Gauging, and Thread Inspection of Casing, Tubing, and Line Pipe Threads	Spec. 5B 14th Ed., 1996	10 422 Pub.: 1993	yes
Gauging and Inspection of Casing, Tubing and Line Pipe Threads	RP 5B1 4th Ed., 1996	15 464	yes
Casing and Tubing	Spec. 5CT 8th Ed., 2004	11 960 Pub.: 2004	yes
Care and Use of Casing and Tubing	RP 5C1 18th Ed., May 1999	10 405 Pub.: 2006	yes
Performance Properties of Casing, Tubing and Drill Pipe	Bull. 5C2 20th Ed., 1987		
Formulas and Calculations for Casing, Tubing, Drill Pipe, and Line Pipe Properties	Bull. 5C3 6th Ed., 1993	10 400 Pub.: 2007	
Round Thread Casing Joint Strength with Combined Internal Pressure and Bending	Bull. 5C4 * 2nd Ed., May 1987		
Evaluation Procedures for Casing & Tubing Connections	RP 5C5 2nd Ed., 1996	13 679 Pub.: 2002	yes
Coiled Tubing Operations in Oil and Gas Well Services	RP 5C7 1st Ed., Dec. 1996		
Line Pipe for pipeline transportation systems	Spec. 5L 44st Ed., 2007	3183 Pub.: 2007	
Internal Coating of Line Pipe for Non-Corrosive Gas Transmission Service	RP 5L2 3rd Ed., May 1987		
Conducting Drop-Weight Tear Tests on Line Pipe	RP 5L3 3rd Ed., 1996		
Unprimed Internal Fusion Bonded Epoxy Coating of Line Pipe	RP 5L7 2nd Ed., 1988/94		
Field Inspection of New Line Pipe	RP 5L8 2nd Ed., 1996		
Corrosion resistant alloy Line Pipe	Spec. 5L C 3rd Ed., 1998		
Corrosion Resistant Alloy Seamless Tubes for Use as Casing, Tubing and Coupling Stock		13 680 Pub.: 2000	yes

* not in 1999 catalog.

Y3 VALVES AND WELLHEAD EQUIPMENT

Title	API	ISO	Agreement CEN
Valves and Wellhead Equipment	Spec. 6A 19th Ed., 2003	10 423 Pub.: 2004	yes
Capabilities of API Flanges Under Combination of Load	Bull. 6AF 2nd Ed., Sept. 1995		
Temperature Derating of API Flanges Under Combination of Loading	Bull. 6AF1 2nd Ed., 1998		
Material Toughness	Bull. 6AM 2nd Ed., Sept. 1995		
Pipeline Valves (Steel, Plug, Ball, Check Valves)	Spec. 6D 22th Ed., Dec. 1999	14 313 Pub.: 2007	yes
Fire Test for Valves	Spec. 6FA 3rd Ed., April 1999		
Fire Test for End Connections	Spec. 6FB 3rd Ed., May 1998		
Fire Test for Valve with Selective Backseats	Spec. 6FC 3rd Ed., April 1999		
Fire Resistance of API/ANSI End Connections	Bull. 6F1 2nd Ed., Febr. 1994		
Improved Fire Resistivity of API Connections	Bull. 6F2 2nd Ed., Febr. 1994		
End Closures, Connectors and Swivels	Spec. 6H 2nd Ed., 1998		
Testing of Oilfield Elastomers–A Tutorial	Bull. 6J 1st Ed., Febr. 1992		
Referenced Standards for Committee 6, Standardization of Valves & Wellhead Equipment	Bull. 6RS * July 1990 + 1992		

* not in 1999 catalog.

Y4 WIRE ROPE

Title	API	ISO	Agreement CEN
Specification for Wire Rope	Spec. 9A 25th Ed., June 2003	10 425 Pub.: 2003	
Application, Care, and Use of Wire Rope for Oil Field Service	RP 9B 9th Ed., May 1986/92		

Y5 PRODUCTION EQUIPMENT

Title	API	ISO	Agreement CEN
Subsurface Sucker Rod Pumps and Fittings	Spec. 11AX 10th Ed., 1996		
Care and Use of Subsurface Pumps	RP 11AR 3rd Ed., 1989		
Sucker Rods	Spec. 11B 26th Ed., 1998	10 428 Pub.: 1993	yes
Care and Handling of Sucker Rods	RP 11BR 8th Ed., 1989		
Reinforced Plastic Sucker Rods	Spec. 11C * 2nd Ed., 1988 + 1991		
Dowhole equisetum – Packers and bridge plugs	Spec 11D1 1st Ed., 2001	14310 Pub.: 2001	yes
Pumping Units	Spec. 11E 17th Ed., 1994	10 431 Pub.: 1993	yes
Guarding of Pumping Units	RP 11ER 2nd Ed., 1990 +1991		
Installation and Lubrication of Pumping Units	RP 11G 4th Ed., 1994		
Data Sheet for the Design of Air Exchange Coolers	Bull. 11K 2nd Ed., 1988/94		
Design Calculations for Sucker Rod Pumping Systems (Conventional Units)	RP 11L 4th Ed., 1988/94		
Sucker Rod Pumping System Design Book	Bull. 11L3 1st Ed., 1970 + 1988/94		
Curves for Selecting Beam Pumping Units	Bull. 11L4 1st Ed., April 1970/94		
Electric Motor Performance Data Request Form	Bull. 11L5 * 1st Ed., Oct. 1990		
Lease Automatic Custody Transfer (LACT) Equipment	Spec. 11N 4th Ed.,1994		
Packaged High Speed Separable Engine-Driven Reciprocating Gas Compressors	Spec. 11P 2nd Ed., 1989	13 631 Pub.: 2002	yes
Electric Submersible Pump Installation	RP 11R * 2nd Ed., May 1986		
Operation, Maintenance and Troubleshooting of Electric Submersible Pump Installations	RP 11S 3rd Ed., Nov. 1994		
Electric Submersible Pump Teardown Report	RP 11S1 3rd Ed., 1997		
Electric Submersible Pump Testing	RP 11S2 2nd Ed., 1997		

* not in 1999 catalog.   *(to be continued)*

Title	API	ISO	Agreement CEN
Electric Submersible Pump Installation	RP 11S3 2nd Ed., March 1999		
Installation and Operation of Wet Steam Generators	RP 11T 2nd Ed., 1999		
Sizing and Selection of Electric Submersible Pumps Installation	RP 11U * 2nd Ed., May 1986		
Gas Lift Mandrels and Associated Subsurface Equipment	Spec. 11V1 2nd Ed., 1995		
Artificial Lift: Side pocket mandrels		17 078-1 Pub.: 2004	yes
Artificial Lift: Side pocket mandrels gas lift valves and flow control devices		17 078-2 Pub.: 2007	yes
Artificial Lift: Latches, seals and interface data for side pocket mandrels and flow control devices		17 078-3	yes
Operation, Maintenance & Trouble-shooting of Gas Lift Installations	RP 11V5 2nd Ed., 1999		
Design of Continuous Flow Gas Lift Installations Using Injection Pressure Operated Valves	RP 11V6 2nd Ed., 1999		
Repair, Testing & Setting Gas Lift Valves	RP 11V7 2nd Ed., 1999		
Downhole Progressive Cavity Pumps - Part 1: Pumps		15 136-1 Pub.: 2001	yes
Downhole Progressive Cavity Pumps - Part 2: Drive heads		15 136-2 Pub.: 2006	yes
Locks and Nipples	Spec. 14L 1st Ed., 2005	16 070 Pub.: 2005	yes
Packers and Bridge Plugs	Spec. 11D1 1st Ed., 2001	14 310 Pub.: 2001	yes
Wire Wrapped Screens		17 824	yes

* not in 1999 catalog.

Y

Y6 LEASE PRODUCTION VESSELS

Title	API	ISO	Agreement CEN
Bolted Tanks for Storage of Production Liquids	Spec. 12B 14th Ed., 1995		
Field Welded Tanks for Storage of Production Liquids	Spec. 12D 10th Ed., 1994		yes
Glycol-Type Gas Deshydration Units	Spec. 12GDU 1990		
Shop Welded Tanks for Storage of Production Liquids	Spec. 12F 11th Ed., 1994		
Oil and Gas Separators	Spec. 12J 7th Ed., 1989		
Indirect-Type Oil Field Heaters	Spec. 12K 7th Ed., 1989		
Vertical and Horizontal Emulsion Treaters	Spec. 12L 4th Ed., 1994		
Operations, Maintenance and Testing of Firebox Flame Arrestors	RP 12N 2nd Ed., 1994		
Fiberglass Reinforced Plastic Tanks	Spec. 12P 2nd Ed., 1995		
Setting, Connecting, Maintenance and Operation of Lease Tanks	RP 12R1 5th Ed., 1997		

Y7 OFFSHORE SAFETY AND ANTI-POLLUTION EQUIPMENT

Title	API	ISO	Agreement CEN
Subsurface Safety Valve Equipment. Functional and technical specifications	Spec. 14A 11th Ed., 2004	10 432 Pub.: 2004	yes
Design, Installation, Repair, and Operation of Subsurface Safety Valve Systems	RP 14B 5th Ed., 2004	10 417 Pub.: 2004	yes
Analysis, Design, Installation and Testing of Basic Surface Safety Systems on Offshore Production Platforms	RP 14C 4th Ed., 1986	10 418 Pub.: 2003	yes
Wellhead Surface Safety Valves and Underwater Safety Valves for Offshore Service	Spec. 14D 8th Ed., 1991	10 433 Pub.: 1994	
Design and Installation of Offshore Production Platform Piping Systems	RP 14E 5th Ed., 1991	13 703 Pub.: 2002	yes
Design and Installation of Electrical Systems for Offshore Production Platforms	RP 14F 4th Ed., 1999		
Fire Prevention and Control on Open Type Offshore Production Platforms	RP 14G 3rd Ed., 1993	13 702 Pub.: 1999	yes
Installation, Maintenance and Repair of Surface Safety Valves Offshore	RP 14H 4th Ed., 1994	10 419 Pub.: 1993	
Tools and Techniques for Identification and Assessment of Hazardous Situations	RP 14J 1st Ed., 1993	17 776 Pub.: 2000	yes
Downhole equisetum – Lock mandrels and landing nipples	Spec 14L 1st Ed., 2005	16070 Pub.: 2005	yes

Y8 PLASTIC PIPE

Title	API	ISO	Agreement CEN
Care and Use of Reinforced Thermosetting Resin Casing and Tubing	RP 15A4 * 1st Ed., March 1976		
High Pressure Fiberglass Line Pipe	Spec. 15HR 2nd Ed., 1995		
Polyethylene Line Pipe	Spec. 15LE 3rd Ed., May 1995		
Thermoplastic Line Pipe (PVC and CPVC)	Spec. 15LP * 2nd Ed., May 1987		
Low Pressure Fiberglass Line Pipe	Spec. 15LR 6th Ed., Sept. 1990	14 692 Pub.: 2002	yes
Care and Use of Reinforced Thermosetting Resin Line Pipe (RTRP)	RP 15L4 * 2nd Ed., March 1976		
Care and Use of Fiberglass Tubulars	RP 15TL4 2nd Ed., March 1999		

* not in 1999 catalog.

Y9 SUBSEA PRODUCTION

Title	API	ISO	Agreement CEN
Design and Operation of Subsea Production Systems:			
General Requirements and Recommendations	RP 17A 4th Ed., 2005	13 628-1 Pub.: 2005	yes
Flexible Pipe Systems for Subsea and Wireline Applications	RP 17J * 1st Ed., 1996	13 628-2 Pub.: 2006	yes
TFL (Through Flowline) Pump Down Systems	RP 17C 1st Ed., July 1991	13 628-3 Pub.: 2000	yes
Subsea Wellhead and Christmas Tree Equipment	Spec. 17D 1st Ed., Oct. 1992.	13 628-4 Pub.: 1999	yes
Design and Operation of Subsea Control Umbilicals	Spec. 17E 2nd Ed., 1998	13 628-5 Pub.: 2002	yes
Subsea Production Control Systems	Spec. 17F * 1st Ed., 1996	13 628-6 Pub.: 2006	yes
Completion/Workover Riser Systems	RP 17G 1st Ed., 1995	13 628-7 Pub.: 2005	yes
Design and Operation of Remote Operated Vehicles (ROV). Interface with subsea production systems	RP 17H * 1st Ed., 1996	13 628-8 Pub.: 2002	yes
ROT Intervention System	RP 17M 1st Ed., 2000	13 628-9 Pub.: 2000	yes
Specification for bonded flexible pipes	Spec 17K 2nd Ed., 2005	13 628-10 Pub.: 2005	yes
Drilling and Production Equipment. Flexible pipe systems for subsea and marine riser applications	RP 17B 1st Ed.	13 628-11 Pub.: 2007	in progress
Dynamic Production Risers	Spec. 17C	16 389	yes

* not in 1999 catalog.

Y10 COMMON PRODUCTION

Title	API	ISO	Agreement CEN
Determining Permeability of Porous Media	RP 27 * 3rd Ed., 1952 + 1956		
Standard Calibration and Format for Nuclear Logs	RP 33 3rd Ed., 1974 + 1990		
Standard Procedures for Evaluation of Hydraulic Fracturing Fluids	RP 39 3rd Ed., May 1998	13 503	
Standard Procedure for Presenting Performance Data on Hydraulic Fracturing Equipment	RP 41 2nd Ed., 1995		

(to be continued)

Title	API	ISO	Agreement CEN
Laboratory Testing of Surface Active Agents for Well Stimulation	RP 42 * 2nd Ed., 1977 + 1990		
Evaluation of Well Perforators	RP 43 5th Ed., Jan. 1991		
Sampling Petroleum Reservoir Fluids	RP 44 * 1st Ed., Jan. 1996		
Analysis of Oil-Field Waters	RP 45 3rd Ed., Aug. 1998		
Drill Stem Test Report Form	RP 48 * 1st Ed., 1972 + 1990		
Occupational Safety for Oil and Gas Drilling and Servicing Operations	RP 54 2nd Ed., Dec. 1991		
Testing Sand Used in Hydraulic Fracturing Operations	RP 56 2nd Ed., Dec. 1995		
Offshore Well Completion, Servicing, Workover, and Plug and Abandonment Operation	RP 57 1st Ed., Jan. 1986		
Testing Sand Used in Gravel Packing Operations	RP 58 2nd Ed., Dec. 1995		
Well Control Operation	RP 59 * 1st Ed., Aug. 1987		
Testing High Strength Proppants Used in Hydraulic Fracturing Operations	RP 60 2nd Ed., Dec. 1995		
Evaluating Short Term Proppant Pack Conductivity	RP 61 * 1st Ed., Oct. 1989		
Evaluation of Polymers Used in Enhanced Oil Recovery Operations	RP 63 1st Ed., June 1990		
Diverter Systems Equipment and Operations	RP 64 1st Ed., July 1991		
Recommended Digital Log Interchange Standard (DLIS)	RP 66 2nd Ed., June 1996		
Measurement of viscous properties of completion fluids	RP 13M 1st Ed., 2003	13 503-1 Pub.: 2003	yes
Measurement of properties used in hydraulic fracture and gravel packing operations		13 503-2 Pub.: 2006	yes
Testing of heavy brines	RP 13J 4th Ed., 2005	13 503-3 Pub.: 2005	yes
Procedures for measuring stimulation and gravel-pack fluid leak-off under static conditions	RP 13mM 1st Ed., 2006	13 503-4 Pub.: 2006	yes
Procedure for measuring the long term conductivity of propants		13 503-5 Pub.: 2006	yes

* not in 1999 catalog.

Y

REFERENCES

API Catalog

ISO:
– ISO/TC 67/SC1: Materials, equipment and offshore structures for petroleum and natural gas industries – Line pipes
– ISO/TC 67/SC2: Materials, equipment and offshore structures for petroleum and natural gas industries – Pipeline transportation systems
– ISO/TC 67/SC3: Materials, equipment and offshore structures for petroleum and natural gas industries – Drilling and completion fluids and well cements
– ISO/TC 67/SC4: Materials, equipment and offshore structures for petroleum and natural gas industries – Drilling and production equipment
– ISO/TC 67/SC5: Materials, equipment and offshore structures for petroleum and natural gas industries – Casing, tubing and drill pipe
– ISO/TC 67/SC6: Materials, equipment and offshore structures for petroleum and natural gas industries – Processing equipment and systems
– ISO/TC 67/SC7: Materials, equipment and offshore structures for petroleum and natural gas industries – Offshore structures.

CEN
– Alain Loppinet – *List of ISO TC67 Standards with the adoption in CEN and in API* – on 2007, December 31st.

Y

Glossary

Abandonment: Converting a drilled well to a condition that can be left indefinitely without further attention and which will not damage fresh water supplies or potential petroleum reservoirs.

Absolute permeability: A measure of the ability of a single fluid (such as water, gas, or oil) to flow through a rock formation when the formation is totally filled (saturated) with that fluid. The permeability measure of a rock filled with a single fluid is different from the permeability measure of the same rock filled with two or more fluids.

Acid inhibitor: Inhibiting additive that acts to top or retard chemical reaction of the acid.

Acid intensifier: Intensifying additive that acts to accelerate or strengthen chemical reaction of the acid.

Acoustic log/logging: Recorded measurement of ultrasonic signal travel through formation rock in order to identify formation rock lithology, porosity, anf fluid saturation.

Asphaltene: Any of the dark, solid constituents of crude oils and other bitumens.

Back-pressure: The pressure maintained on equipment or systems through which a fluid flows.

Backwash: To reverse the flow of fluid from a water injection well in order to get rid of sediment that has clogged the wellbore.

Basic sediment and water (BS&W): Sediments and water found in crude oil.

Battery: Equipment to process or store crude oil from one or more wells.

Bitumen: Petroleum in semi-solid or solid forms.

Blanking plug: A plug used to cut off flow of liquid.

Blowout preventer: A safety device installed immediately above the casing or in the drill-stem that can close the borehole in an emergency.

Bubble point: 1. The temperature and pressure at which part of a liquid begins to convert to gas. For example, if a certain volume of liquid is held at constant pressure, but its temperature is increased, a point is reached when bubbles of gas begin to form in the liquid. That is the bubble point. Similarly, if a certain volume of liquid is held at a constant temperature but the pressure is reduced, the point at which gas begins to form is the bubble point. **2.** The temperature and pressure at which gas, held in solution in crude oil, breaks out of solution as free gas.

Bullheading: Overpowering a well by pumping a kill fluid down the tubing or casing and killing the well.

Caliper log: A record showing variations in wellbore diameter by depth, indicating undue enlargement due to caving in, washout, or other causes. The caliper log also reveals corrosion, scaling, or pitting inside tubular goods.

Centrifugal pump: A pump with an impeller or rotor, an impeller shaft, and a casing, which discharges fluid by centrifugal force.

Check valve: Valve which operates to allow flow in only one direction.

Coalbed methane (CBM): Natural gas trapped in coal seams.

Compressibility factor (gas-deviation factor, supercompressibility factor) is a multiplying factor introduced into the ideal-gas law to account for the departure of true gases from ideal behavior ($pV = nzRT$; z is the compressibility factor).

Condensate: A light hydrocarbon liquid obtained by condensation of hydrocarbon vapors. It consists of varying proportions of butane, propane, pentane, and heavier fractions, with little or no methane or ethane.

Connate water: Water retained in the pore spaces, or interstices, of a formation from the time the formation was created.

Conventional crude oil: Petroleum found in liquid form, flowing naturally or capable of being pumped without further processing or dilution.

Critical state is the term used to identify the unique condition of pressure, temperature, and composition in which all properties of coexisting vapor and liquid become identical.

Demulsifier/demulsifying: agents/chemical products used to break down crude oil/water emulsions by reducing surface of the oil film surrounding water droplets.

Density: The gravity of crude oil, indicating the proportion of large, carbon-rich molecules, generally measured in kilograms per cubic meter or degrees on the American Petroleum Institute (API) gravity scale.

Discover well: An exploratory well that encounters a previously untapped oil or gas deposit.

Dissolved gas (solution gas) identifies material ordinarily gaseous at atmospheric conditions but which is part of a liquid phase at elevated pressure and temperature.

Drawdown: 1. The difference between static and flowing bottomhole pressures. **2.** The distance between the static level and the pumping level of the fluid in the annulus of a pumping well.

Emulsion: Colloidal mixture of two immiscible fluids, one being dispersed in the other in the form of fine droplets.

Enhanced oil recovery: Usually refers to tertiary recovery methods which alter oil properties in the reservoir in order to improve recovery.

Exploratory well: A well in an area where petroleum has not been previously found or one targeted for formations above or below known reservoirs.

Field: The surface area above one or more underground petroleum pools sharing the same or related infrastructure.

Flow line: Pipe, usually buried, through which oil or gas travels from the well to a processing facility.

Flow regime: The physical geometry exhibited by a multiphase flow in a conduit; for example, liquid occupying the bottom of the conduit with the gas phase flowing above, or a liquid phase with bubbles of gas.

Flow test: Determination of productivity of a well by measuring total pressure drop and pressure drop per unit of formation section open to a well during flow at a given production rate.

Flowing bottomhole pressure: Pressure at the bottom of the wellbore during normal oil production.

Fluid: A substance readily assuming the shape of the container in which it is placed; e.g. oil, gas, water or mixtures of these.

Formation volume factor (FVF): The factor that is used to convert stock tank barrels of oil to reservoir barrels. It is the ratio between the space occupied by a barrel of oil containing solution gas at reservoir conditions and a barrel of dead oil at surface conditions. Also called reservoir volume factor.

Gamma ray log/logging: Recorded measurement of natural formation radiation in order to identify formation rocks.

Gas: Hydrocarbons in the gaseous state at the prevailing temperature and pressure.

Gas cap: A free-gas phase overlying an oil zone and occuring within the same producing formation as the oil.

Gas cap drive: Reservoir drive provided by the expansion of compressed gas in a free state above the reservoir fluid being produced.

Gas-liquid ratio (GLR): The gas volume flow rate, relative to the total liquid volume flow rate (oil and water), all volumes converted to volumes at standard pressure and temperature.

Gas-oil ratio (GOR): The gas volume flow rate, relative to the oil volume flow rate, both converted to volumes at standard pressure and temperature.

Gas volume fraction (GVF): The gas volume flow rate, relative to the multiphase volume flow rate, at the pressure and temperature prevailing in that section. The GVF is normally expressed as a percentage.

Z

Heavy crude oil: Oil with a gravity below 28 degrees API (or d = 0.887 g/cm^3).

Hold-up: The cross-sectional area locally occupied by one of the phases of a multiphase flow, relative to the cross-sectional area of the conduit at the same local position.

Homogeneous multiphase flow: A multiphase flow in which all phases are evenly distributed over the cross-section of a closed conduit; i.e. the composition is the same at all points.

Huff and puff: Cyclic steam/hot water injection into a well in order to stimulate oil production.

Hydrate/gas hydrate: Icy lattice containing gas molecules, causing blockage of flowlines and pipelines transporting natural gas.

Hydrostatic pressure: The force exerted by a body of fluid at rest, which increases directly with the density and the depth of the fluid and expressed in Pa (or psi). The hydrostatic pressure of fresh water is 10 kPa per meter (or 0.433 psi per foot) of depth. In a water drive field, the term refers to the pressure that may furnish the primary energy for production.

Inflow performance relationship: The relation between the midpoint pressure of the producing interval and the liquid inflow rate of a producing well.

Injection well: A well used for injecting fluids (air, steam, water, natural gas, gas liquids, surfactants, alkalines, polymers, etc.) into an underground formation for the purpose of increasing recovery).

Interfacial tension: Surface tension occurring at the interface of two liquids.

Laterolog: Logging instrument in which electric current is forced to flow radially through the formation.

Lease: Legal document giving an operator the right to drill for or produce oil or gas, also, the land on which a lease has been obtained.

Light crude oil: Liquid petroleum with a gravity of 28 degrees API (or d = 0.887 g/cm^3) or higher.

Limestone and dolomite: Calcium carbonate-rich sedimentary rocks in which oil or gas reservoirs are often found.

Mass flow rate: The mass of fluid flowing through the cross-section of a conduit in unit time.

Methane: The principal constituent of natural gas; the simplest hydrocarbon molecule, containing one carbon atom and four hydrogen atoms.

Microlaterolog: A resistivity logging instrument with one center electrode and three circular ring electrodes around the center electrode.

Microlog: A resistivity logging instrument with electrodes mounted at short spacing in an insulating pad.

Z

Middle distillates: Medium-density refined petroleum products, including kerosene, stove oil, jet fuel and light fuel oil.

Mobility ratio: Ratio of mobility of a driving fluid to that of a driven fluid.

Mud: Fluid circulated down the drill pipe and up the annulus during drilling to remove cuttings, cool and lubricate the bit, and maintain desired pressure in the well.

Multiphase flow: Two or more phases flowing simultaneously in a conduit; this document deals in particular with multiphase flows of oil, gas and water.

Multiphase flow rate: The total amount of the two or three phases of a multiphase flow flowing through the cross-section of a conduit in unit time. The multiphase flow rate should be specified as multiphase volume rate or multiphase mass flow rate.

Multiphase flow rate meter: A device for measuring the flow rate of a multiphase flow through a cross-section of a conduit. It is necessary to specify whether the multiphase flow rate meter measures the multiphase volume or mass flow rate.

Multiphase flow velocity: The flow velocity of a multiphase flow. It may also be defined by the relationship (Multiphase volume flow rate/Pipe cross-section).

Multiphase fraction meter: A device for measuring the phase area fractions of oil, gas and water of a multiphase flow through a cross-section of a conduit.

Multiphase meter: A device for measuring the phase area fractions and flow rates of oil, gas and water of a multiphase flow through a cross-section of a conduit. It is necessary to specify whether the multiphase meter measures volume or mass flow rates.

Natural gas liquids: Liquids obtained during natural gas production, including ethane, propane, butane, and condensate.

Neutron log/logging: Recorded measurement of artificially produced radiation within a well in order to measure formation fluids.

Newtonian fluid: A simple fluid in which the state of stress at any point is proportional to the time rate of strain at that point; the proportionality factor is the viscosity coefficient.

Oil: Hydrocarbon in the liquid state at the prevailing temperature and pressure conditions.

Oil-continuous multiphase flow: A multiphase flow of oil/gas/water characterized in that the water is distributed as water droplets surrounded by oil. Electrically, the mixture acts as an insulator.

Operator: The company or individual responsible for managing an exploration, development or production operation.

Packer method: Controlled acidizing treatment in which the packer prevents acid from traveling further up the well annulus than necessary.

Pay section: Producing formation.

Z

Pay zone: The producing formation.

Permeability: The capacity of a reservoir rock to transmit fluids.

Petroleum: A naturally occurring mixture of hydrocarbons in gaseous, liquid or solid form.

Phase: In multiphase metering, "phase" is used in the sense of one constituent in a mixture of several fluids. In particular, the term refers to either oil, gas or water in a mixture of any number of the three.

Phase area fraction: The cross-sectional area locally occupied by one of the phases of a multiphase flow, relative to the cross-sectional area of a conduit at the same local position.

Phase flow rate: The amount of one phase of a multiphase flow flowing through the cross-section of a conduit in unit time. The phase flow rate may be specified as phase volume flow rate or as phase mass flow rate.

Phase mass fraction: The phase mass flow rate of one of the phases of a multiphase flow, relative to the multiphase mass flow rate.

Phase velocity: The velocity of one phase of a multiphase flow at a cross-section of a conduit. It may also be defined by the relationship: superficial phase velocity × phase area fraction.

Phase volume fraction: The phase volume flow rate of one of the phases of a multiphase flow, relative to the multiphase volume flow rate.

Pilot flood: A pilot-scale waterflooding project conducted in order to evaluate operation procedures and to assess and predict performance of the waterflood.

Pinnacle reef: A conical formation, usually composed of limestone, in which hydrocarbons might be trapped.

Plugback: Sealing well casing to separate producing intervals in the wellbore from depleted intervals.

Pool: A natural underground reservoir containing, or appearing to contain, an accumulation of petroleum.

Porosity: The open or void space within rock.

Radioactive log: Measurement of natural and artificially produced radiation within a well in order to identify formation rocks and fluids.

Radioactive tracer log/logging: Record of travel of a radioactive tracer substance within a formation or within the borehole.

Reservoir drive: Energy or force in a reservoir that causes reservoir fluid to move toward a well and up to the surface.

Residue gas: Gas remaining after natural gas is processed and liquids are removed.

Retarded acid: Acidizing solution whose reactivity is slowed so that the acid can penetrate deeper into a formation before being spent.

Royalty: The "owner's share" of production or revenues retained by government or freehold mineral rights holders.

Sand-exclusion completion: Completing a well in which sand production is discouraged with the installation of slotted or screen well liners or by gravel packing the borehole.

Sandstone: Compacted sedimentary rock composed mainly of quartz or feldspar; a common rock in which oil/or water condensate.

Screen-out: Well congested with sand which has fallen out of produced fluids in the well.

Sedimentary basin: A geographical area in which much of the rock is sedimentary (as opposed to igneous or metamorphic) and therefore likely to contain hydrocarbons.

Shale: Rock formed from clay.

Shrinkage: refers to the decrease in volume of a liquid phase caused by release of solution gas and/or by the thermal contraction of the liquid. Shrinkage may be expressed (1) as a percentage of the final resulting stock-tank oil or (2) as a percentage of the original volume of the liquid. Shrinkage factor is the reciprocal of FVF expressed as barrels (or m^3) of stock-tank oil per barrel (or m^3) of reservoir oil. A reservoir oil that resulted in 0.75 bbl of stock-tank oil per 1 bbl of reservoir oil would have a shrinkage of 0.25/0.75 = 33% under Definition 1, a shrinkage of 0.25/1.00 = 25% under Definition 2, a shrinkage factor of 0.75, and an FVF of 1.00/0.75 = 1.33.

Shut-in bottomhole pressure: The pressure at the bottom of a well when the surface valves on the well are completely closed, caused by formation fluids at the bottom of the well.

Sink: Pressure gradient surrounding the borehole within its drainage radius.

Slip: Term used to describe the flow conditions that exist when the phases have different velocities at a cross-section of a conduit. The slip may be quantitatively expressed by the phase velocity difference between the phases.

Slip ratio: The ratio between two phase velocities.

Slip velocity: The phase velocity difference between two phases.

Slug: A measured amount of liquid used to displace or force fluid flow in the reservoir.

Solution gas-oil ratio, R_s, expresses the amount of gas in solution, or dissolved, in a liquid. The reference oil may be stock-tank oil or residual oil. On occasion, reservoir saturated oil is used as a reference.

Sour gas: Natural gas containing relatively large amounts of sulfur/sulfur compounds.

Spent acid: Acidizing solution sludge or residue.

Standard conditions (surface) are 1 bar and 15.6°C (14.7 psia and 60°F). Gas volumes may be specified on occasion at pressures slightly removed from 1 bar (14.7 psia).

Static bottomhole pressure: Pressure at the bottom of the wellbore when there is no flow of oil.

Steam soak: Steam/hot water injection into a well in order to stimulate oil production.

Stock-tank oil is the liquid that results from production of reservoir material through surface equipment that separates normally gaseous components. Stock-tank oil may be caused to vary in composition and properties by varying the conditions of gas/liquid separation. Stock-tank oil is normally reported at 1 bar and 15.6°C (14.7 psia and 60°F) but may be measured under other conditions and corrected to the standard condition.

Z

Stripper/stripper well: An oil well that produces a limited amount of oil, e.g., no more than 10 bbl/D or 1.6 m³/d.

Superficial phase velocity: The flow velocity of a multiphase flow, assuming that the phase occupies the whole conduit by itself. It may also be defined by the relationship (phase volume flow rate/pipe cross-section).

Surfactant/surface active agent: Chemical additives that lower the surface tension of a solution.

Swabbing: Cleaning out a well with a special tool connected to a wireline.

Sweep efficiency: The efficiency with which water displaces oil or gas in a water drive oil or gas field. Water flowing in from the aquifer does not displace the oil or gas uniformly but channels past certain areas due to variations in porosity and permeability.

Sweet oil and gas: Petroleum containing little or no hydrogen sulphide.

Thermal recovery: Tertiary recovery methods in which the oil the reservoir is affected by thermal treatments, including in situ combustion, steam injection, and other methods of reservoir heating.

Torque: The turning force that is supplied to a shaft or other rotary mechanism to cause it to rotate or tend to do so. Torque is measured in units of length and force (newton-meters, foot-pounds).

Tubingless completion: A method of producing a well in which only small-diameter production casing is set through the pay zone, with no tubing or inner production string used to bring formation fluids to the surface. This type of completion has its best application in low-pressure, dry-gas reservoirs.

Uncontrolled acidizing treatment: Pumping acid solution down a well, followed by displacement fluid to force the acid out into the formation.

Unitization: Process whereby owners of adjoining properties pool reserves from a single unit operated by one of the owners; production is divided among the owners according to the unitization agreement.

Velocity profile: The mean velocity distribution of a fluid at a cross-section of a conduit. The velocity profile may be visualized by means of a two- or three-dimensional graph.

Viscosity: The resistance to flow, or "stickness", of a fluid.

Void fraction: The cross-sectional area locally occupied by the gas phase of a multiphase flow, relative to the cross-sectional area of the conduit at the same local position.

Volatile organic compounds (VOCs): gas and vapors, such as benzene, released by petroleum refineries, petrochemical plants, plastics manufacturing and the distribution and use of gasoline, among other sources; VOCs include carcinogens and chemicals which react with sunlight and nitrogen oxides to form ground-level ozone or "smog".

Z

Volume flow rate: The volume of fluid flowing through the cross-section of a conduit in unit time at the pressure and temperature prevailing in that section.

Water-continuous multiphase flow: A multiphase flow of oil/gas/water characterized in that the oil is distributed as oil droplets surrounded by water. Electrically, the mixture acts as a conductor.

Water cut (WC): The water volume flow rate, relative to the total liquid volume flow rate (oil and water), both converted to volumes at standard pressure and temperature. The WC is normally expressed as a percentage.

Water drive: Reservoir drive provided by the force of water under pressure below the reservoir fluid being produced.

Water-in-liquid ratio (WLR): The water volume flow rate, relative to the total liquid volume flow rate (oil and water), at the pressure and temperature prevailing in that section.

Wet/rich gas: Natural gas containing significantly large amounts of associated petroleum liquids.

Wildcat: A well drilled in an area where no oil or gas production exists.

REFERENCES

1   Hall LW (1986) *Petroleum Production Operations*. The University of Texas
2   Petroleum Communication Foundation (1998) *Glossary of Industry Terms*. PanCanadian
3   *Handbook of Multiphase Metering* (1995) produced for The Norwegian Society for Oil and Gas Measurement, NFOGM Report No. 1, 1995
4   Bradley HB, Ed. (1992) *Petroleum Engineering Handbook*. Chapter 22, Oil system correlation. SPE, Richardson, TX
5   Gray F (1995) *Petroleum Production in Nontechnical language*. PennWell Books, Tulsa, OK.

Index

www.ingramcontent.com/pod-product-compliance
Lightning Source LLC
Chambersburg PA
CBHW081216220326
41598CB00037B/6794